Student Solution Manual for McKeague's
Intermediate Algebra: Concepts and Graphs

Prepared by
Ross Rueger
Department of Mathematics
College of the Sequoias
Visalia, California

Student Solutions Manual for McKeague's Intermediate Algebra: Concepts and Graphs

Ross Rueger

Publisher: XYZ Textbooks

Sales: Amy Jacobs, Richard Jones, Bruce Spears, Rachael Hillman

Cover Design: Kyle Schoenberger

Printed in the United States of America

ISBN-13: 978-1-936368-30-3 / ISBN-10: 1-936368-30-7

XYZ Textbooks

For product information and technology assistance, contact us at
XYZ Textbooks, 1-877-745-3499

For permission to use material from this text or product,
e-mail: **info@mathtv.com**

1339 Marsh Street
San Luis Obispo, CA 93401
USA

For your course and learning solutions, visit **www.xyztextbooks.com**

Contents

Preface

This *Student Solutions Manual* contains complete solutions to all odd-numbered exercises of *Intermediate Algebra, Concepts and Graphs* by Charles P. McKeague. Every attempt has been made to format solutions for readability and accuracy, and I apologize for any errors that you may encounter. If you have any comments, suggestions, error corrections, or alternative solutions please feel free to drop me a note or send an email (address below).

Please use this manual with some degree of caution. Be sure that you have attempted a solution, and re-attempted it, before you look it up in this manual. Algebra can only be learned by *doing*, and not by observing! As you use this manual, do not just read the solution but work it along with the manual, using my solution to check your work. If you use this manual in that fashion then it should be helpful to you as you do homework and study for tests.

I wish to express my appreciation to Pat McKeague at MathTV.com and Patrick McKeague at XYZ Textbooks for asking me to be involved with this textbook. This book provides a complete and affordable course in intermediate algebra, and you will find the text very easy to read and understand. Good luck!

Ross Rueger
College of the Sequoias
rossrueger@gmail.com

January 2011

Chapter 1
Numbers, Variables, and Expressions

1.1 Recognizing Patterns

1. The pattern is to add 1, so the next term is 5. **3.** The pattern is to add 2, so the next term is 10.
5. These numbers are squares, so the next term is $5^2 = 25$.
7. The pattern is to add 7, so the next term is 29.
9. These numbers are cubes, so the next term is $5^3 = 125$.
11. One possibility is: Δ **13.** One possibility is: ⊙
15. The pattern is to add 4, so the next two numbers are 17 and 21.
17. The pattern is to add –1, so the next two numbers are –2 and –3.
19. The pattern is to add –3, so the next two numbers are –4 and –7.
21. The pattern is to add $-\frac{1}{4}$, so the next two numbers are $-\frac{1}{2}$ and $-\frac{3}{4}$.
23. The pattern is to add $\frac{1}{2}$, so the next two numbers are $\frac{5}{2}$ and 3.
25. The pattern is to multiply by 3, so the next number is 27.
27. The pattern is to multiply by –3, so the next number is –270.
29. The pattern is to multiply by $\frac{1}{2}$, so the next number is $\frac{1}{8}$.
31. The pattern is to multiply by $\frac{1}{2}$, so the next number is $\frac{5}{2}$.
33. The pattern is to multiply by –5, so the next number is –625.
35. The pattern is to multiply by $-\frac{1}{5}$, so the next number is $-\frac{1}{125}$.
37. **a**. The pattern is to add 4, so the next number is 12.
b. The pattern is to multiply by 2, so the next number is 16.

39. Completing the table:

Two Numbers a and b	Their Product ab	Their Sum $a+b$
1,–24	–24	–23
–1,24	–24	23
2,–12	–24	–10
–2,12	–24	10
3,–8	–24	–5
–3,8	–24	5
4,–6	–24	–2
–4,6	–24	2

41. The even numbers are 2, 8, and 34.

43. **a.** Using the rule for order of operations: $3 \bullet 5 + 4 = 15 + 4 = 19$

b. Using the rule for order of operations: $3(5+4) = 3 \bullet 9 = 27$

c. Using the rule for order of operations: $3 \bullet 5 + 3 \bullet 4 = 15 + 12 = 27$

45. **a.** Using the rule for order of operations: $6 + 3 \bullet 4 - 2 = 6 + 12 - 2 = 16$

b. Using the rule for order of operations: $6 + 3(4-2) = 6 + 3 \bullet 2 = 6 + 6 = 12$

c. Using the rule for order of operations: $(6+3)(4-2) = 9 \bullet 2 = 18$

47. **a.** Using the rule for order of operations: $(7-4)(7+4) = 3 \bullet 11 = 33$

b. Using the rule for order of operations: $7^2 - 4^2 = 49 - 16 = 33$

49. **a.** Using the rule for order of operations: $(5+7)^2 = 12^2 = 144$

b. Using the rule for order of operations: $5^2 + 7^2 = 25 + 49 = 74$

c. Using the rule for order of operations: $5^2 + 2 \bullet 5 \bullet 7 + 7^2 = 25 + 70 + 49 = 144$

51. **a.** Using the rule for order of operations: $2 + 3 \bullet 2^2 + 3^2 = 2 + 3 \bullet 4 + 9 = 2 + 12 + 9 = 23$

b. Using the rule for order of operations: $2 + 3(2^2 + 3^2) = 2 + 3(4+9) = 2 + 3(13) = 2 + 39 = 41$

c. Using the rule for order of operations: $(2+3)(2^2 + 3^2) = (2+3)(4+9) = (5)(13) = 65$

53. **a.** Using the rule for order of operations: $40 - 10 \div 5 + 1 = 40 - 2 + 1 = 38 + 1 = 39$

b. Using the rule for order of operations: $(40-10) \div 5 + 1 = 30 \div 5 + 1 = 6 + 1 = 7$

c. Using the rule for order of operations: $(40-10) \div (5+1) = 30 \div 6 = 5$

55. **a.** Using the rule for order of operations: $40 + [10 - (4-2)] = 40 + (10-2) = 40 + 8 = 48$

b. Using the rule for order of operations: $40 - 10 - 4 - 2 = 30 - 4 - 2 = 26 - 2 = 24$

57. **a.** Using the rule for order of operations:

$$3 + 2(2 \bullet 3^2 + 1) = 3 + 2(2 \bullet 9 + 1) = 3 + 2(18+1) = 3 + 2 \bullet 19 = 3 + 38 = 41$$

b. Using the rule for order of operations:

$$(3+2)(2 \bullet 3^2 + 1) = 5(2 \bullet 9 + 1) = 5(18+1) = 5 \bullet 19 = 95$$

59. The counting numbers are: 1,2

61. The rational numbers are: $-6, -5.2, 0, 1, 2, 2.3, \frac{9}{2}$

63. The irrational numbers are: $-\sqrt{7}, -\pi, \sqrt{17}$

65. The nonnegative integers are: 0,1,2

67. The sequence of air temperatures is: 41°,37.5°,34°,30.5°,27°,23.5°. This is an arithmetic sequence.

69. Completing the table:

Elevation (ft)	Boiling Point (°F)
–2,000	215.6
–1,000	213.8
0	212
1,000	210.2
2,000	208.4
3,000	206.6

1.2 Products

1. Using the associative property: $5(3y)=(5\bullet 3)y=15y$

3. Using the associative property: $4\left(\frac{1}{4}a\right)=\left(4\bullet\frac{1}{4}\right)a=a$

5. Using the associative property: $10(0.3x)=(10\bullet 0.3)x=3x$

7. Using the associative property: $\frac{2}{3}\left(\frac{3}{2}x\right)=\left(\frac{2}{3}\bullet\frac{3}{2}\right)x=x$

9. Using the associative property: $15\left(\frac{2}{3}x\right)=\left(\frac{15}{1}\bullet\frac{2}{3}\right)x=10x$

11. Using the associative property: $-15\left(\frac{x}{5}\right)=\left(-\frac{15}{1}\bullet\frac{1}{5}\right)x=-3x$

13. Using the associative property: $x\left(\frac{5}{x}\right)=\frac{x}{1}\bullet\frac{5}{x}=5$

15. Applying the distributive property: $5(3a+2)=5\bullet 3a+5\bullet 2=15a+10$

17. Applying the distributive property: $(5t+1)8=5t\bullet 8+1\bullet 8=40t+8$

19. Applying the distributive property: $\frac{1}{3}(4x+6)=\frac{1}{3}\bullet 4x+\frac{1}{3}\bullet 6=\frac{4}{3}x+2$

21. Applying the distributive property: $\frac{1}{5}(10+5y)=\frac{1}{5}\bullet 10+\frac{1}{5}\bullet 5y=2+y$

23. Applying the distributive property: $\frac{3}{4}(8x-4)=\frac{3}{4}\bullet 8x-\frac{3}{4}\bullet 4=6x-3$

25. Applying the distributive property: $\frac{5}{6}(12x-18)=\frac{5}{6}\bullet 12x-\frac{5}{6}\bullet 18=10x-15$

27. Applying the distributive property: $8\left(\frac{1}{8}x+3\right)=8\bullet\frac{1}{8}x+8\bullet 3=x+24$

29. Applying the distributive property: $6\left(\frac{1}{2}x-\frac{1}{3}y\right)=6\bullet\frac{1}{2}x-6\bullet\frac{1}{3}y=3x-2y$

31. Applying the distributive property: $20\left(\frac{2}{5}x+\frac{1}{4}y\right)=20\bullet\frac{2}{5}x+20\bullet\frac{1}{4}y=8x+5y$

33. Applying the distributive property: $8\left(\frac{x}{8}+\frac{y}{2}\right)=8\bullet\frac{1}{8}x+8\bullet\frac{1}{2}y=x+4y$

35. Applying the distributive property: $12\left(\frac{a}{4}+\frac{1}{2}\right)=12\bullet\frac{1}{4}a+12\bullet\frac{1}{2}=3a+6$

37. Applying the distributive property:
$$12\left(\frac{y}{2}+\frac{y}{4}+\frac{y}{6}\right)=12\bullet\frac{1}{2}y+12\bullet\frac{1}{4}y+12\bullet\frac{1}{6}y=6y+3y+2y=11y$$

39. Applying the distributive property: $10(0.3x+0.7y)=10\bullet 0.3x+10\bullet 0.7y=3x+7y$

41. Applying the distributive property: $100(0.06x+0.07y)=100\bullet 0.06x+100\bullet 0.07y=6x+7y$

43. Applying the distributive property: $0.05(x+2{,}000)=0.05\bullet x+0.05\bullet 2{,}000=0.05x+100$

45. Applying the distributive property: $0.12(x+500)=0.12\bullet x+0.12\bullet 500=0.12x+60$

47. Applying the distributive property: $a\left(1+\frac{1}{a}\right)=a\bullet 1+a\bullet\frac{1}{a}=a+1$

49. Applying the distributive property: $3\left(x+\frac{1}{3}\right)=3\bullet x+3\bullet\frac{1}{3}=3x+1$

51. Applying the distributive property: $x\left(1+\frac{2}{x}\right)=x\bullet 1+x\bullet\frac{2}{x}=x+2$

53. Applying the distributive property: $-5(2x-3)=-5\bullet 2x+5\bullet 3=-10x+15$

55. Applying the distributive property: $-4(2x-1)=-4\bullet 2x+4\bullet 1=-8x+4$

57. Applying the distributive property: $-1(5-x)=-1\bullet 5+1\bullet x=-5+x=x-5$

59. Applying the distributive property: $-1(7-x)=-1\bullet 7+1\bullet x=-7+x=x-7$

61. Using properties of exponents: $(5x^3)(7x^4)=(5\bullet 7)x^{3+4}=35x^7$

63. Using properties of exponents: $(-4x)(7x^3)=(-4\bullet 7)x^{1+3}=-28x^4$

65. Using properties of exponents: $(2x^2)^3(x^4)^5=(2)^3(x^2)^3(x^4)^5=8x^6\bullet x^{20}=8x^{6+20}=8x^{26}$

67. Using the distributive property:

$$2x^2(3x^2-2x+1)=2x^2(3x^2)-2x^2(2x)+2x^2(1)=6x^4-4x^3+2x^2$$

69. Using the distributive property:

$$2ab(a^2-ab+1)=2ab(a^2)-2ab(ab)+2ab(1)=2a^3b-2a^2b^2+2ab$$

71. Multiplying using the FOIL method: $(2x-3)(x-4)=2x^2-3x-8x+12=2x^2-11x+12$

73. Multiplying using the FOIL method: $(a+2)(2a-1)=2a^2+4a-a-2=2a^2+3a-2$

75. Multiplying using the FOIL method: $(2x-5)(3x-2)=6x^2-15x-4x+10=6x^2-19x+10$

77. Multiplying using the FOIL method: $(2x+3)(a+4)=2ax+3a+8x+12$

79. Multiplying using the FOIL method: $(5x-4)(5x+4)=25x^2-20x+20x-16=25x^2-16$

81. Multiplying using the FOIL method: $(x-3)^2=(x-3)(x-3)=x^2-3x-3x+9=x^2-6x+9$

83. Multiplying using the FOIL method:

$$(5x+1)^2=(5x+1)(5x+1)=25x^2+5x+5x+1=25x^2+10x+1$$

85. Multiplying using the FOIL method: $\left(2x-\frac{1}{2}\right)\left(x+\frac{3}{2}\right)=2x^2-\frac{1}{2}x+3x-\frac{3}{4}=2x^2+\frac{5}{2}x-\frac{3}{4}$

87. **a.** Finding the product: $(x-1)(x^2+x+1)=x^3+x^2+x-x^2-x-1=x^3-1$

b. Finding the product: $(x-2)(x^2+2x+4)=x^3+2x^2+4x-2x^2-4x-8=x^3-8$

c. Finding the product: $(x-3)(x^2+3x+9)=x^3+3x^2+9x-3x^2-9x-27=x^3-27$

d. Finding the product: $(x-4)(x^2+4x+16)=x^3+4x^2+16x-4x^2-16x-64=x^3-64$

89. Multiplying using the vertical format:

$$\begin{array}{rrr} x & -5 & \\ x & +3 & \\ \hline x^2 & -5x & \\ & +3x & -15 \\ \hline x^2 & -2x & -15 \end{array}$$

The product is $x^2 - 2x - 15$.

91. Multiplying using the vertical format:

$$\begin{array}{rrr} 2x^2 & -3 & \\ 3x^2 & -5 & \\ \hline 6x^4 & -9x^2 & \\ & -10x^2 & +15 \\ \hline 6x^4 & -19x^2 & +15 \end{array}$$

The product is $6x^4 - 19x^2 + 15$.

93. Multiplying using the vertical format:

$$\begin{array}{rrrr} x^2 & +6x & +5 & \\ & x & +3 & \\ \hline x^3 & +6x^2 & +5x & \\ & +3x^2 & +18x & +15 \\ \hline x^3 & +9x^2 & +23x & +15 \end{array}$$

The product is $x^3 + 9x^2 + 23x + 15$.

95. Multiplying using the vertical format:

$$\begin{array}{rrrr} a^2 & +ab & +b^2 & \\ & a & -b & \\ \hline a^3 & +a^2b & +ab^2 & \\ & -a^2b & -ab^2 & -b^3 \\ \hline a^3 & & & -b^3 \end{array}$$

The product is $a^3 - b^3$.

97. Multiplying using the vertical format:

$$\begin{array}{rrrr} 4x^2 & -2xy & +y^2 & \\ & 2x & +y & \\ \hline 8x^3 & -4x^2y & +2xy^2 & \\ & +4x^2y & -2xy^2 & +y^3 \\ \hline 8x^3 & & & +y^3 \end{array}$$

The product is $8x^3 + y^3$.

99. Multiplying using the vertical format:

$$\begin{array}{rrrr} a^2 & +ab & +b^2 & \\ & 2a & -3b & \\ \hline 2a^3 & +2a^2b & +2ab^2 & \\ & -3a^2b & -3ab^2 & -3b^3 \\ \hline 2a^3 & -a^2b & -ab^2 & -3b^3 \end{array}$$

The product is $2a^3 - a^2b - ab^2 - 3b^3$.

101. The revenue is: $R = xp = (1{,}200 - 100p)p = 1{,}200p - 100p^2$

1.3 Sums and Differences

1. Simplifying: $5a + 7 + 8a + a = (5a + 8a + a) + 7 = 14a + 7$
3. Simplifying: $2(5x+1) + 2x = 10x + 2 + 2x = (10x + 2x) + 2 = 12x + 2$
5. Simplifying: $3 + 4(5a+3) + 4a = 3 + 20a + 12 + 4a = (20a + 4a) + (3 + 12) = 24a + 15$
7. Simplifying: $5x + 3(x+2) + 7 = 5x + 3x + 6 + 7 = 8x + 13$
9. Simplifying: $5(x+2y) + 4(3x+y) = 5x + 10y + 12x + 4y = (5x + 12x) + (10y + 4y) = 17x + 14y$
11. Simplifying: $5b + 3(4b+a) + 6a = 5b + 12b + 3a + 6a = 17b + 9a$
13. Simplifying: $3(5x+4) - x = 15x + 12 - x = 14x + 12$
15. Simplifying: $6 - 7(3-m) = 6 - 21 + 7m = 7m - 15$
17. Simplifying: $7 - 2(3x-1) + 4x = 7 - 6x + 2 + 4x = -2x + 9$
19. Simplifying: $5(3y+1) - (8y-5) = 15y + 5 - 8y + 5 = 7y + 10$
21. Simplifying: $4(2-6x) - (3-4x) = 8 - 24x - 3 + 4x = -20x + 5$
23. Simplifying: $10 - 4(2x+1) - (3x-4) = 10 - 8x - 4 - 3x + 4 = -11x + 10$

25. Simplifying: $3x-5(x-3)-2(1-3x)=3x-5x+15-2+6x=4x+13$
27. Simplifying: $0.06x+0.05(10{,}000-x)=0.06x+500-0.05x=0.01x+500$
29. Simplifying: $0.12x+0.10(15{,}000-x)=0.12x+1{,}500-0.10x=0.02x+1{,}500$
31. Simplifying: $-(a+1)-4a=-a-1-4a=-5a-1$
33. Simplifying: $(x-3)(x-2)+2=x^2-3x-2x+6+2=x^2-5x+8$
35. Simplifying: $(2x-3)(4x+3)+4=8x^2+6x-12x-9+4=8x^2-6x-5$
37. Simplifying: $(x+4)(x-5)+(-5)(2)=x^2-5x+4x-20-10=x^2-x-30$
39. Simplifying: $2(x-3)+x(x+2)=2x-6+x^2+2x=x^2+4x-6$
41. Simplifying: $3x(x+1)-2x(x-5)=3x^2+3x-2x^2+10x=x^2+13x$
43. Simplifying: $x(x+2)-3=x^2+2x-3$
45. Simplifying: $a(a-3)+6=a^2-3a+6$
47. Simplifying:
$$(6x^3-4x^2+2x)+(9x^2-6x+3)=6x^3-4x^2+2x+9x^2-6x+3=6x^3+5x^2-4x+3$$
49. Simplifying: $(a^2-a-1)-(-a^2+a+1)=a^2-a-1+a^2-a-1=2a^2-2a-2$
51. Simplifying:
$$(x^3+4x^2+4x)+(2x^2+8x+8)=x^3+4x^2+4x+2x^2+8x+8=x^3+6x^2+12x+8$$
53. a. Substituting $x=0$: $-\frac{1}{3}x+2=-\frac{1}{3}(0)+2=0+2=2$
    b. Substituting $x=3$: $-\frac{1}{3}x+2=-\frac{1}{3}(3)+2=-1+2=1$
    c. Substituting $x=-3$: $-\frac{1}{3}x+2=-\frac{1}{3}(-3)+2=1+2=3$
55. a. Substituting $x=2$ and $y=-1$: $2x+y=2(2)+(-1)=4-1=3$
    b. Substituting $x=0$ and $y=3$: $2x+y=2(0)+(0)=0+3=3$
    c. Substituting $x=\frac{3}{2}$ and $y=-7$: $2x+y=2\left(\frac{3}{2}\right)+(-7)=3+(-7)=-4$
57. a. Substituting $y=4$: $y(2y+3)=4[2(4)+3]=4(8+3)=4(11)=44$
    b. Substituting $y=-\frac{11}{2}$: $y(2y+3)=-\frac{11}{2}\left[2\left(-\frac{11}{2}\right)+3\right]=-\frac{11}{2}(-11+3)=-\frac{11}{2}(-8)=\frac{88}{2}=44$
59. Substituting $x=7{,}000$ and $y=8{,}000$:
$$0.06x+0.07y=0.06(7{,}000)+0.07(8{,}000)=420+560=980$$
61. Substituting $x=10$ and $y=12$: $0.05x+0.10y=0.05(10)+0.10(12)=0.5+1.2=1.7$

63. a. Substituting $a = 3$, $b = -2$, and $c = 4$: $b^2 - 4ac = (-2)^2 - 4(3)(4) = 4 - 48 = -44$

b. Substituting $a = 1$, $b = -3$, and $c = -28$: $b^2 - 4ac = (-3)^2 - 4(1)(-28) = 9 + 112 = 121$

c. Substituting $a = 1$, $b = -6$, and $c = 9$: $b^2 - 4ac = (-6)^2 - 4(1)(9) = 36 - 36 = 0$

d. Substituting $a = 0.1$, $b = -27$, and $c = 1700$:

$$b^2 - 4ac = (-27)^2 - 4(0.1)(1700) = 729 - 680 = 49$$

65. Calculating: $-500 + 27(100) - 0.1(100)^2 = -500 + 2700 - 1000 = 1200$

67. Calculating: $-0.05(130)^2 + 9.5(130) - 200 = -845 + 1235 - 200 = 190$

69. For these calculations, remember that 1 minute = 60 seconds.

a. Calculating the difference: 4:22:14 – 4:18:44 = 4:21:74 – 4:18:44 = 0:03:30

b. Calculating the difference: 4:24:42 – 4:24:32 = 0:0:10

c. Calculating the difference: 4:25:57 – 4:18:44 = 0:07:13

1.4 Factorizations

1. Factoring the number: $288 = 12 \bullet 24 = (3 \bullet 4) \bullet (4 \bullet 6) = (3 \bullet 2 \bullet 2) \bullet (2 \bullet 2 \bullet 3 \bullet 2) = 2^5 \bullet 3^2$

3. Factoring the number: $210 = 10 \bullet 21 = (2 \bullet 5) \bullet (3 \bullet 7) = 2 \bullet 3 \bullet 5 \bullet 7$

5. Factoring the number: $1,925 = 25 \bullet 77 = (5 \bullet 5) \bullet (7 \bullet 11) = 5^2 \bullet 7 \bullet 11$

7. Factoring the number: $598 = 2 \bullet 299 = 2 \bullet (13 \bullet 23) = 2 \bullet 13 \bullet 23$

9. Reducing the fraction: $\dfrac{165}{385} = \dfrac{3 \bullet 5 \bullet 11}{5 \bullet 7 \bullet 11} = \dfrac{3}{7}$

11. Reducing the fraction: $\dfrac{385}{735} = \dfrac{5 \bullet 7 \bullet 11}{3 \bullet 5 \bullet 7 \bullet 7} = \dfrac{11}{3 \bullet 7} = \dfrac{11}{21}$

13. Factoring the trinomial: $x^2 - 2x - 24 = (x - 6)(x + 4)$

15. Factoring the trinomial: $x^2 - 10x + 25 = (x - 5)(x - 5) = (x - 5)^2$

17. Factoring the trinomial: $21x^2 - 23x + 6 = (7x - 3)(3x - 2)$

19. Factoring the difference of squares: $x^2 - 16 = (x + 4)(x - 4)$

21. Factoring the difference of squares: $a^2 - 1 = (a + 1)(a - 1)$

23. Factoring the difference of squares: $a^2 - 16b^2 = (a + 4b)(a - 4b)$

25. Factoring the difference of squares: $9x^2 - 49 = (3x + 7)(3x - 7)$

27. Factoring the difference of squares: $16x^4 - 49 = (4x^2 + 7)(4x^2 - 7)$

29. Factoring the difference of squares: $t^4 - 81 = (t^2 + 9)(t^2 - 9) = (t^2 + 9)(t + 3)(t - 3)$

31. Factoring the sum of cubes: $x^3 + y^3 = (x + y)(x^2 - xy + y^2)$

33. Factoring the difference of cubes: $8x^3 - 27y^3 = (2x - 3y)(4x^2 + 6xy + 9y^2)$

35. Factoring the sum of cubes: $t^3 + \dfrac{1}{27} = \left(t + \dfrac{1}{3}\right)\left(t^2 - \dfrac{1}{3}t + \dfrac{1}{9}\right)$

37. Factoring the sum of cubes: $64a^3 + 125b^3 = (4a + 5b)(16a^2 - 20ab + 25b^2)$

39. Factoring the polynomial: $2x^3-5x^2-3x=x(2x^2-5x-3)=x(2x+1)(x-3)$

41. Factoring the polynomial: $x^3-2x^2-24x=x(x^2-2x-24)=x(x-6)(x+4)$

43. Factoring the polynomial: $100x^2-300x=100x(x-3)$

45. Factoring the polynomial: $20a^2-45=5(4a^2-9)=5(2a+3)(2a-3)$

47. Factoring the polynomial: $9a^3-16a=a(9a^2-16)=a(3a+4)(3a-4)$

49. Factoring the polynomial: $12y-2xy-2x^2y=-2y(x^2+x-6)=-2y(x+3)(x-2)$

51. Factoring by grouping: $ax+2x+3a+6=x(a+2)+3(a+2)=(a+2)(x+3)$

53. Factoring by grouping: $x^2-3ax-2x+6a=x(x-3a)-2(x-3a)=(x-3a)(x-2)$

55. Factoring by grouping:
$$4x^3+12x^2-9x-27=4x^2(x+3)-9(x+3)=(x+3)(4x^2-9)=(x+3)(2x+3)(2x-3)$$

57. Factoring by grouping:
$$2x^3+x^2-18x-9=x^2(2x+1)-9(2x+1)=(2x+1)(x^2-9)=(2x+1)(x+3)(x-3)$$

59. Factoring the polynomial: $4x^2-31x-8=(4x+1)(x-8)$

61. The polynomial x^2+49 does not factor (prime).

63. Factoring the polynomial: $150x^3+65x^2-280x=5x(30x^2+13x-56)=5x(5x+8)(6x-7)$

65. Factoring the polynomial: $24x^2+2x-5=(12x-5)(2x+1)$

67. Factoring the polynomial: $x^6-1=(x^3+1)(x^3-1)=(x+1)(x^2-x+1)(x-1)(x^2+x+1)$

69. Factoring the polynomial:
$$12a^2(x-7)-75(x-7)=3(x-7)(4a^2-25)=3(x-7)(2a+5)(2a-5)$$

71. Factoring the polynomial: $15t^2+t-16=(15t+16)(t-1)$

73. Factoring the polynomial: $100x^2-100x-600=100(x^2-x-6)=100(x-3)(x+2)$

75. Factoring the polynomial: $4x^3+16xy^2=4x(x^2+4y^2)$

77. Factoring the polynomial: $30x^2+97x+77=(5x+7)(6x+11)$

79. Factoring the right side: $h=96+80t-16t^2=-16(t^2-5t-6)=-16(t-6)(t+1)$
Evaluating when $t=6$: $h=-16(6-6)(6+1)=-16(0)(7)=0$ feet
Evaluating when $t=3$: $h=-16(3-6)(3+1)=-16(-3)(4)=192$ feet

81. Using factoring by grouping:
$$P+Pr+(P+Pr)r=(P+Pr)+(P+Pr)r=(P+Pr)(1+r)=P(1+r)(1+r)=P(1+r)^2$$

1.5 Quotients

1. Simplifying the expression: $\frac{0-4}{0-2}=\frac{-4}{-2}=2$

3. Simplifying the expression: $\frac{-4-4}{-4-2}=\frac{-8}{-6}=\frac{4}{3}$

5. Simplifying the expression: $\frac{-6+6}{-6-3}=\frac{0}{-9}=0$

7. Simplifying the expression: $\frac{-2-4}{2-2}=\frac{-6}{0}$, which is undefined

9. Simplifying the expression: $\frac{3-(-1)}{-3-3}=\frac{3+1}{-6}=\frac{4}{-6}=-\frac{2}{3}$

11. Simplifying the expression: $\frac{-7}{0}$ is undefined

13. Simplifying the expression: $\frac{-3+9}{2 \bullet 5-10}=\frac{6}{10-10}=\frac{6}{0}$, which is undefined

15. Simplifying the expression: $\frac{15(-5)-25}{2(-10)}=\frac{-75-25}{-20}=\frac{-100}{-20}=5$

17. Simplifying the expression: $\frac{3(-1)-4(-2)}{8-5}=\frac{-3+8}{3}=\frac{5}{3}$

19. Simplifying the expression: $\frac{5^2-2^2}{-5+2}=\frac{25-4}{-3}=\frac{21}{-3}=-7$

21. Simplifying the expression: $\frac{(8-4)^2}{8^2-4^2}=\frac{4^2}{64-16}=\frac{16}{48}=\frac{1}{3}$

23. Simplifying the expression: $\frac{3 \bullet 10^2+4 \bullet 10+5}{345}=\frac{300+40+5}{345}=\frac{345}{345}=1$

25. Simplifying the expression: $\frac{6(-4)-2(5-8)}{-6-3-5}=\frac{-24-2(-3)}{-14}=\frac{-24+6}{-14}=\frac{-18}{-14}=\frac{9}{7}$

27. Simplifying the expression:

$$\frac{1}{2}\left(\frac{1.2}{1.4}-1\right)=\frac{1}{2}\left(\frac{12}{14}-1\right)=\frac{1}{2}\left(\frac{6}{7}-1\right)=\frac{1}{2}\left(\frac{6}{7}-\frac{7}{7}\right)=\frac{1}{2}\left(-\frac{1}{7}\right)=-\frac{1}{14}\approx -0.07$$

29. Simplifying the expression: $\frac{(6.8)(3.9)}{7.8}=\frac{26.52}{7.8}=3.4$

31. Simplifying the expression: $\frac{0.0005(200)}{(0.25)^2}=\frac{0.1}{0.0625}=1.6$

33. a. Adding: $50+(-80)=-30$

    b. Subtracting: $50-(-80)=50+80=130$

    c. Multiplying: $50(-80)=-4{,}000$

    d. Dividing: $\frac{50}{-80}=-\frac{5}{8}=-0.625$

35. a. Adding: $\frac{3}{4}+\left(-\frac{1}{2}\right)=\frac{3}{4}+\left(-\frac{2}{4}\right)=\frac{1}{4}$

b. Subtracting: $\frac{3}{4}-\left(-\frac{1}{2}\right)=\frac{3}{4}+\frac{1}{2}=\frac{3}{4}+\frac{2}{4}=\frac{5}{4}$

c. Multiplying: $\frac{3}{4}\left(-\frac{1}{2}\right)=-\frac{3}{8}$

d. Dividing: $\frac{3}{4}\div\left(-\frac{1}{2}\right)=\frac{3}{4}\left(-\frac{2}{1}\right)=-\frac{3}{2}$

37. Completing the table:

		Sum	Difference	Product	Quotient
a	b	$a+b$	$a-b$	ab	$\frac{a}{b}$
3	12	15	−9	36	$\frac{1}{4}$
−3	12	9	−15	−36	$-\frac{1}{4}$
3	−12	−9	15	−36	$-\frac{1}{4}$
−3	−12	−15	9	36	$\frac{1}{4}$

39. Writing with positive exponents: $3^{-2}=\frac{1}{3^2}=\frac{1}{9}$

41. Writing with positive exponents: $(-2)^{-5}=\frac{1}{(-2)^5}=-\frac{1}{32}$

43. Writing with positive exponents: $\left(\frac{3}{4}\right)^{-2}=\left(\frac{4}{3}\right)^{2}=\frac{16}{9}$

45. Writing with positive exponents: $\left(\frac{1}{3}\right)^{-2}+\left(\frac{1}{2}\right)^{-3}=3^2+2^3=9+8=17$

47. Simplifying: $x^{-4}x^{7}=x^{-4+7}=x^{3}$

49. Simplifying: $\left(a^2b^{-5}\right)^3=a^6b^{-15}=\frac{a^6}{b^{15}}$

51. Simplifying: $\left(5y^4\right)^{-3}\left(2y^{-2}\right)^3=5^{-3}y^{-12}2^3y^{-6}=\frac{2^3}{5^3}y^{-18}=\frac{8}{125y^{18}}$

53. Simplifying: $\frac{x^{-1}}{x^9}=x^{-1-9}=x^{-10}=\frac{1}{x^{10}}$

55. Simplifying: $\frac{a^4}{a^{-6}}=a^{4-(-6)}=a^{4+6}=a^{10}$

57. Simplifying: $\frac{t^{-10}}{t^{-4}}=t^{-10-(-4)}=t^{-10+4}=t^{-6}=\frac{1}{t^6}$

59. Simplifying: $\left(\frac{x^5}{x^3}\right)^6=\left(x^{5-3}\right)^6=\left(x^2\right)^6=x^{12}$

61. Simplifying: $\frac{\left(x^5\right)^6}{\left(x^3\right)^4}=\frac{x^{30}}{x^{12}}=x^{30-12}=x^{18}$

63. Simplifying: $\dfrac{(x^{-2})^3(x^3)^{-2}}{x^{10}} = \dfrac{x^{-6}x^{-6}}{x^{10}} = \dfrac{x^{-12}}{x^{10}} = x^{-12-10} = x^{-22} = \dfrac{1}{x^{22}}$

65. Simplifying: $\dfrac{5a^8b^3}{20a^5b^{-4}} = \frac{5}{20}a^{8-5}b^{3-(-4)} = \frac{1}{4}a^3b^7 = \dfrac{a^3b^7}{4}$

67. Simplifying: $\dfrac{(3x^{-2}y^8)^4}{(9x^4y^{-3})^2} = \dfrac{81x^{-8}y^{32}}{81x^8y^{-6}} = x^{-8-8}y^{32+6} = x^{-16}y^{38} = \dfrac{y^{38}}{x^{16}}$

69. Simplifying: $\left(\dfrac{8x^2y}{4x^4y^{-3}}\right)^4 = (2x^{2-4}y^{1+3})^4 = (2x^{-2}y^4)^4 = 16x^{-8}y^{16} = \dfrac{16y^{16}}{x^8}$

71. Simplifying: $\left(\dfrac{x^{-5}y^2}{x^{-3}y^5}\right)^{-2} = (x^{-5+3}y^{2-5})^{-2} = (x^{-2}y^{-3})^{-2} = x^4y^6$

73. Simplifying: $\dfrac{(3x-5)-(3a-5)}{x-a} = \dfrac{3x-5-3a+5}{x-a} = \dfrac{3x-3a}{x-a} = \dfrac{3(x-a)}{x-a} = 3$

75. Simplifying: $\dfrac{(x^2-4)-(a^2-4)}{x-a} = \dfrac{x^2-4-a^2+4}{x-a} = \dfrac{x^2-a^2}{x-a} = \dfrac{(x+a)(x-a)}{x-a} = x+a$

77. Completing the table:

x	$\dfrac{x-3}{3-x}$
–2	–1
–1	–1
0	–1
1	–1
2	–1

79. Completing the table:

x	$\dfrac{x-5}{x^2-25}$	$\dfrac{1}{x+5}$
0	$\frac{1}{5}$	$\frac{1}{5}$
2	$\frac{1}{7}$	$\frac{1}{7}$
–2	$\frac{1}{3}$	$\frac{1}{3}$
5	undefined	$\frac{1}{10}$
–5	undefined	undefined

1.6 Unit Analysis and Scientific Notation

1. Converting to miles: $14,494 \text{ feet} \bullet \dfrac{1 \text{ mile}}{5,280 \text{ feet}} \approx 2.7 \text{ miles}$

3. Converting to miles per hour: $\dfrac{1088 \text{ feet}}{1 \text{ second}} \bullet \dfrac{1 \text{ mile}}{5280 \text{ feet}} \bullet \dfrac{3600 \text{ seconds}}{1 \text{ hour}} \approx 742 \text{ miles per hour}$

5. Finding the speed: $\dfrac{785 \text{ feet}}{20 \text{ minutes}} \bullet \dfrac{1 \text{ mile}}{5280 \text{ feet}} \bullet \dfrac{60 \text{ minutes}}{1 \text{ hour}} \approx 0.45 \text{ miles per hour}$

7. Converting to miles per hour:

$$\frac{100 \text{ yards}}{10.5 \text{ seconds}} \bullet \frac{3 \text{ feet}}{1 \text{ yard}} \bullet \frac{1 \text{ mile}}{5280 \text{ feet}} \bullet \frac{3600 \text{ seconds}}{1 \text{ hour}} \approx 19.5 \text{ miles per hour}$$

9. Converting the units: $10,000 \text{ steps} \bullet \dfrac{2.5 \text{ feet}}{1 \text{ step}} \bullet \dfrac{1 \text{ mile}}{5,280 \text{ feet}} \approx 4.7 \text{ miles}$

Yes, the facts are essentially correct.

11. Converting each area to acres:

Smallest: $55 \bullet 100 \text{ square yards} \bullet \dfrac{9 \text{ square feet}}{1 \text{ square yards}} \bullet \dfrac{1 \text{ acre}}{43,560 \text{ square feet}} \approx 1.1 \text{ acres}$

Rose Bowl: $72 \bullet 116 \text{ square yards} \bullet \dfrac{9 \text{ square feet}}{1 \text{ square yards}} \bullet \dfrac{1 \text{ acre}}{43,560 \text{ square feet}} \approx 1.7 \text{ acres}$

Largest: $75 \bullet 120 \text{ square yards} \bullet \dfrac{9 \text{ square feet}}{1 \text{ square yards}} \bullet \dfrac{1 \text{ acre}}{43,560 \text{ square feet}} \approx 1.9 \text{ acres}$

13. Finding the total grams: $80 \text{ vitamins} \bullet \dfrac{30 \text{ mg}}{1 \text{ vitamin}} \bullet \dfrac{1 \text{ g}}{1000 \text{ mg}} = 2.4 \text{ g}$

15. Finding the total grams: $240 \text{ tablets} \bullet \dfrac{500 \text{ mg}}{1 \text{ tablet}} \bullet \dfrac{1 \text{ g}}{1000 \text{ mg}} = 120 \text{ g}$

17. Completing the table:

Expanded Form	Scientific Notation $(n \times 10^r)$
0.000357	3.57×10^{-4}
0.00357	3.57×10^{-3}
0.0357	3.57×10^{-2}
0.357	3.57×10^{-1}
3.57	3.57×10^{0}
35.7	3.57×10^{1}
357	3.57×10^{2}
3,570	3.57×10^{3}
35,700	3.57×10^{4}

19. Completing the table:

Jupiter's Moon	Period (seconds)	
Io	153,000	1.53×10^5
Europa	307,000	3.07×10^5
Ganymede	618,000	6.18×10^5
Callisto	1,440,000	1.44×10^6

21. Writing in scientific notation: $630{,}000{,}000 = 6.3 \times 10^8$ seconds

23. **a.** For Google, the number is: $350{,}000{,}000 = 3.5 \times 10^8$
    b. For Yahoo, the number is: $200{,}000{,}000 = 2.0 \times 10^8$
    c. For LookSmart, the number is: $75{,}000{,}000 = 7.5 \times 10^7$

25. Multiplying to find the distance:
$$(1.7 \times 10^6 \text{ light-years})(5.9 \times 10^{12} \text{ miles/light-year}) \approx 1.0 \times 10^{19} \text{ miles}$$

27. **a.** Writing in scientific notation: $\$4.22 \times 10^{11}$
    b. Dividing: $\dfrac{\$4.22 \times 10^{11}}{6 \times 10^7} \approx \$7.033 \times 10^3$.
    The average credit card debt was $7,033 per household.

29. Finding the product: $(3 \times 10^3)(2 \times 10^5) = 6 \times 10^8$

31. Finding the product: $(3.5 \times 10^4)(5 \times 10^{-6}) = 17.5 \times 10^{-2} = 1.75 \times 10^{-1}$

33. Finding the product: $(5.5 \times 10^{-3})(2.2 \times 10^{-4}) = 12.1 \times 10^{-7} = 1.21 \times 10^{-6}$

35. Finding the quotient: $\dfrac{8.4 \times 10^5}{2 \times 10^2} = 4.2 \times 10^3$

37. Finding the quotient: $\dfrac{6 \times 10^8}{2 \times 10^{-2}} = 3 \times 10^{10}$

39. Finding the quotient: $\dfrac{2.5 \times 10^{-6}}{5 \times 10^{-4}} = 0.5 \times 10^{-2} = 5 \times 10^{-3}$

41. Simplifying: $\dfrac{(6 \times 10^8)(3 \times 10^5)}{9 \times 10^7} = \dfrac{18 \times 10^{13}}{9 \times 10^7} = 2 \times 10^6$

43. Simplifying: $\dfrac{(5 \times 10^3)(4 \times 10^{-5})}{2 \times 10^{-2}} = \dfrac{20 \times 10^{-2}}{2 \times 10^{-2}} = 10 = 1 \times 10^1$

45. Simplifying: $\dfrac{(2.8 \times 10^{-7})(3.6 \times 10^4)}{2.4 \times 10^3} = \dfrac{10.08 \times 10^{-3}}{2.4 \times 10^3} = 4.2 \times 10^{-6}$

Chapter 1 Test

See www.mathtv.com for video solutions to all problems in this chapter test.

Chapter 2
Equations and Inequalities in One Variable

2.1 Linear and Quadratic Equations

1. Solving the equation:

$$\begin{aligned} 7y-4&=2y+11\\ 5y-4&=11\\ 5y&=15\\ y&=3 \end{aligned}$$

3. Solving the equation:

$$\begin{aligned} -\frac{2}{5}x+\frac{2}{15}&=\frac{2}{3}\\ 15\left(-\frac{2}{5}x+\frac{2}{15}\right)&=15\left(\frac{2}{3}\right)\\ -6x+2&=10\\ -6x&=8\\ x&=-\frac{4}{3} \end{aligned}$$

5. Solving the equation:

$$\begin{aligned} 0.14x+0.08(10{,}000-x)&=1220\\ 0.14x+800-0.08x&=1220\\ 0.06x+800&=1220\\ 0.06x&=420\\ x&=7{,}000 \end{aligned}$$

7. Solving the equation:

$$\begin{aligned} 5(y+2)-4(y+1)&=3\\ 5y+10-4y-4&=3\\ y+6&=3\\ y&=-3 \end{aligned}$$

9. Solving the equation:

$$\begin{aligned} x^2-5x-6&=0\\ (x+1)(x-6)&=0\\ x&=-1,6 \end{aligned}$$

11. Solving the equation:

$$\begin{aligned} 9a^3&=16a\\ 9a^3-16a&=0\\ a(9a^2-16)&=0\\ a(3a+4)(3a-4)&=0\\ a&=-\frac{4}{3},0,\frac{4}{3} \end{aligned}$$

13. Solving the equation:

$$\begin{aligned} (x+6)(x-2)&=-7\\ x^2+4x-12&=-7\\ x^2+4x-5&=0\\ (x+5)(x-1)&=0\\ x&=-5,1 \end{aligned}$$

15. Solving the equation:

$$\begin{aligned} 2y^3-9y&=-3y^2\\ 2y^3+3y^2-9y&=0\\ y(2y^2+3y-9)&=0\\ y(2y-3)(y+3)&=0\\ y&=-3,0,\frac{3}{2} \end{aligned}$$

17. Solving the equation:

$$4x^3+12x^2-9x-27=0$$
$$4x^2(x+3)-9(x+3)=0$$
$$(x+3)(4x^2-9)=0$$
$$(x+3)(2x+3)(2x-3)=0$$
$$x=-3,-\frac{3}{2},\frac{3}{2}$$

19. **a.** Solving the equation:

$$8x-5=0$$
$$8x=5$$
$$x=\frac{5}{8}$$

b. Adding: $(8x-5)+(2x-3)=10x-8$

c. Multiplying: $(8x-5)(2x-3)=16x^2-24x-10x+15=16x^2-34x+15$

d. Solving the equation:

$$16x^2-34x+15=0$$
$$(8x-5)(2x-3)=0$$
$$x=\frac{5}{8},\frac{3}{2}$$

21. **a.** Solving the equation:

$$9x-25=0$$
$$9x=25$$
$$x=\frac{25}{9}$$

b. Solving the equation:

$$9x^2-25=0$$
$$(3x+5)(3x-5)=0$$
$$x=-\frac{5}{3},\frac{5}{3}$$

c. Solving the equation:

$$9x^2-25=56$$
$$9x^2-81=0$$
$$9(x+3)(x-3)=0$$
$$x=-3,3$$

d. Solving the equation:

$$9x^2-25=30x-50$$
$$9x^2-30x+25=0$$
$$(3x-5)^2=0$$
$$3x-5=0$$
$$3x=5$$
$$x=\frac{5}{3}$$

23. Solving the equation:

$$-3-4x=15$$
$$-4x=18$$
$$x=-\frac{9}{2}$$

25. Solving the equation:

$$x^3-5x^2+6x=0$$
$$x(x^2-5x+6)=0$$
$$x(x-2)(x-3)=0$$
$$x=0,2,3$$

27. Solving the equation:

$$\begin{aligned} 0 &= 6{,}400a + 70 \\ -70 &= 6{,}400a \\ a &= -\frac{70}{6{,}400} = -\frac{7}{640} \end{aligned}$$

29. Solving the equation:

$$\begin{aligned} 5(2x+1) &= 12 \\ 10x + 5 &= 12 \\ 10x &= 7 \\ x &= \frac{7}{10} \end{aligned}$$

31. Solving the equation:

$$\begin{aligned} 100P &= 2{,}400 \\ P &= \frac{2{,}400}{100} = 24 \end{aligned}$$

33. Solving the equation:

$$\begin{aligned} 5\left(-\frac{19}{15}\right) + 5y &= 9 \\ -\frac{19}{3} + 5y &= 9 \\ 5y &= \frac{27}{3} + \frac{19}{3} = \frac{46}{3} \\ y &= \frac{46}{15} \end{aligned}$$

35. Solving the equation:

$$\begin{aligned} 3x^2 + x &= 10 \\ 3x^2 + x - 10 &= 0 \\ (3x-5)(x+2) &= 0 \\ x &= -2, \frac{5}{3} \end{aligned}$$

37. Solving the equation:

$$\begin{aligned} (y+3)^2 + y^2 &= 9 \\ y^2 + 6y + 9 + y^2 &= 9 \\ 2y^2 + 6y &= 0 \\ 2y(y+3) &= 0 \\ y &= -3, 0 \end{aligned}$$

39. Solving the equation:

$$\begin{aligned} 15 - 3(x-1) &= x - 2 \\ 15 - 3x + 3 &= x - 2 \\ -3x + 18 &= x - 2 \\ -4x + 18 &= -2 \\ -4x &= -20 \\ x &= 5 \end{aligned}$$

41. Solving the equation:

$$\begin{aligned} 2(20+x) &= 3(20-x) \\ 40 + 2x &= 60 - 3x \\ 40 + 5x &= 60 \\ 5x &= 20 \\ x &= 4 \end{aligned}$$

43. Solving the equation:

$$\begin{aligned} 0.08x + 0.09(9{,}000 - x) &= 750 \\ 0.08x + 810 - 0.09x &= 750 \\ -0.01x + 810 &= 750 \\ -0.01x &= -60 \\ x &= 6{,}000 \end{aligned}$$

45. Solving the equation:

$$\begin{aligned} (x+3)^2 + 1^2 &= 2 \\ x^2 + 6x + 9 + 1 &= 2 \\ x^2 + 6x + 8 &= 0 \\ (x+4)(x+2) &= 0 \\ x &= -4, -2 \end{aligned}$$

47. Solving the equation:

$$\begin{aligned} 3x - 6 &= 3(x+4) \\ 3x - 6 &= 3x + 12 \\ -6 &= 12 \end{aligned}$$

Since this statement is false, there is no solution.

49. Solving the equation:

$$\begin{aligned} 2(4t-1)+3 &= 5t+4+3t \\ 8t-2+3 &= 8t+4 \\ 8t+1 &= 8t+4 \\ 1 &= 4 \end{aligned}$$

Since this statement is false, there is no solution.

51. Solving the equation:

$$\begin{aligned} 7(x+2)-4(2x-1) &= 18-x \\ 7x+14-8x+4 &= 18-x \\ -x+18 &= -x+18 \\ 18 &= 18 \end{aligned}$$

Since this statement is true, the solution is all real numbers.

53. Solving the equation:

$$\begin{aligned} -35 &= -0.0035A+70 \\ -105 &= -0.0035A \\ A &= \frac{-105}{-0.0035} = 30{,}000 \end{aligned}$$

The altitude is 30,000 feet.

55. Solving the equation:

$$\begin{aligned} x \bullet 42 &= 21 \\ x &= \frac{21}{42} = \frac{1}{2} \end{aligned}$$

57. Solving the equation:

$$\begin{aligned} 25 &= 0.4x \\ x &= \frac{25}{0.4} = 62.5 \end{aligned}$$

59. Solving the equation:

$$\begin{aligned} 12-4y &= 12 \\ -4y &= 0 \\ y &= 0 \end{aligned}$$

61. Solving the equation:

$$\begin{aligned} 525 &= 900-300p \\ -375 &= -300p \\ p &= \frac{-375}{-300} = \frac{5}{4} \end{aligned}$$

63. Solving the equation:

$$\begin{aligned} 48 &= 64t-16t^2 \\ 16t^2-64t+48 &= 0 \\ 16(t^2-4t+3) &= 0 \\ 16(t-1)(t-3) &= 0 \\ t &= 1, 3 \end{aligned}$$

65. Solving the equation:

$$\begin{aligned} 486.7 &= 78.5+31.4h \\ 408.2 &= 31.4h \\ h &= \frac{408.2}{31.4} = 13 \end{aligned}$$

2.2 Formulas

1. Substituting $x = 0$:

$$3(0) - 4y = 12$$
$$-4y = 12$$
$$y = -3$$

3. Substituting $x = 4$:

$$3(4) - 4y = 12$$
$$12 - 4y = 12$$
$$-4y = 0$$
$$y = 0$$

5. Substituting $y = 0$:

$$2x - 3 = 0$$
$$2x = 3$$
$$x = \frac{3}{2}$$

7. Substituting $y = 5$:

$$2x - 3 = 5$$
$$2x = 8$$
$$x = 4$$

9. Substituting $y = -\frac{6}{5}$:

$$x - 2\left(-\frac{6}{5}\right) = 4$$
$$x + \frac{12}{5} = 4$$
$$x = \frac{8}{5}$$

11. Substituting $x = 160$ and $y = 0$:

$$0 = a(160 - 80)^2 + 70$$
$$-70 = a(80)^2$$
$$6{,}400a = -70$$
$$a = -\frac{70}{6{,}400} = -\frac{7}{640}$$

13. Substituting $p = 1.5$: $R = (900 - 300 \bullet 1.5)(1.5) = (450)(1.5) = 675$

15. **a.** Substituting $x = 100$: $P = -0.1(100)^2 + 27(100) + 1{,}700 = -1{,}000 + 2{,}700 + 1{,}700 = 3{,}400$

b. Substituting $x = 170$: $P = -0.1(170)^2 + 27(170) - 1{,}700 = -2{,}890 + 4{,}590 + 1{,}700 = 3{,}400$

17. **a.** Substituting $t = \frac{1}{4}$: $h = 16 + 32\left(\frac{1}{4}\right) - 16\left(\frac{1}{4}\right)^2 = 16 + 8 - 1 = 23$

b. Substituting $t = \frac{7}{4}$: $h = 16 + 32\left(\frac{7}{4}\right) - 16\left(\frac{7}{4}\right)^2 = 16 + 56 - 49 = 23$

19. Substituting $d = 30$, $r = 12$, and $t = 3$:

$$30 = (12 - c) \bullet 3$$
$$10 = 12 - c$$
$$c = 2$$

21. Substituting $x = 5$ and $y = 15$:

$$15 = K(5)$$
$$K = 3$$

23. Substituting $P = 48$ and $V = 50$:

$$50 = \frac{K}{48}$$
$$K = 50 \bullet 48 = 2{,}400$$

25. Substituting $x = 2$:

$$5(2) - 3y = -15$$
$$10 - 3y = -15$$
$$-3y = -25$$
$$y = \frac{25}{3}$$

27. Substituting $x=-\frac{1}{5}$:

$$\begin{aligned} 5\left(-\frac{1}{5}\right)-3y&=-15\\ -1-3y&=-15\\ -3y&=-14\\ y&=\frac{14}{3} \end{aligned}$$

29. Solving for r:

$$\begin{aligned} d&=rt\\ r&=\frac{d}{t} \end{aligned}$$

31. Solving for t:

$$\begin{aligned} d&=(r+c)t\\ t&=\frac{d}{r+c} \end{aligned}$$

33. Solving for l:

$$\begin{aligned} A&=lw\\ l&=\frac{A}{w} \end{aligned}$$

35. Solving for t:

$$\begin{aligned} I&=prt\\ t&=\frac{I}{pr} \end{aligned}$$

37. Solving for T:

$$\begin{aligned} PV&=nRT\\ T&=\frac{PV}{nR} \end{aligned}$$

39. Solving for x:

$$\begin{aligned} y&=mx+b\\ y-b&=mx\\ x&=\frac{y-b}{m} \end{aligned}$$

41. Solving for F:

$$\begin{aligned} C&=\frac{5}{9}(F-32)\\ \frac{9}{5}C&=F-32\\ F&=\frac{9}{5}C+32 \end{aligned}$$

43. Solving for v:

$$\begin{aligned} h&=vt+16t^2\\ h-16t^2&=vt\\ v&=\frac{h-16t^2}{t} \end{aligned}$$

45. Solving for d:

$$\begin{aligned} A&=a+(n-1)d\\ A-a&=(n-1)d\\ d&=\frac{A-a}{n-1} \end{aligned}$$

47. Solving for y:

$$\begin{aligned} 2x+3y&=6\\ 3y&=-2x+6\\ y&=\frac{-2x+6}{3}\\ y&=-\frac{2}{3}x+2 \end{aligned}$$

49. Solving for y:

$$\begin{aligned} -3x+5y&=15\\ 5y&=3x+15\\ y&=\frac{3x+15}{5}\\ y&=\frac{3}{5}x+3 \end{aligned}$$

51. Solving for y:

$$\begin{aligned} 2x-6y+12&=0\\ -6y&=-2x-12\\ y&=\frac{-2x-12}{-6}\\ y&=\frac{1}{3}x+2 \end{aligned}$$

53. Solving for x:

$$\begin{aligned} ax+4&=bx+9\\ ax-bx+4&=9\\ ax-bx&=5\\ x(a-b)&=5\\ x&=\frac{5}{a-b} \end{aligned}$$

55. Solving for h:

$$\begin{aligned} S&=\pi r^2+2\pi rh\\ S-\pi r^2&=2\pi rh\\ h&=\frac{S-\pi r^2}{2\pi r} \end{aligned}$$

57. Solving for x:

$$\begin{aligned} -3x+4y&=12\\ -3x&=-4y+12\\ x&=\frac{-4y+12}{-3}\\ x&=\frac{4}{3}y-4 \end{aligned}$$

59. Solving for x:

$$\begin{aligned} ax+3&=cx-7\\ ax-cx+3&=-7\\ ax-cx&=-10\\ x(a-c)&=-10\\ x&=\frac{-10}{a-c}=\frac{10}{c-a} \end{aligned}$$

61. Solving for y:

$$\begin{aligned} x&=2y-3\\ 2y&=x+3\\ y&=\frac{x+3}{2}=\frac{1}{2}x+\frac{3}{2} \end{aligned}$$

63. Solving for y:

$$\begin{aligned} y-3&=-2(x+4)\\ y-3&=-2x-8\\ y&=-2x-5 \end{aligned}$$

65. Solving for y:

$$\begin{aligned} y-3&=-\frac{2}{3}(x+3)\\ y-3&=-\frac{2}{3}x-2\\ y&=-\frac{2}{3}x+1 \end{aligned}$$

67. Solving for y:

$$\begin{aligned} y-4&=-\frac{1}{2}(x+1)\\ y-4&=-\frac{1}{2}x-\frac{1}{2}\\ y&=-\frac{1}{2}x+\frac{7}{2} \end{aligned}$$

69. **a.** Solving for y:

$$\frac{y+1}{x-0}=4$$
$$y+1=4(x-0)$$
$$y+1=4x$$
$$y=4x-1$$

b. Solving for y:

$$\frac{y+2}{x-4}=-\frac{1}{2}$$
$$y+2=-\frac{1}{2}(x-4)$$
$$y+2=-\frac{1}{2}x+2$$
$$y=-\frac{1}{2}x$$

c. Solving for y:

$$\frac{y+3}{x-7}=0$$
$$y+3=0(x-7)$$
$$y+3=0$$
$$y=-3$$

71. Solving for y:

$$\frac{x}{8}+\frac{y}{2}=1$$
$$8\left(\frac{x}{8}+\frac{y}{2}\right)=8(1)$$
$$x+4y=8$$
$$4y=-x+8$$
$$y=-\frac{1}{4}x+2$$

73. Solving for y:

$$\frac{x}{5}+\frac{y}{-3}=1$$
$$15\left(\frac{x}{5}+\frac{y}{-3}\right)=15(1)$$
$$3x-5y=15$$
$$-5y=-3x+15$$
$$y=\frac{3}{5}x-3$$

75. **a.** Solving the equation:

$$-4x+5=20$$
$$-4x=15$$
$$x=-\frac{15}{4}=-3.75$$

b. Substituting $x=3$: $-4x+5=-4(3)+5=-7$

c. Solving for y:

$$-4x+5y=20$$
$$5y=4x+20$$
$$y=\frac{4}{5}x+4$$

d. Solving for x:

$$-4x+5y=20$$
$$-4x=-5y+20$$
$$x=\frac{5}{4}y-5$$

77. Substituting $A=30$, $P=30$, and $N=4$: $W=\frac{APN}{2000}=\frac{(30)(30)4}{2000}=\frac{9}{5}=1.8$ tons

79. Substituting $h = 192$:

$$\begin{aligned}192 &= 96 + 80t - 16t^2 \\ 0 &= -16t^2 + 80t - 96 \\ 0 &= -16(t^2 - 5t + 6) \\ 0 &= -16(t-2)(t-3) \\ t &= 2, 3\end{aligned}$$

The bullet will be 192 feet in the air after 2 sec and 3 sec.

81. Let c represent the rate of the current. The equation is:

$$\begin{aligned}2(15 - c) &= 18 \\ 30 - 2c &= 18 \\ -2c &= -12 \\ c &= 6\end{aligned}$$

The speed of the current is 6 mph.

83. Let w represent the rate of the wind. The equation is:

$$\begin{aligned}4(258 - w) &= 864 \\ 1032 - 4w &= 864 \\ -4w &= -168 \\ w &= 42\end{aligned}$$

The speed of the wind is 42 mph.

85. The distance traveled by the rider is the circumference: $C = \pi(65) \approx (3.14)(65) \approx 204.1$ feet

Finding the rate: $\dfrac{204.1 \text{ feet}}{30 \text{ seconds}} \approx 6.8$ feet per second

87. Substituting $R = \$7{,}000$:

$$\begin{aligned}R &= xp \\ 7000 &= (1700 - 100p)p \\ 7000 &= 1700p - 100p^2 \\ 0 &= -100p^2 + 1700p - 7000 \\ 0 &= -100(p^2 - 17p + 70) \\ 0 &= -100(p-7)(p-10) \\ p &= 7, 10\end{aligned}$$

The calculators should be sold for either \$7 or \$10.

89. Substituting the values: $S = \dfrac{480 \bullet 216 \bullet 30 \bullet 150}{35000} = 13{,}330$ kilobytes

91. Substituting $n = 1$, $y = 7$, and $z = 15$:

$$\begin{aligned}x^1 + 7^1 &= 15^1 \\ x + 7 &= 15 \\ x &= 8\end{aligned}$$

93. For Shar, $M = 220 - 46 = 174$ and $R = 60$:

$$T = R + 0.6(M - R) = 60 + 0.6(174 - 60) = 128.4 \text{ beats per minute}$$

For Sara, $M = 220 - 26 = 194$ and $R = 60$:

$$T = R + 0.6(M - R) = 60 + 0.6(194 - 60) = 140.4 \text{ beats per minute}$$

95. Translating into symbols: $2x - 3$

97. Translating into symbols: $x + y = 180$

99. Solving the equation:

$$\begin{aligned} x + 2x &= 90 \\ 3x &= 90 \\ x &= 30 \end{aligned}$$

101. Solving the equation:

$$\begin{aligned} 2(2x - 3) + 2x &= 45 \\ 4x - 6 + 2x &= 45 \\ 6x - 6 &= 45 \\ 6x &= 51 \\ x &= \frac{51}{6} = 8.5 \end{aligned}$$

103. Solving the equation:

$$\begin{aligned} 6x + 5(10{,}000 - x) &= 56{,}000 \\ 6x + 50{,}000 - 5x &= 56{,}000 \\ x + 50{,}000 &= 56{,}000 \\ x &= 6{,}000 \end{aligned}$$

2.3 Applications

1. Let w represent the width and $2w$ represent the length. Using the perimeter formula:

$$\begin{aligned} 2w + 2(2w) &= 60 \\ 2w + 4w &= 60 \\ 6w &= 60 \\ w &= 10 \end{aligned}$$

The dimensions are 10 feet by 20 feet.

3. Let s represent the side of the square. Using the perimeter formula:

$$\begin{aligned} 4s &= 28 \\ s &= 7 \end{aligned}$$

The length of each side is 7 feet.

5. Let x represent the shortest side, $x + 3$ represent the medium side, and $2x$ represent the longest side. Using the perimeter formula:

$$\begin{aligned} x + x + 3 + 2x &= 23 \\ 4x + 3 &= 23 \\ 4x &= 20 \\ x &= 5 \end{aligned}$$

The shortest side is 5 inches.

7. Let w represent the width and $2w-3$ represent the length. Using the perimeter formula:

$$2w+2(2w-3)=18$$
$$2w+4w-6=18$$
$$6w-6=18$$
$$6w=24$$
$$w=4$$

The width is 4 meters.

9. Let w represent the width and $2w$ represent the length. Using the perimeter formula:

$$2w+2(2w)=48$$
$$2w+4w=48$$
$$6w=48$$
$$w=8$$

The width is 8 feet and the length is 16 feet. Finding the cost:

$$C=1.75(32)+2.25(16)=56+36=92$$

The cost to build the pen is \$92.00.

11. Let b represent the amount of money Shane had at the beginning of the trip. Using the percent increase:

$$b+0.50b=300$$
$$1.5b=300$$
$$b=200$$

Shane had \$200.00 at the beginning of the trip.

13. Let c represent the cost for the bookstore. Using the markup equation:

$$c+0.33c=115$$
$$1.33c=115$$
$$c\approx 86.47$$

The cost to the bookstore was approximately \$86.47.

15. Let R represent the total box office receipts. The equation is:

$$0.53R=52.8$$
$$R\approx 99.6$$

The total receipts were approximately \$99.6 million.

17. Let x represent one angle and $8x$ represent the other angle. Since the angles are supplementary:

$$x+8x=180$$
$$9x=180$$
$$x=20$$

The two angles are 20° and 160°.

19. **a.** Let x represent one angle and $4x - 12$ represent the other angle. Since the angles are complementary:

$$x + 4x - 12 = 90$$
$$5x - 12 = 90$$
$$5x = 102$$
$$x = 20.4$$
$$4x - 12 = 4(20.4) - 12 = 69.6$$

The two angles are 20.4° and 69.6°.

b. Let x represent one angle and $4x - 12$ represent the other angle. Since the angles are supplementary:

$$x + 4x - 12 = 180$$
$$5x - 12 = 180$$
$$5x = 192$$
$$x = 38.4$$
$$4x - 12 = 4(38.4) - 12 = 141.6$$

The two angles are 38.4° and 141.6°.

21. Let x represent the smallest angle, $3x$ represent the largest angle, and $3x - 9$ represent the third angle. The equation is:

$$x + 3x + 3x - 9 = 180$$
$$7x - 9 = 180$$
$$7x = 189$$
$$x = 27$$

The three angles are 27°, 72° and 81°.

23. Let x represent the largest angle, $\frac{1}{3}x$ represent the smallest angle, and $\frac{1}{3}x + 10$ represent the third angle. The equation is:

$$\frac{1}{3}x + x + \frac{1}{3}x + 10 = 180$$
$$\frac{5}{3}x + 10 = 180$$
$$\frac{5}{3}x = 170$$
$$x = 102$$

The three angles are 34°, 44° and 102°.

25. Let x represent the measure of the two base angles, and $2x + 8$ represent the third angle. The equation is:

$$x + x + 2x + 8 = 180$$
$$4x + 8 = 180$$
$$4x = 172$$
$$x = 43$$

The three angles are 43°, 43° and 94°.

27. Let h represent the height the ladder makes with the building. Using the Pythagorean theorem:

$$7^2 + h^2 = 25^2$$
$$49 + h^2 = 625$$
$$h^2 = 576$$
$$h = 24$$

The ladder reaches a height of 24 feet along the building.

29. Let x, $x + 2$, and $x + 4$ represent the three sides. Using the Pythagorean theorem:

$$x^2 + (x+2)^2 = (x+4)^2$$
$$x^2 + x^2 + 4x + 4 = x^2 + 8x + 16$$
$$2x^2 + 4x + 4 = x^2 + 8x + 16$$
$$x^2 - 4x - 12 = 0$$
$$(x-6)(x+2) = 0$$
$$x = 6 \quad (x = -2 \text{ is impossible})$$

The lengths of the three sides are 6, 8, and 10.

31. Let w represent the width and $3w + 2$ represent the length. Using the area formula:

$$w(3w+2) = 16$$
$$3w^2 + 2w = 16$$
$$3w^2 + 2w - 16 = 0$$
$$(3w+8)(w-2) = 0$$
$$w = 2 \quad (w = -\frac{8}{3} \text{ is impossible})$$

The dimensions are 2 feet by 8 feet.

33. Let h represent the height and $4h + 2$ represent the base. Using the area formula:

$$\frac{1}{2}(4h+2)(h) = 36$$
$$4h^2 + 2h = 72$$
$$4h^2 + 2h - 72 = 0$$
$$2(2h^2 + h - 36) = 0$$
$$2(2h+9)(h-4) = 0$$
$$h = 4 \quad (h = -\frac{9}{2} \text{ is impossible})$$

The base is 18 inches and the height is 4 inches.

35. Let x represent the amount invested at 8% and $9000 - x$ represent the amount invested at 9%. The equation is:

$$\begin{aligned} 0.08x + 0.09(9000 - x) &= 750 \\ 0.08x + 810 - 0.09x &= 750 \\ -0.01x + 810 &= 750 \\ -0.01x &= -60 \\ x &= 6000 \end{aligned}$$

She invested \$6,000 at 8% and \$3,000 at 9%.

37. Let x represent the amount invested at 12% and $15000 - x$ represent the amount invested at 10%. The equation is:

$$\begin{aligned} 0.12x + 0.10(15000 - x) &= 1600 \\ 0.12x + 1500 - 0.10x &= 1600 \\ 0.02x + 1500 &= 1600 \\ 0.02x &= 100 \\ x &= 5000 \end{aligned}$$

The investment was \$5,000 at 12% and \$10,000 at 10%.

39. Let x represent the amount invested at 8% and $6000 - x$ represent the amount invested at 9%. The equation is:

$$\begin{aligned} 0.08x + 0.09(6000 - x) &= 500 \\ 0.08x + 540 - 0.09x &= 500 \\ -0.01x + 540 &= 500 \\ -0.01x &= -40 \\ x &= 4000 \end{aligned}$$

Stacey invested \$4,000 at 8% and \$2,000 at 9%.

41. Let t represent the required time. Using the Pythagorean theorem:

$$\begin{aligned} (15t)^2 + (20t)^2 &= 75^2 \\ 225t^2 + 400t^2 &= 5625 \\ 625t^2 &= 5625 \\ 625t^2 - 5625 &= 0 \\ 625(t^2 - 9) &= 0 \\ t &= -3, 3 \end{aligned}$$

Since the time must be positive, the time is 3 hours.

43. Completing the table:

Speed (miles per hour)	Distance (miles)
20	10
30	15
40	20
50	25
60	30
70	35

45. Completing the table:

Time (hours)	Distance Upstream (miles)	Distance Downstream (miles)
1	6	14
2	12	28
3	18	42
4	24	56
5	30	70
6	36	84

47. Let x represent the number of father tickets and $75 - x$ represent the number of son tickets. The equation is:

$$\begin{aligned} 2x + 1.5(75 - x) &= 127.5 \\ 2x + 112.5 - 1.5x &= 127.5 \\ 0.5x + 112.5 &= 127.5 \\ 0.5x &= 15 \\ x &= 30 \end{aligned}$$

So 30 fathers and 45 sons attended the breakfast.

49. The total money collected is: $\$1204 - \$250 = \$954$

Let x represent the amount of her sales (not including tax). Since this amount includes the tax collected, the equation is:

$$\begin{aligned} x + 0.06x &= 954 \\ 1.06x &= 954 \\ x &= 900 \end{aligned}$$

Her sales were \$900, so the sales tax is: $0.06(900) = \$54$

51. Let x represent the length of Patrick's call. He talks 1 minute at 40 cents and $x - 1$ minutes at 30 cents, so the equation is:

$$\begin{aligned} 40(1) + 30(x - 1) + 50 &= 1380 \\ 40 + 30x - 30 + 50 &= 1380 \\ 30x + 60 &= 1380 \\ 30x &= 1320 \\ x &= 44 \end{aligned}$$

Patrick talked for 44 minutes.

53. Completing the table:

t	0	$\frac{1}{4}$	1	$\frac{7}{4}$	2
h	0	7	16	7	0

55. Completing the table:

Year	Sales (billions of dollars)
2005	7
2006	7.5
2007	8
2008	8.6
2009	9.2

57. Completing the table:

w	l	A
2	22	44
4	20	80
6	18	108
8	16	128
10	14	140
12	12	144

59. Completing the table:

Age (years)	Maximum Heart Rate (beats per minute)
18	202
19	201
20	200
21	199
22	198
23	197

61. Completing the table:

Resting Heart Rate (beats per minute)	Training Heart Rate (beats per minute)
60	144
62	144.8
64	145.6
68	147.2
70	148
72	148.8

63. Graphing the inequality:

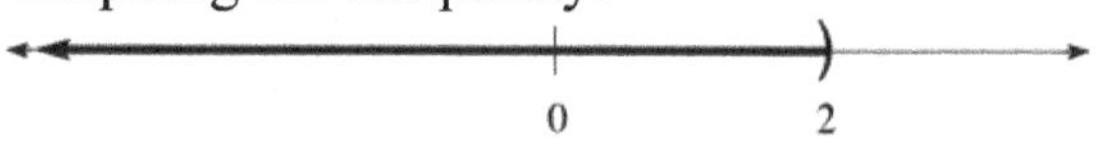

65. Graphing the inequality:

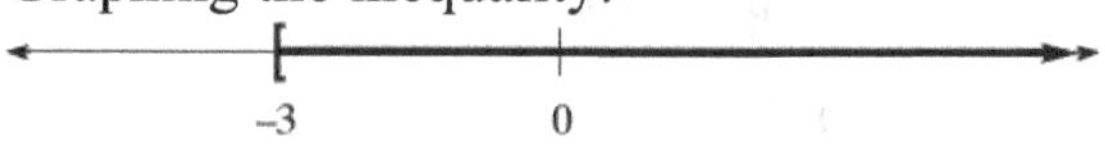

67. Solving the equation:

$$\begin{aligned} -2x-3&=7 \\ -2x&=10 \\ x&=-5 \end{aligned}$$

69. Solving the equation:

$$\begin{aligned} 3(2x-4)-7x&=-3x \\ 6x-12-7x&=-3x \\ -x-12&=-3x \\ -12&=-2x \\ x&=6 \end{aligned}$$

2.4 Linear Inequalities in One Variable

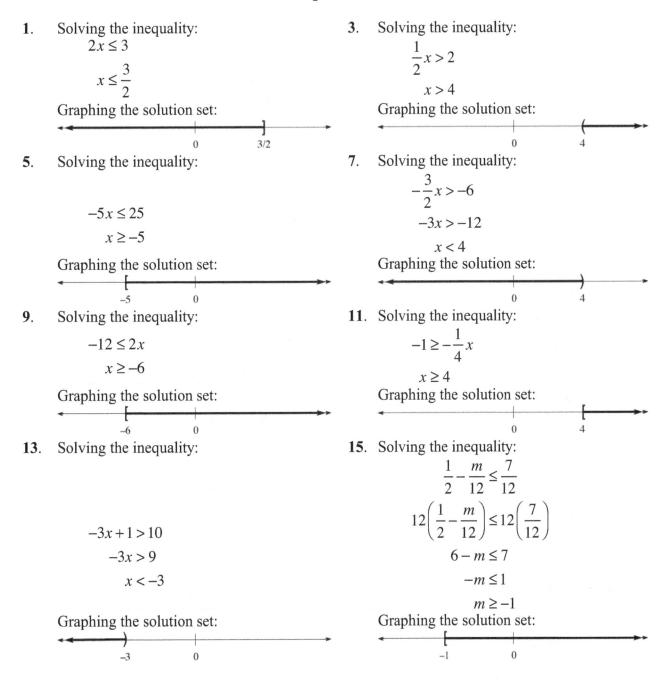

1. Solving the inequality:

$$2x \le 3$$
$$x \le \frac{3}{2}$$

Graphing the solution set:

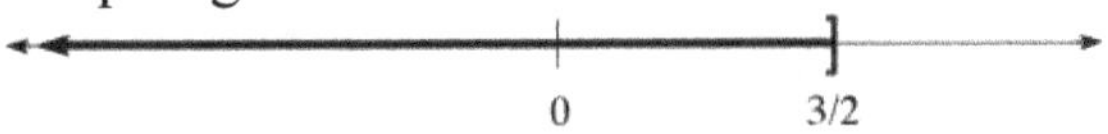

3. Solving the inequality:

$$\frac{1}{2}x > 2$$
$$x > 4$$

Graphing the solution set:

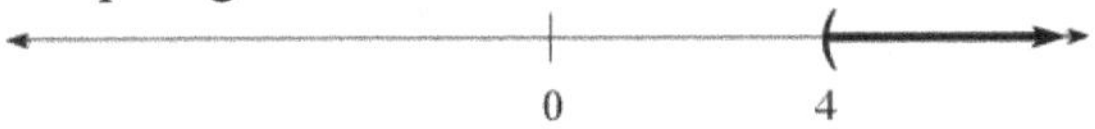

5. Solving the inequality:

$$-5x \le 25$$
$$x \ge -5$$

Graphing the solution set:

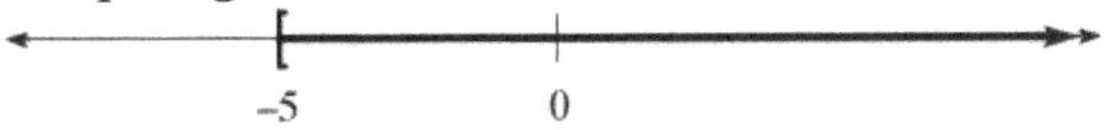

7. Solving the inequality:

$$-\frac{3}{2}x > -6$$
$$-3x > -12$$
$$x < 4$$

Graphing the solution set:

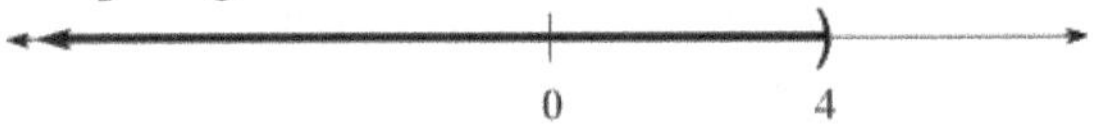

9. Solving the inequality:

$$-12 \le 2x$$
$$x \ge -6$$

Graphing the solution set:

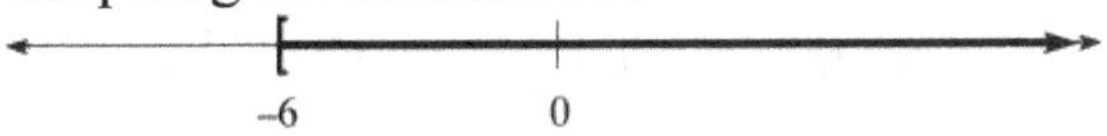

11. Solving the inequality:

$$-1 \ge -\frac{1}{4}x$$
$$x \ge 4$$

Graphing the solution set:

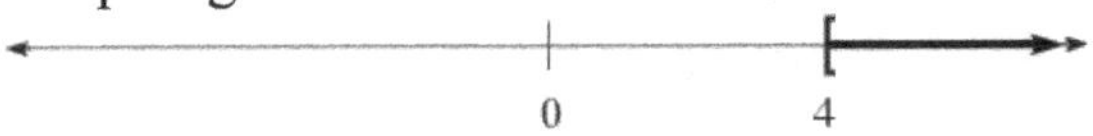

13. Solving the inequality:

$$-3x + 1 > 10$$
$$-3x > 9$$
$$x < -3$$

Graphing the solution set:

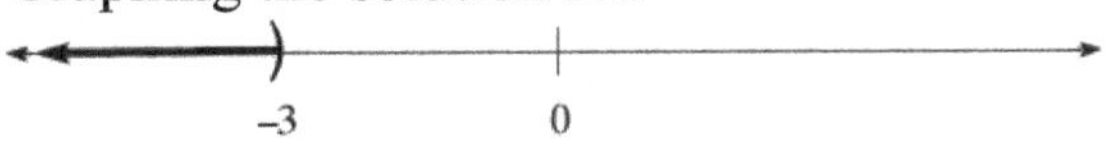

15. Solving the inequality:

$$\frac{1}{2} - \frac{m}{12} \le \frac{7}{12}$$
$$12\left(\frac{1}{2} - \frac{m}{12}\right) \le 12\left(\frac{7}{12}\right)$$
$$6 - m \le 7$$
$$-m \le 1$$
$$m \ge -1$$

Graphing the solution set:

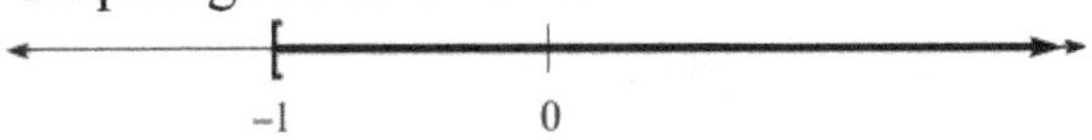

17. Solving the inequality:

$$\frac{1}{2} \geq -\frac{1}{6} - \frac{2}{9}x$$
$$18\left(\frac{1}{2}\right) \geq 18\left(-\frac{1}{6} - \frac{2}{9}x\right)$$
$$9 \geq -3 - 4x$$
$$12 \geq -4x$$
$$x \geq -3$$

Graphing the solution set:

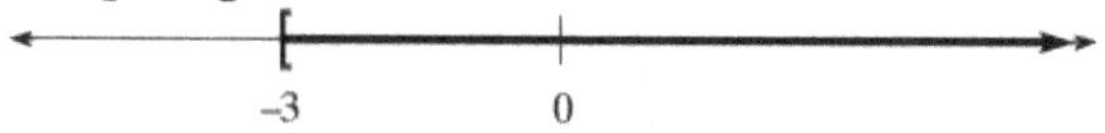

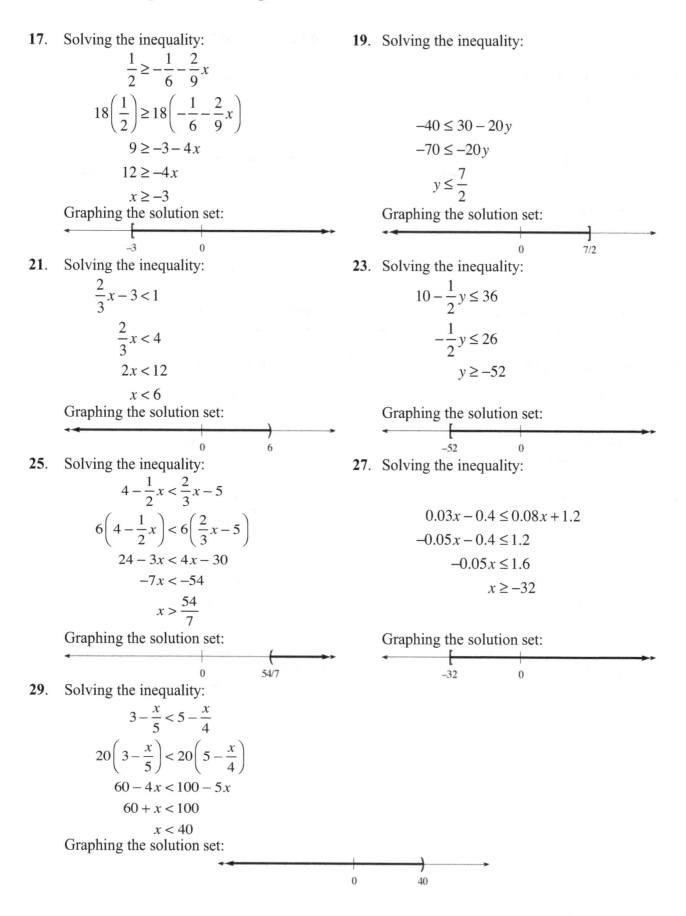

19. Solving the inequality:

$$-40 \leq 30 - 20y$$
$$-70 \leq -20y$$
$$y \leq \frac{7}{2}$$

Graphing the solution set:

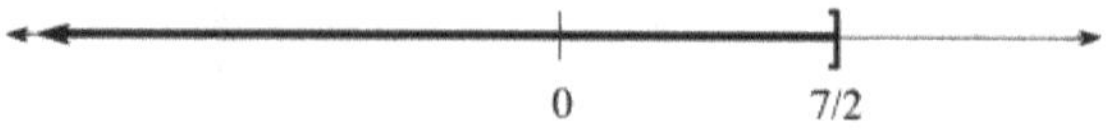

21. Solving the inequality:

$$\frac{2}{3}x - 3 < 1$$
$$\frac{2}{3}x < 4$$
$$2x < 12$$
$$x < 6$$

Graphing the solution set:

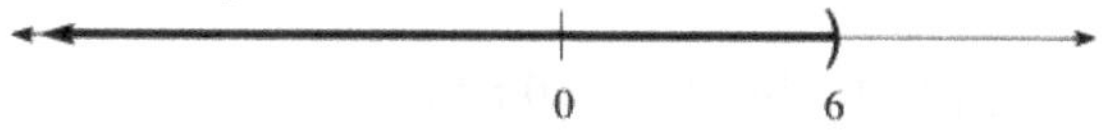

23. Solving the inequality:

$$10 - \frac{1}{2}y \leq 36$$
$$-\frac{1}{2}y \leq 26$$
$$y \geq -52$$

Graphing the solution set:

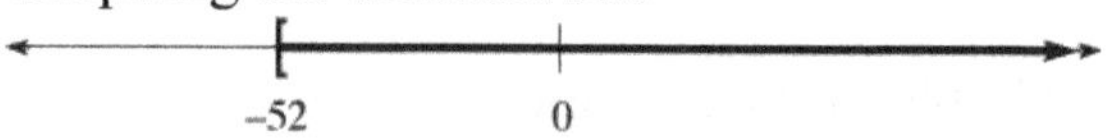

25. Solving the inequality:

$$4 - \frac{1}{2}x < \frac{2}{3}x - 5$$
$$6\left(4 - \frac{1}{2}x\right) < 6\left(\frac{2}{3}x - 5\right)$$
$$24 - 3x < 4x - 30$$
$$-7x < -54$$
$$x > \frac{54}{7}$$

Graphing the solution set:

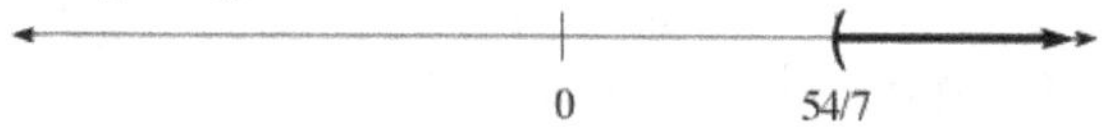

27. Solving the inequality:

$$0.03x - 0.4 \leq 0.08x + 1.2$$
$$-0.05x - 0.4 \leq 1.2$$
$$-0.05x \leq 1.6$$
$$x \geq -32$$

Graphing the solution set:

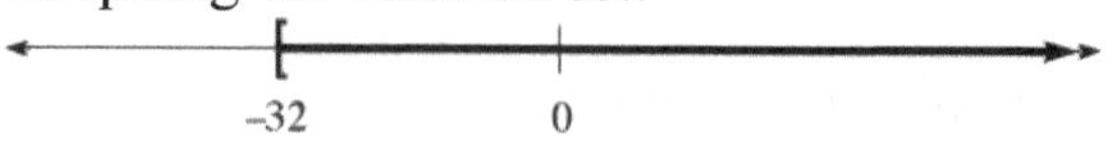

29. Solving the inequality:

$$3 - \frac{x}{5} < 5 - \frac{x}{4}$$
$$20\left(3 - \frac{x}{5}\right) < 20\left(5 - \frac{x}{4}\right)$$
$$60 - 4x < 100 - 5x$$
$$60 + x < 100$$
$$x < 40$$

Graphing the solution set:

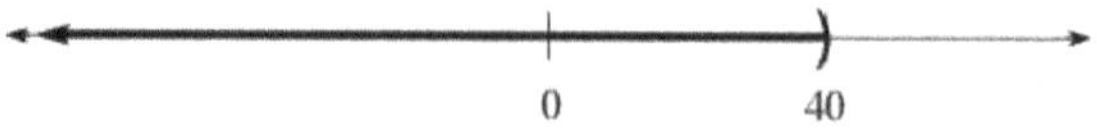

31. Solving the inequality:

$$\begin{aligned} 2(3y+1) &\le -10 \\ 6y+2 &\le -10 \\ 6y &\le -12 \\ y &\le -2 \end{aligned}$$

The solution set is $(-\infty,-2]$.

33. Solving the inequality:

$$\begin{aligned} -(a+1)-4a &\le 2a-8 \\ -a-1-4a &\le 2a-8 \\ -5a-1 &\le 2a-8 \\ -7a &\le -7 \\ a &\ge 1 \end{aligned}$$

The solution set is $[1,\infty)$.

35. Solving the inequality:

$$\begin{aligned} \frac{1}{3}t-\frac{1}{2}(5-t) &< 0 \\ 6\left(\frac{1}{3}t-\frac{1}{2}(5-t)\right) &< 6(0) \\ 2t-3(5-t) &< 0 \\ 2t-15+3t &< 0 \\ 5t-15 &< 0 \\ 5t &< 15 \\ t &< 3 \end{aligned}$$

The solution set is $(-\infty,3)$.

37. Solving the inequality:

$$\begin{aligned} -2 &\le 5-7(2a+3) \\ -2 &\le 5-14a-21 \\ -2 &\le -16-14a \\ 14 &\le -14a \\ a &\le -1 \end{aligned}$$

The solution set is $(-\infty,-1]$.

39. Solving the inequality:

$$\begin{aligned} -\frac{1}{3}(x+5) &\le -\frac{2}{9}(x-1) \\ 9\left[-\frac{1}{3}(x+5)\right] &\le 9\left[-\frac{2}{9}(x-1)\right] \\ -3(x+5) &\le -2(x-1) \\ -3x-15 &\le -2x+2 \\ -x-15 &\le 2 \\ -x &\le 17 \\ x &\ge -17 \end{aligned}$$

The solution set is $[-17,\infty)$.

41. Solving the inequality:

$$\begin{aligned} 5(x-2)-7(x+1) &\le -4x+3 \\ 5x-10-7x-7 &\le -4x+3 \\ -2x-17 &\le -4x+3 \\ 2x-17 &\le 3 \\ 2x &\le 20 \\ x &\le 10 \end{aligned}$$

The solution set is $(-\infty,10]$.

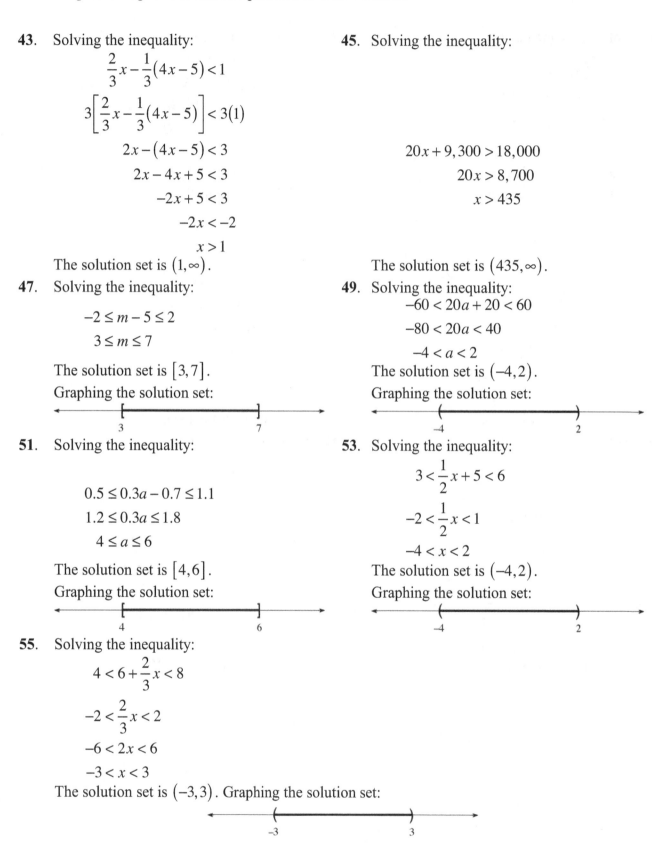

43. Solving the inequality:

$$\frac{2}{3}x-\frac{1}{3}(4x-5)<1$$
$$3\left[\frac{2}{3}x-\frac{1}{3}(4x-5)\right]<3(1)$$
$$2x-(4x-5)<3$$
$$2x-4x+5<3$$
$$-2x+5<3$$
$$-2x<-2$$
$$x>1$$

The solution set is $(1,\infty)$.

45. Solving the inequality:

$$20x+9,300>18,000$$
$$20x>8,700$$
$$x>435$$

The solution set is $(435,\infty)$.

47. Solving the inequality:

$$-2\le m-5\le 2$$
$$3\le m\le 7$$

The solution set is $[3,7]$.

Graphing the solution set:

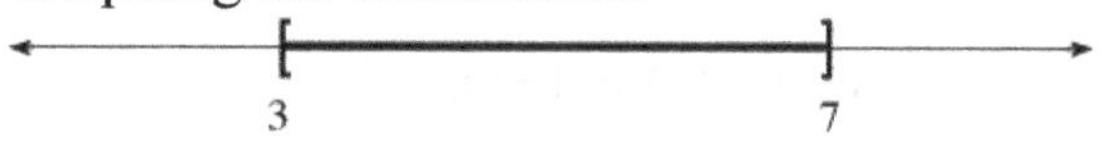

49. Solving the inequality:

$$-60<20a+20<60$$
$$-80<20a<40$$
$$-4<a<2$$

The solution set is $(-4,2)$.

Graphing the solution set:

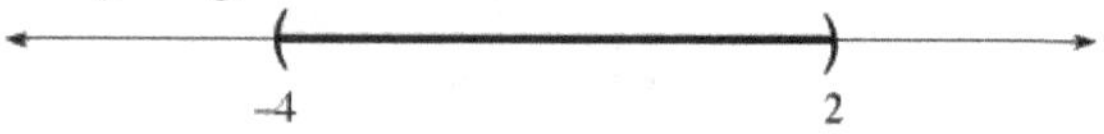

51. Solving the inequality:

$$0.5\le 0.3a-0.7\le 1.1$$
$$1.2\le 0.3a\le 1.8$$
$$4\le a\le 6$$

The solution set is $[4,6]$.

Graphing the solution set:

53. Solving the inequality:

$$3<\frac{1}{2}x+5<6$$
$$-2<\frac{1}{2}x<1$$
$$-4<x<2$$

The solution set is $(-4,2)$.

Graphing the solution set:

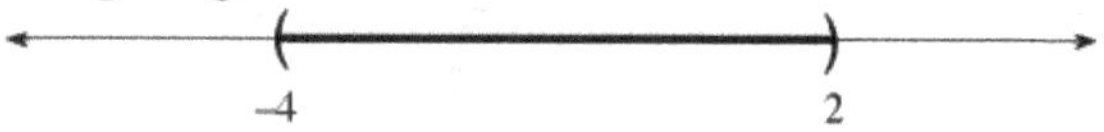

55. Solving the inequality:

$$4<6+\frac{2}{3}x<8$$
$$-2<\frac{2}{3}x<2$$
$$-6<2x<6$$
$$-3<x<3$$

The solution set is $(-3,3)$. Graphing the solution set:

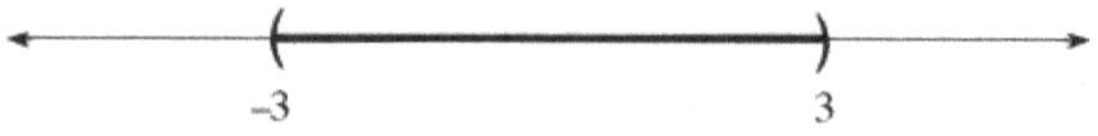

57. Solving the inequality:

$$\begin{aligned} x+5 \le -2 \quad &\text{or} \quad x+5 \ge 2 \\ x \le -7 \quad &\text{or} \quad x \ge -3 \end{aligned}$$

The solution set is $(-\infty,-7]\cup[-3,\infty)$. Graphing the solution set:

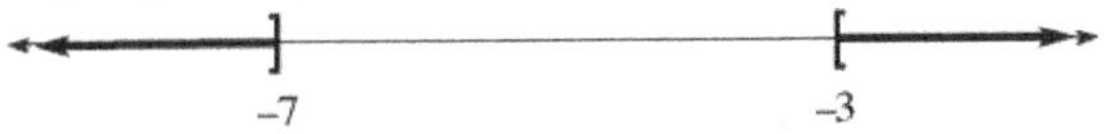

59. Solving the inequality:

$$\begin{aligned} 5y+1 \le -4 \quad &\text{or} \quad 5y+1 \ge 4 \\ 5y \le -5 \quad &\text{or} \quad 5y \ge 3 \\ y \le -1 \quad &\text{or} \quad y \ge \frac{3}{5} \end{aligned}$$

The solution set is $(-\infty,-1]\cup\left[\frac{3}{5},\infty\right)$. Graphing the solution set:

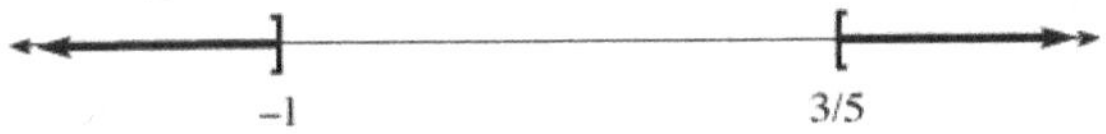

61. Solving the inequality:

$$\begin{aligned} 2x+5 < 3x-1 \quad &\text{or} \quad x-4 > 2x+6 \\ -x+5 < -1 \quad &\text{or} \quad -x-4 > 6 \\ -x < -6 \quad &\text{or} \quad -x > 10 \\ x > 6 \quad &\text{or} \quad x < -10 \end{aligned}$$

The solution set is $(-\infty,-10)\cup(6,\infty)$. Graphing the solution set:

63. Solving the inequality:

$$\begin{aligned} 3x+1 < -8 \quad &\text{or} \quad -2x+1 \le -3 \\ 3x < -9 \quad &\text{or} \quad -2x \le -4 \\ x < -3 \quad &\text{or} \quad x \ge 2 \end{aligned}$$

The solution set is $(-\infty,-3)\cup[2,\infty)$. Graphing the solution set:

65. **a.** Evaluating when $x=0$: $-\frac{1}{2}x+1=-\frac{1}{2}(0)+1=1$

b. Solving the equation:

$$\begin{aligned} -\frac{1}{2}x+1&=-7 \\ -\frac{1}{2}x&=-8 \\ x&=16 \end{aligned}$$

c. Substituting $x=0$: $-\frac{1}{2}x+1=-\frac{1}{2}(0)+1=1$

No, 0 is not a solution to the inequality.

d. Solving the inequality:

$$-\frac{1}{2}x+1<-7$$
$$-\frac{1}{2}x<-8$$
$$x>16$$

67. Translating into an inequality statement: $-2<x\le 4$

69. Translating into an inequality statement: $x<-4$ or $x\ge 1$

71. **a.** Solving the inequality:

$$900-300p\ge 300$$
$$-300p\ge -600$$
$$p\le 2$$

They should charge \$2.00 per pad or less.

b. Solving the inequality:

$$900-300p>600$$
$$-300p>-300$$
$$p<1$$

They should charge less than \$1.00 per pad.

c. Solving the inequality:

$$900-300p<525$$
$$-300p<-375$$
$$p>1.25$$

They should charge more than \$1.25 per pad.

d. Solving the inequality:

$$900-300p\le 375$$
$$-300p\le -525$$
$$p\ge 1.75$$

They should charge \$1.75 per pad or more.

73. **a.** Solving the inequality:

$$0.36x+15.9<17$$
$$0.36x<1.1$$
$$x<3.06$$

In the years between 1980 and 1983 the average fuel efficiency was less than 17 mpg.

b. Solving the inequality:

$$0.36x+15.9>20$$
$$0.36x>4.1$$
$$x>11.39$$

In the years 1991 (partway through) and after the average fuel efficiency was more than 20 mpg.

75. Solving the inequality:

$$x\le 0.08I$$
$$x\le 0.08\left(\frac{24{,}000}{12}\right)$$
$$x\le 160$$

The amount of monthly debt should be less than or equal to \$160.

77. For eggs to hatching, the inequality is $0.7\le r\le 0.8$. For hatched chicks to fledglings, the inequality is $0.5\le r\le 0.7$. For fledglings to age of first breeding, the inequality is $r<0.5$.

79. Solving the equation:

$$3x-6=9$$
$$3x=15$$
$$x=5$$

81. Solving the equation:

$$\frac{2}{3}x-3=-7$$
$$\frac{2}{3}x=-4$$
$$x=-6$$

83. Solving the equation:

$$\begin{aligned} x+3&=x+8 \\ 3&=8 \end{aligned}$$

The equation has no solution $(\varnothing)$.

85. Solving the equation:

$$\begin{aligned} x+3&=-x+8 \\ 2x+3&=8 \\ 2x&=5 \\ x&=\frac{5}{2} \end{aligned}$$

2.5 Equations with Absolute Value

1. Solving the equation:

$$\begin{aligned} |x|&=4 \\ x&=-4,4 \end{aligned}$$

3. Solving the equation:

$$\begin{aligned} 2&=|a| \\ a&=-2,2 \end{aligned}$$

5. The equation $|x|=-3$ has no solution, or $\varnothing$.

7. Solving the equation:

$$\begin{aligned} |a|+2&=3 \\ |a|&=1 \\ a&=-1,1 \end{aligned}$$

9. Solving the equation:

$$\begin{aligned} |y|+4&=3 \\ |y|&=-1 \end{aligned}$$

The equation $|y|=-1$ has no solution, or $\varnothing$.

11. Solving the equation:

$$\begin{aligned} |a-4|&=\frac{5}{3} \\ a-4&=-\frac{5}{3},\frac{5}{3} \\ a&=\frac{7}{3},\frac{17}{3} \end{aligned}$$

13. Solving the equation:

$$\begin{aligned} \left|\frac{3}{5}a+\frac{1}{2}\right|&=1 \\ \frac{3}{5}a+\frac{1}{2}&=-1,1 \\ \frac{3}{5}a&=-\frac{3}{2},\frac{1}{2} \\ a&=-\frac{5}{2},\frac{5}{6} \end{aligned}$$

15. Solving the equation:

$$\begin{aligned} 60&=|20x-40| \\ 20x-40&=-60,60 \\ 20x&=-20,100 \\ x&=-1,5 \end{aligned}$$

17. Since $|2x+1|=-3$ is impossible, there is no solution, or $\varnothing$.

19. Solving the equation:

$$\left|\frac{3}{4}x-6\right|=9$$
$$\frac{3}{4}x-6=-9,9$$
$$\frac{3}{4}x=-3,15$$
$$3x=-12,60$$
$$x=-4,20$$

21. Solving the equation:

$$\left|1-\frac{1}{2}a\right|=3$$
$$1-\frac{1}{2}a=-3,3$$
$$-\frac{1}{2}a=-4,2$$
$$a=-4,8$$

23. Solving the equation:

$$|2x-5|=3$$
$$2x-5=-3,3$$
$$2x=2,8$$
$$x=1,4$$

25. Solving the equation:

$$|4-7x|=5$$
$$4-7x=-5,5$$
$$-7x=-9,1$$
$$x=-\frac{1}{7},\frac{9}{7}$$

27. Solving the equation:

$$\left|3-\frac{2}{3}y\right|=5$$
$$3-\frac{2}{3}y=-5,5$$
$$-\frac{2}{3}y=-8,2$$
$$-2y=-24,6$$
$$y=-3,12$$

29. Solving the equation:

$$|3x+4|+1=7$$
$$|3x+4|=6$$
$$3x+4=-6,6$$
$$3x=-10,2$$
$$x=-\frac{10}{3},\frac{2}{3}$$

31. Solving the equation:

$$|3-2y|+4=3$$
$$|3-2y|=-1$$

Since this equation is impossible, there is no solution, or $\varnothing$.

33. Solving the equation:

$$\begin{aligned} 3+|4t-1| &= 8 \\ |4t-1| &= 5 \\ 4t-1 &= -5,5 \\ 4t &= -4,6 \\ t &= -1,\frac{3}{2} \end{aligned}$$

35. Solving the equation:

$$\begin{aligned} \left|9-\frac{3}{5}x\right|+6 &= 12 \\ \left|9-\frac{3}{5}x\right| &= 6 \\ 9-\frac{3}{5}x &= -6,6 \\ -\frac{3}{5}x &= -15,-3 \\ -3x &= -75,-15 \\ x &= 5,25 \end{aligned}$$

37. Solving the equation:

$$\begin{aligned} 5 &= \left|\frac{2}{7}x+\frac{4}{7}\right|-3 \\ \left|\frac{2}{7}x+\frac{4}{7}\right| &= 8 \\ \frac{2}{7}x+\frac{4}{7} &= -8,8 \\ 2x+4 &= -56,56 \\ 2x &= -60,52 \\ x &= -30,26 \end{aligned}$$

39. Solving the equation:

$$\begin{aligned} 2 &= -8+\left|4-\frac{1}{2}y\right| \\ \left|4-\frac{1}{2}y\right| &= 10 \\ 4-\frac{1}{2}y &= -10,10 \\ -\frac{1}{2}y &= -14,6 \\ y &= -12,28 \end{aligned}$$

41. Solving the equation:

$$\begin{aligned} |3(x+1)|-4 &= -1 \\ |3x+3| &= 3 \\ 3x+3 &= -3,3 \\ 3x &= -6,0 \\ x &= -2,0 \end{aligned}$$

43. Solving the equation:

$$\begin{aligned} |1+3(2x-1)| &= 5 \\ |1+6x-3| &= 5 \\ |6x-2| &= 5 \\ 6x-2 &= -5,5 \\ 6x &= -3,7 \\ x &= -\frac{1}{2},\frac{7}{6} \end{aligned}$$

45. Solving the equation:

$$3=-2+\left|5-\frac{2}{3}a\right|$$

$$\left|5-\frac{2}{3}a\right|=5$$

$$5-\frac{2}{3}a=-5,5$$

$$-\frac{2}{3}a=-10,0$$

$$-2a=-30,0$$

$$a=0,15$$

47. Solving the equation:

$$6=|7(k+3)-4|$$

$$|7k+21-4|=6$$

$$|7k+17|=6$$

$$7k+17=-6,6$$

$$7k=-23,-11$$

$$k=-\frac{23}{7},-\frac{11}{7}$$

49. Solving the equation:

$$|3a+1|=|2a-4|$$

$3a+1=2a-4$ or $3a+1=-2a+4$

$a+1=-4$ $\quad$ $5a=3$

$a=-5$ $\quad$ $a=\frac{3}{5}$

51. Solving the equation:

$$\left|x-\frac{1}{3}\right|=\left|\frac{1}{2}x+\frac{1}{6}\right|$$

$x-\frac{1}{3}=\frac{1}{2}x+\frac{1}{6}$ or $x-\frac{1}{3}=-\frac{1}{2}x-\frac{1}{6}$

$6x-2=3x+1$ $\quad$ $6x-2=-3x-1$

$3x-2=1$ $\quad$ $9x-2=-1$

$3x=3$ $\quad$ $9x=1$

$x=1$ $\quad$ $x=\frac{1}{9}$

53. Solving the equation:

$$|y-2|=|y+3|$$

$y-2=y+3$ or $y-2=-y-3$

$-2=-3$ $\quad$ $2y=-1$

y=impossible $\quad$ $y=-\frac{1}{2}$

55. Solving the equation:

$$|3x-1|=|3x+1|$$

$3x-1=3x+1$ or $3x-1=-3x-1$

$-1=1$ $\quad$ $6x=0$

x=impossible $\quad$ $x=0$

57. Solving the equation:

$$|0.03-0.01x|=|0.04+0.05x|$$

$$0.03-0.01x=0.04+0.05x \quad \text{or} \quad 0.03-0.01x=-0.04-0.05x$$

$$-0.06x=0.01 \qquad\qquad 0.04x=-0.07$$

$$x=-\frac{1}{6} \qquad\qquad x=-\frac{7}{4}$$

59. Since $|x-2|=|2-x|$ is always true, the solution set is all real numbers.

61. Since $\left|\frac{x}{5}-1\right|=\left|1-\frac{x}{5}\right|$ is always true, the solution set is all real numbers.

63. Solving the equation:

$$\left|\frac{2}{3}b-\frac{1}{4}\right|=\left|\frac{1}{6}b+\frac{1}{2}\right|$$

$$\frac{2}{3}b-\frac{1}{4}=\frac{1}{6}b+\frac{1}{2} \quad \text{or} \quad \frac{2}{3}b-\frac{1}{4}=-\frac{1}{6}b-\frac{1}{2}$$

$$8b-3=2b+6 \qquad\qquad 8b-3=-2b-6$$

$$6b=9 \qquad\qquad 10b=-3$$

$$b=\frac{3}{2} \qquad\qquad b=-\frac{3}{10}$$

65. Solving the equation:

$$|0.1a-0.04|=|0.3a+0.08|$$

$$0.1a-0.04=0.3a+0.08 \quad \text{or} \quad 0.1a-0.04=-0.3a-0.08$$

$$-0.2a=0.12 \qquad\qquad 0.4a=-0.04$$

$$a=-\frac{3}{5} \qquad\qquad a=-\frac{1}{10}$$

67. **a.** Solving the equation:

$$4x-5=0$$

$$4x=5$$

$$x=\frac{5}{4}=1.25$$

b. Solving the equation:

$$|4x-5|=0$$

$$4x-5=0$$

$$4x=5$$

$$x=\frac{5}{4}=1.25$$

c. Solving the equation:

$$4x-5=3$$

$$4x=8$$

$$x=2$$

d. Solving the equation:

$$|4x-5|=3$$

$$4x-5=-3,3$$

$$4x=2,8$$

$$x=\frac{1}{2},2$$

e. Solving the equation:

$$|4x-5| = |2x+3|$$

$$4x-5 = 2x+3 \quad \text{or} \quad 4x-5 = -2x-3$$

$$2x-5 = 3 \qquad\qquad 6x-5 = -3$$

$$2x = 8 \qquad\qquad 6x = 2$$

$$x = 4 \qquad\qquad x = \frac{1}{3}$$

69. Setting $R = 722$:

$$-60|x-11| + 962 = 722$$
$$-60|x-11| = -240$$
$$|x-11| = 4$$
$$x-11 = -4, 4$$
$$x = 7, 15$$

The revenue was 722 million dollars in the years 1987 and 1995.

71. Solving the inequality:

$$2x-5 < 3$$
$$2x < 8$$
$$x < 4$$

73. Solving the inequality:

$$-4 \le 3a+7$$
$$-11 \le 3a$$
$$a \ge -\frac{11}{3}$$

75. Solving the inequality:

$$4t-3 \le -9$$
$$4t \le -6$$
$$t \le -\frac{3}{2}$$

2.6 Inequalities Involving Absolute Value

1. Solving the inequality:

$$|x| < 3$$
$$-3 < x < 3$$

Graphing the solution set:

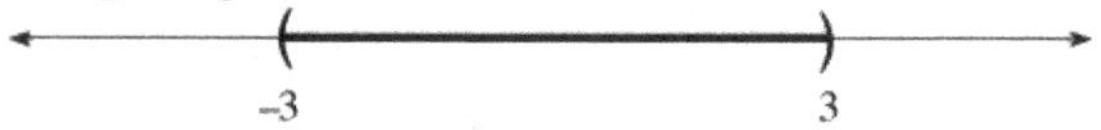

3. Solving the inequality:

$$|x| \ge 2$$
$$x \le -2 \text{ or } x \ge 2$$

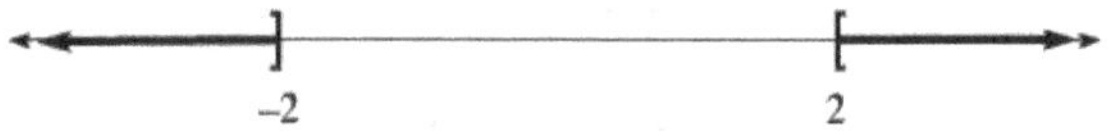

5. Solving the inequality:

$$|x| + 2 < 5$$
$$|x| < 3$$
$$-3 < x < 3$$

Graphing the solution set:

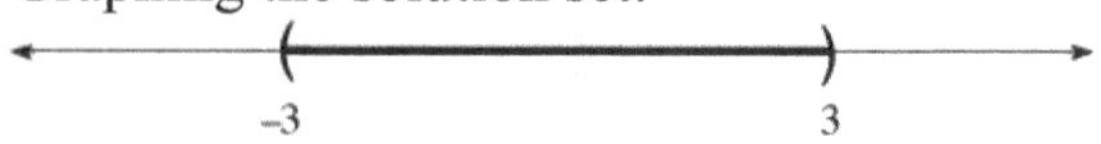

7. Solving the inequality:

$$|t| - 3 > 4$$
$$|t| > 7$$
$$t < -7 \text{ or } t > 7$$

Graphing the solution set:

9. Since the inequality $|y| < -5$ is never true, there is no solution, or $\varnothing$. Graphing the solution set:

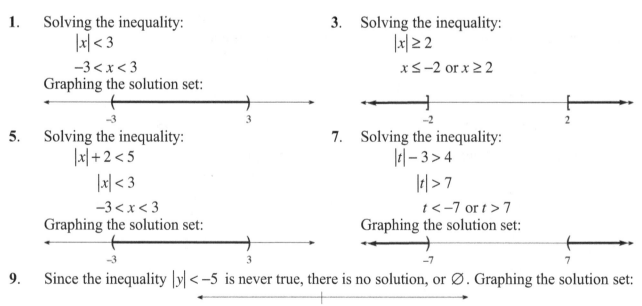

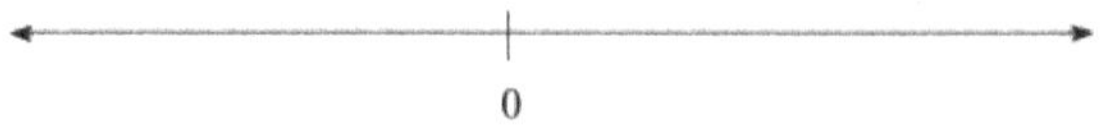

11. Since the inequality $|x| \ge -2$ is always true, the solution set is all real numbers. Graphing the solution set:

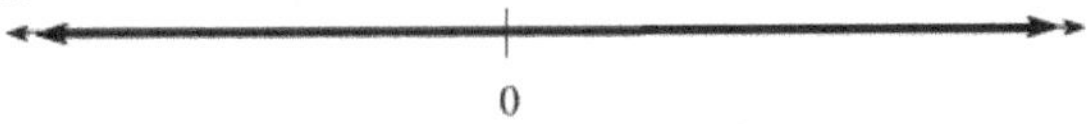

13. Solving the inequality:

$$|x-3| < 7$$
$$-7 < x-3 < 7$$
$$-4 < x < 10$$

Graphing the solution set:

15. Solving the inequality:

$$|a+5| \ge 4$$
$$a+5 \le -4 \text{ or } a+5 \ge 4$$
$$a \le -9 \text{ or } a \ge -1$$

Graphing the solution set:

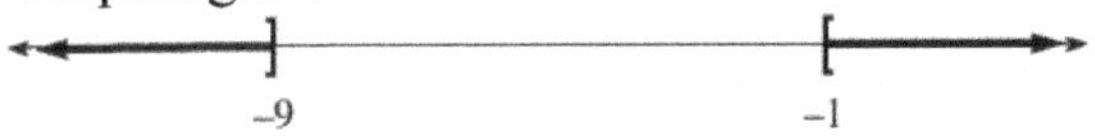

17. Since the inequality $|a-1| < -3$ is never true, there is no solution, or $\varnothing$. Graphing the solution set:

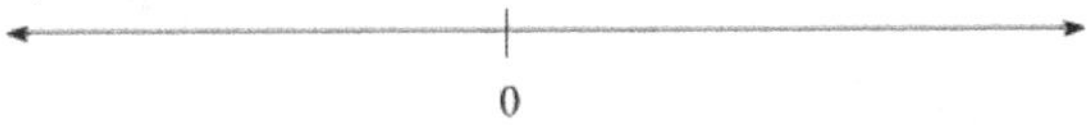

19. Solving the inequality:

$$|2x-4| < 6$$
$$-6 < 2x-4 < 6$$
$$-2 < 2x < 10$$
$$-1 < x < 5$$

Graphing the solution set:

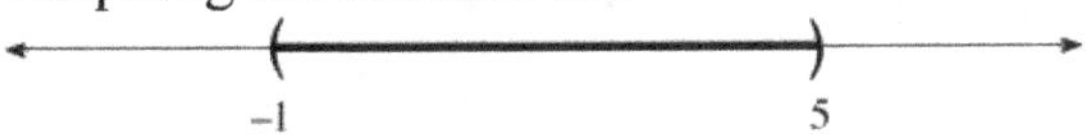

21. Solving the inequality:

$$|3y+9| \ge 6$$
$$3y+9 \le -6 \quad \text{or} \quad 3y+9 \ge 6$$
$$3y \le -15 \qquad 3y \ge -3$$
$$y \le -5 \qquad y \ge -1$$

Graphing the solution set:

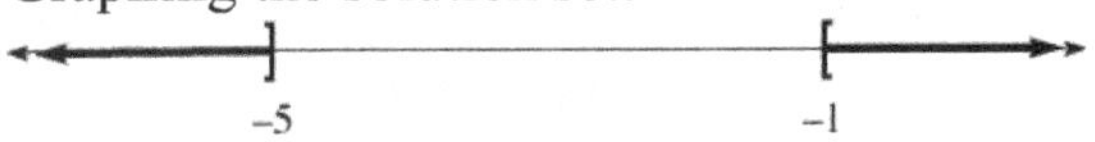

23. Solving the inequality:

$$|2k+3| \ge 7$$
$$2k+3 \le -7 \quad \text{or} \quad 2k+3 \ge 7$$
$$2k \le -10 \qquad 2k \ge 4$$
$$k \le -5 \qquad k \ge 2$$

Graphing the solution set:

25. Solving the inequality:

$$|x-3|+2 < 6$$
$$|x-3| < 4$$
$$-4 < x-3 < 4$$
$$-1 < x < 7$$

Graphing the solution set:

27. Solving the inequality:

$$|2a+1|+4 \ge 7$$
$$|2a+1| \ge 3$$
$$2a+1 \le -3 \quad \text{or} \quad 2a+1 \ge 3$$
$$2a \le -4 \qquad 2a \ge 2$$
$$a \le -2 \qquad a \ge 1$$

Graphing the solution set:

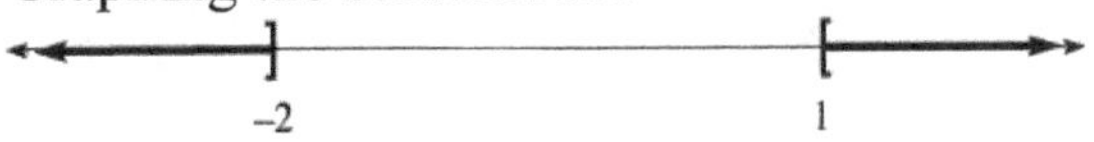

29. Solving the inequality:

$$|3x+5|-8 < 5$$
$$|3x+5| < 13$$
$$-13 < 3x+5 < 13$$
$$-18 < 3x < 8$$
$$-6 < x < \frac{8}{3}$$

Graphing the solution set:

31. Solving the inequality:

$$\begin{aligned} |x-3| &\le 5 \\ -5 \le x-3 &\le 5 \\ -2 \le x &\le 8 \end{aligned}$$

The solution set is $[-2,8]$.

33. Solving the inequality:

$$\begin{aligned} |3y+1| &< 5 \\ -5 < 3y+1 &< 5 \\ -6 < 3y &< 4 \\ -2 < y &< \frac{4}{3} \end{aligned}$$

The solution set is $\left(-2,\frac{4}{3}\right)$.

35. Solving the inequality:

$$|a+4| \ge 1$$

$$\begin{aligned} a+4 &\le -1 \quad \text{or} \quad & a+4 &\ge 1 \\ a &\le -5 & a &\ge -3 \end{aligned}$$

The solution set is $(-\infty,-5]\cup[-3,\infty)$.

37. Solving the inequality:

$$|2x+5| > 2$$

$$\begin{aligned} 2x+5 &< -2 \quad \text{or} \quad & 2x+5 &> 2 \\ 2x &< -7 & 2x &> -3 \\ x &< -\frac{7}{2} & x &> -\frac{3}{2} \end{aligned}$$

The solution set is $\left(-\infty,-\frac{7}{2}\right)\cup\left(-\frac{3}{2},\infty\right)$.

39. Solving the inequality:

$$\begin{aligned} |-5x+3| &\le 8 \\ -8 \le -5x+3 &\le 8 \\ -11 \le -5x &\le 5 \\ \frac{11}{5} \ge x &\ge -1 \end{aligned}$$

The solution set is $\left[-1,\frac{11}{5}\right]$.

41. Solving the inequality:

$$\begin{aligned} |-3x+7| &< 2 \\ -2 < -3x+7 &< 2 \\ -9 < -3x &< -5 \\ 3 > x &> \frac{5}{3} \end{aligned}$$

The solution set is $\left(\frac{5}{3},3\right)$.

43. Solving the inequality:

$$|5-x| > 3$$

$$\begin{aligned} 5-x &< -3 \quad \text{or} \quad & 5-x &> 3 \\ -x &< -8 & -x &> -2 \\ x &> 8 & x &< 2 \end{aligned}$$

Graphing the solution set:

45. Solving the inequality:

$$\left|3-\frac{2}{3}x\right| \ge 5$$

$$\begin{aligned} 3-\frac{2}{3}x &\le -5 \quad \text{or} \quad & 3-\frac{2}{3}x &\ge 5 \\ -\frac{2}{3}x &\le -8 & -\frac{2}{3}x &\ge 2 \\ -2x &\le -24 & -2x &\ge 6 \\ x &\ge 12 & x &\le -3 \end{aligned}$$

Graphing the solution set:

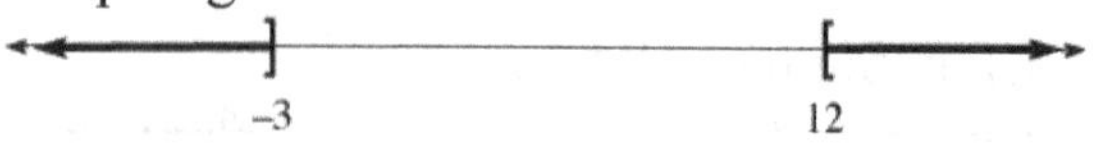

47. Solving the inequality:

$$\left|2-\frac{1}{2}x\right|>1$$

$$2-\frac{1}{2}x<-1 \quad \text{or} \quad 2-\frac{1}{2}x>1$$

$$-\frac{1}{2}x<-3 \qquad\qquad -\frac{1}{2}x>-1$$

$$x>6 \qquad\qquad x<2$$

Graphing the solution set:

49. Solving the inequality:

$$|x-1|<0.01$$

$$-0.01<x-1<0.01$$

$$0.99<x<1.01$$

51. Solving the inequality:

$$|2x+1|\geq\frac{1}{5}$$

$$2x+1\leq-\frac{1}{5} \quad \text{or} \quad 2x+1\geq\frac{1}{5}$$

$$2x\leq-\frac{6}{5} \qquad\qquad 2x\geq-\frac{4}{5}$$

$$x\leq-\frac{3}{5} \qquad\qquad x\geq-\frac{2}{5}$$

53. Solving the inequality:

$$|3x-2|\leq\frac{1}{3}$$

$$-\frac{1}{3}\leq 3x-2\leq\frac{1}{3}$$

$$\frac{5}{3}\leq 3x\leq\frac{7}{3}$$

$$\frac{5}{9}\leq x\leq\frac{7}{9}$$

55. Solving the inequality:

$$\left|\frac{3x+1}{2}\right|>\frac{1}{2}$$

$$\frac{3x+1}{2}<-\frac{1}{2} \quad \text{or} \quad \frac{3x+1}{2}>\frac{1}{2}$$

$$3x+1<-1 \qquad\qquad 3x+1>1$$

$$3x<-2 \qquad\qquad 3x>0$$

$$x<-\frac{2}{3} \qquad\qquad x>0$$

57. Solving the inequality:

$$\left|\frac{4-3x}{2}\right|\geq 1$$

$$\frac{4-3x}{2}\leq-1 \quad \text{or} \quad \frac{4-3x}{2}\geq 1$$

$$4-3x\leq-2 \qquad\qquad 4-3x\geq 2$$

$$-3x\leq-6 \qquad\qquad -3x\geq-2$$

$$x\geq 2 \qquad\qquad x\leq\frac{2}{3}$$

59. Solving the inequality:

$$\left|\frac{3x-2}{5}\right|\leq\frac{1}{2}$$

$$-\frac{1}{2}\leq\frac{3x-2}{5}\leq\frac{1}{2}$$

$$-\frac{5}{2}\leq 3x-2\leq\frac{5}{2}$$

$$-\frac{1}{2}\leq 3x\leq\frac{9}{2}$$

$$-\frac{1}{6}\leq x\leq\frac{3}{2}$$

61. Solving the inequality:

$$\left|2x-\frac{1}{5}\right|<0.3$$

$$-0.3<2x-0.2<0.3$$

$$-0.1<2x<0.5$$

$$-0.05<x<0.25$$

63. Writing as an absolute value inequality: $|x| \le 4$

65. Writing as an absolute value inequality: $|x-5| \le 1$

67. **a.** Evaluating when $x=0$: $|5x+3| = |5(0)+3| = |3| = 3$

b. Solving the equation:

$$\begin{aligned} |5x+3| &= 7 \\ 5x+3 &= -7, 7 \\ 5x &= -10, 4 \\ x &= -2, \frac{4}{5} \end{aligned}$$

c. Substituting $x=0$: $|5x+3| = |5(0)+3| = |3| = 3$

No, 0 is not a solution to the inequality.

d. Solving the inequality:

$$|5x+3| > 7$$

$$\begin{aligned} 5x+3 &< -7 \\ 5x &< -10 \\ x &< -2 \end{aligned} \quad \text{or} \quad \begin{aligned} 5x+3 &> 7 \\ 5x &> 4 \\ x &> \frac{4}{5} \end{aligned}$$

69. The absolute value inequality is: $|x-65| \le 10$

71. Simplifying: $3^{-2} = \frac{1}{3^2} = \frac{1}{9}$

73. Simplifying: $\frac{15x^3y^8}{5xy^{10}} = \frac{15}{5}x^{3-1}y^{8-10} = 3x^2y^{-2} = \frac{3x^2}{y^2}$

75. Simplifying: $\frac{\left(3x^{-3}y^5\right)^{-2}}{\left(9xy^{-2}\right)^{-1}} = \frac{3^{-2}x^6y^{-10}}{9^{-1}x^{-1}y^2} = \frac{9^1}{3^2}x^{6+1}y^{-10-2} = x^7y^{-12} = \frac{x^7}{y^{12}}$

77. Writing in scientific notation: $54{,}000 = 5.4 \times 10^4$

79. Writing in expanded form: $6.44 \times 10^3 = 6{,}440$

81. Simplifying: $\left(3 \times 10^8\right)\left(4 \times 10^{-5}\right) = 12 \times 10^3 = 1.2 \times 10^4$

Chapter 2 Test

See www.mathtv.com for video solutions to all problems in this chapter test.

Chapter 3
Equations and Inequalities in Two Variables

3.1 Paired Data and the Rectangular Coordinate System

1. Plotting the points:

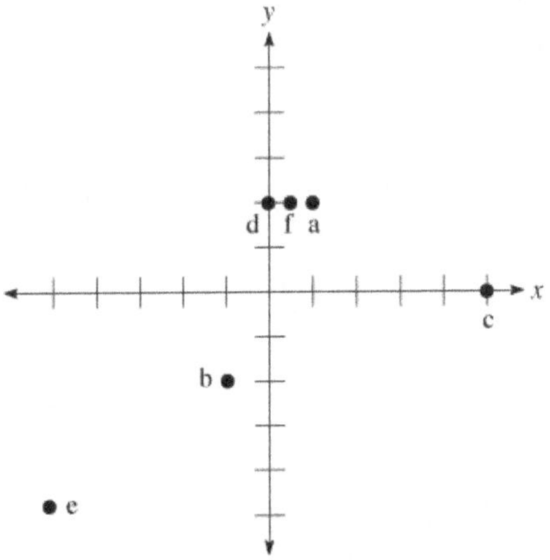

3. A. (4,1) B. (–4,3)
C. (–2,–5) D. (2,–2)
E. (0,5) F. (–4,0)
G. (1,0)

5. Table b, since its values match the equation.

7. Since the y-intercept is –2 and the slope is $\frac{2}{3}$, this is the graph of **b**.

9. Since the graph is translated up 2 units, the equation is $y = x + 2$.

11. Since the graph is translated down 3 units, the equation is $y = |x| - 3$.

13. **a.** Graphing the line:

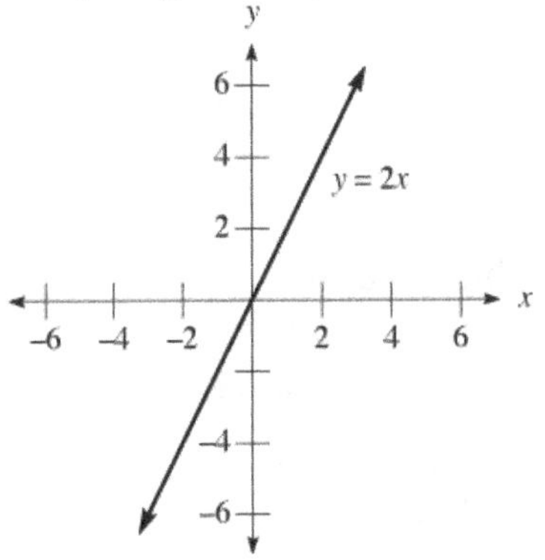

15. **a.** Graphing the curve:

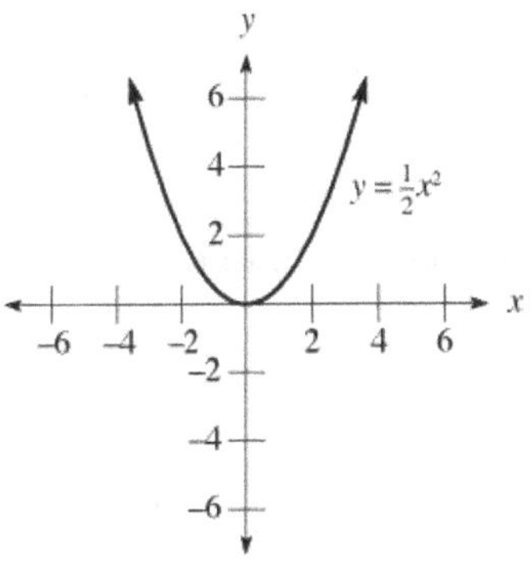

b. Graphing the line:

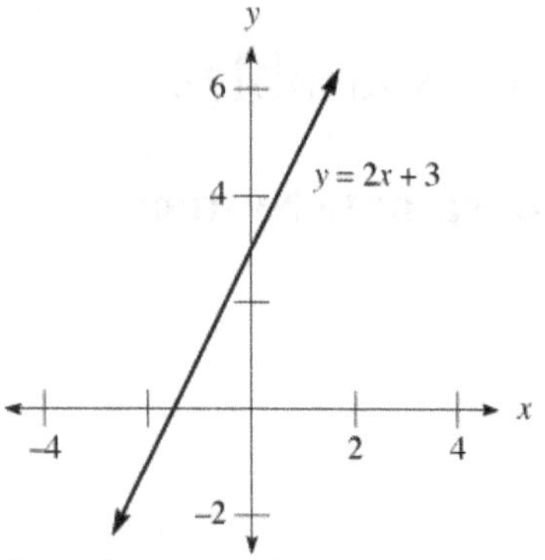

c. Graphing the line:

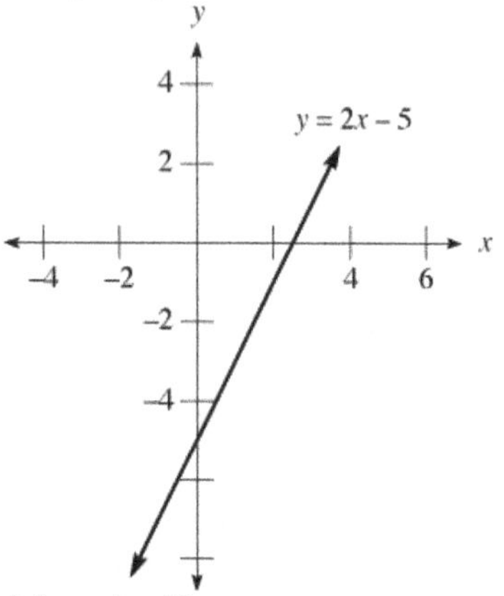

b. Graphing the curve:

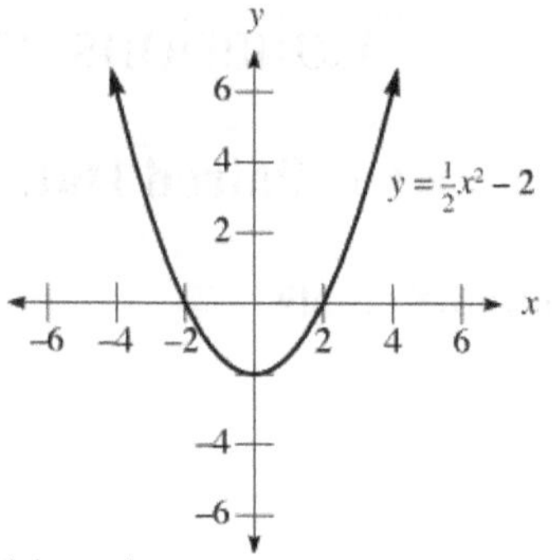

c. Graphing the curve:

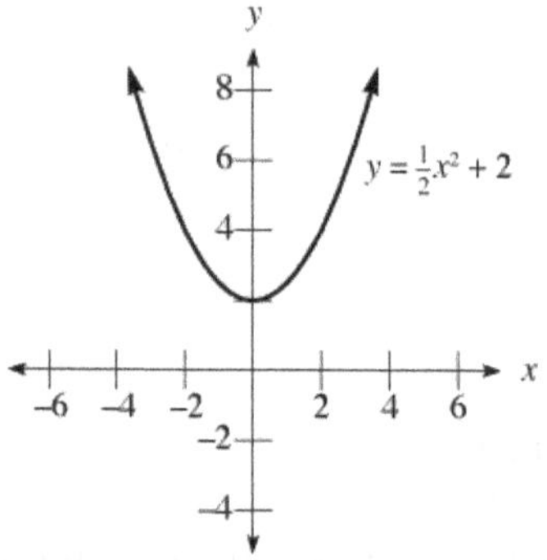

17. Graphing the line:

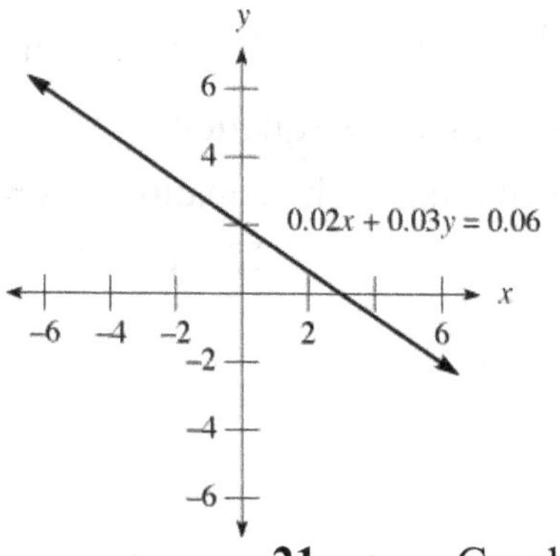

19. **a.** Graphing the line:

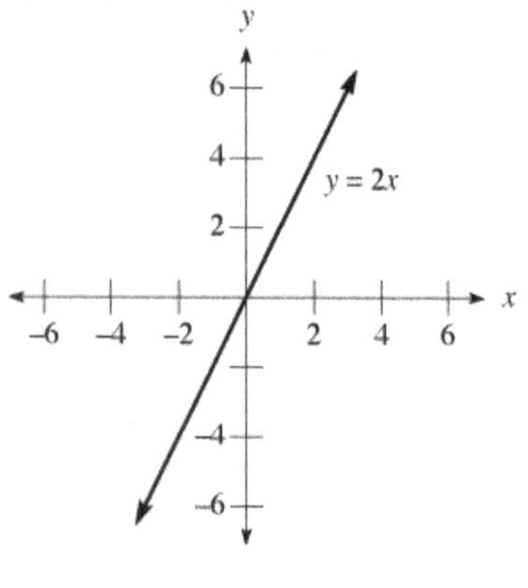

21. **a.** Graphing the line:

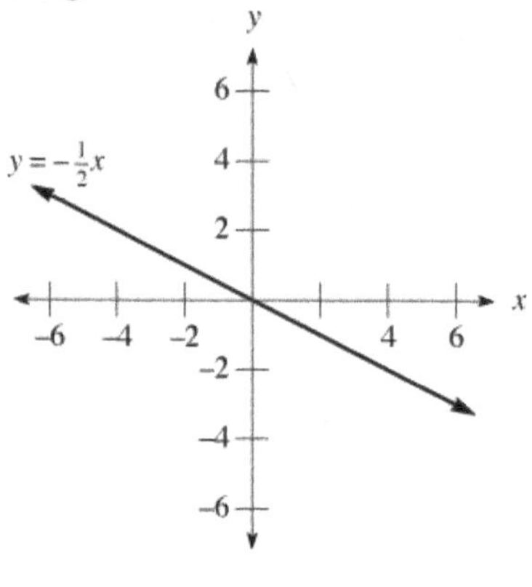

b. Graphing the line:

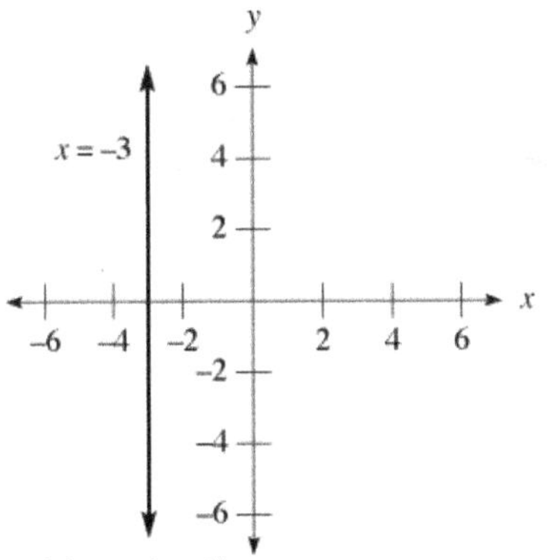

c. Graphing the line:

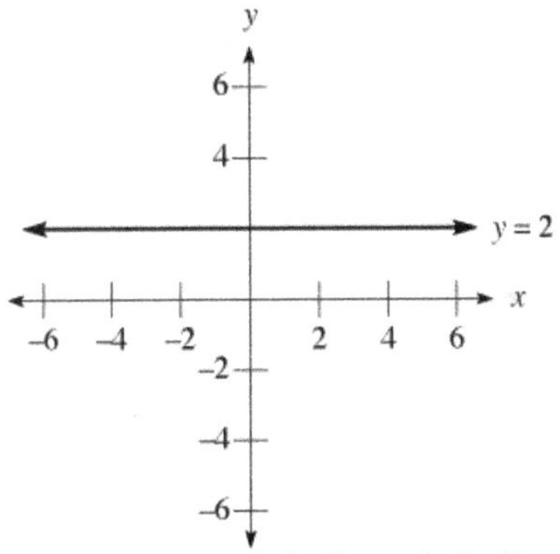

b. Graphing the line:

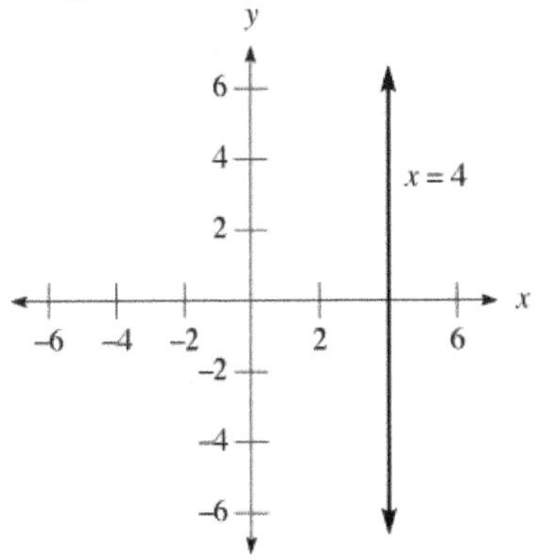

c. Graphing the line:

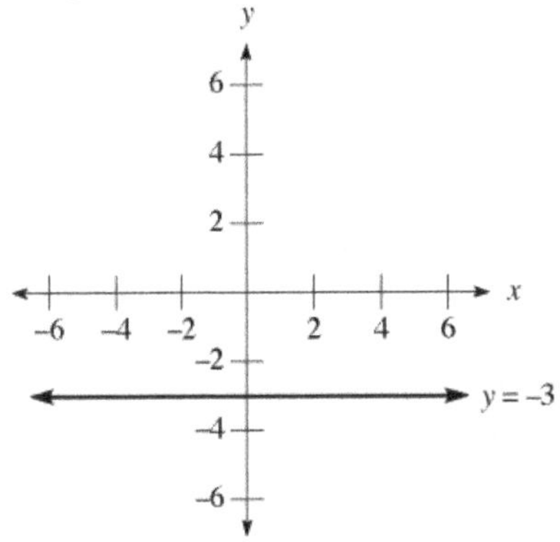

23. The x-intercepts are $(-3,0)$ and $(3,0)$, and the y-intercept is $(0,-9)$:

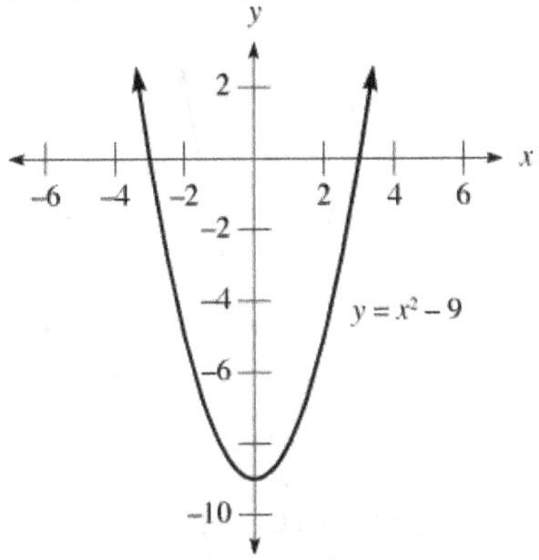

25. The x-intercept is $(2,0)$ and the y-intercept is $(0,-4)$:

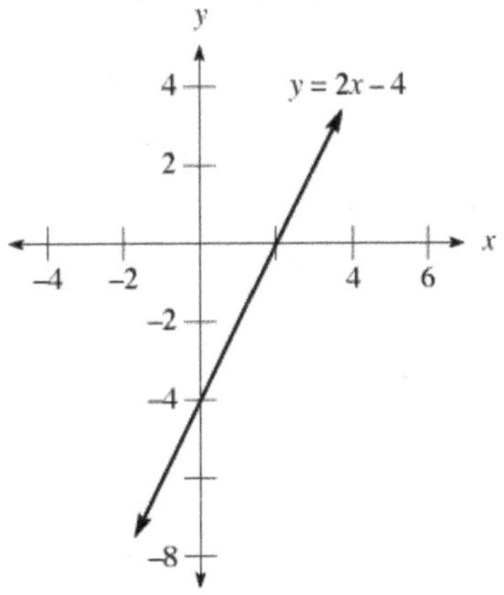

27. The x-intercept is $(-2,0)$ and the y-intercept is $(0,1)$:

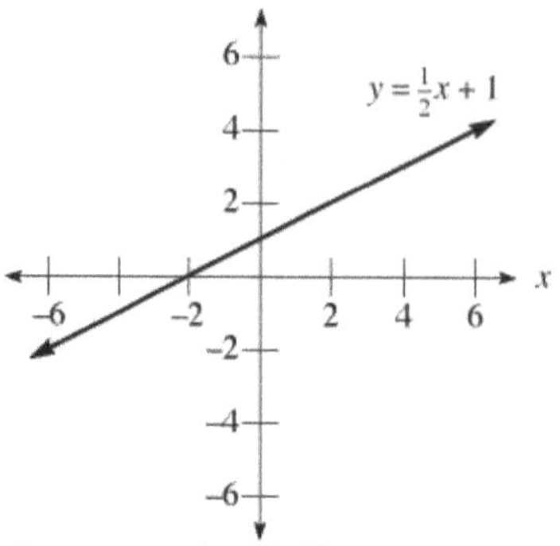

29. The x-intercept is $(0,0)$ and the y-intercept is $(0,0)$:

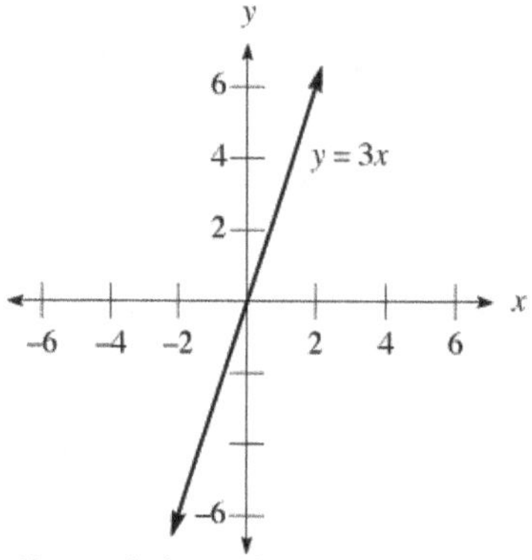

31. The x-intercepts are $(0,0)$ and $(1,0)$, and the y-intercept is $(0,0)$:

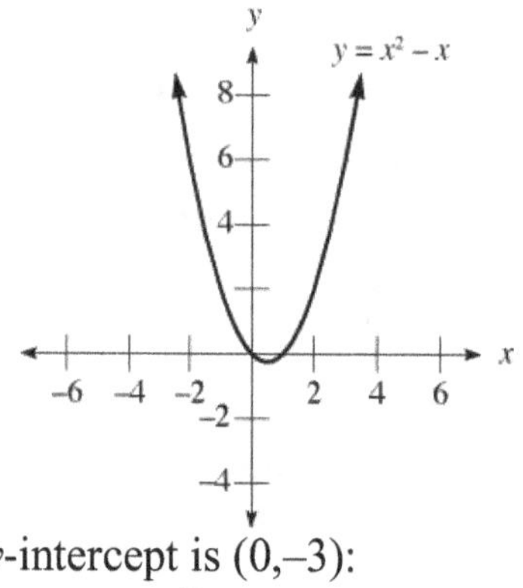

33. The x-intercept is $(3,0)$ and the y-intercept is $(0,-3)$:

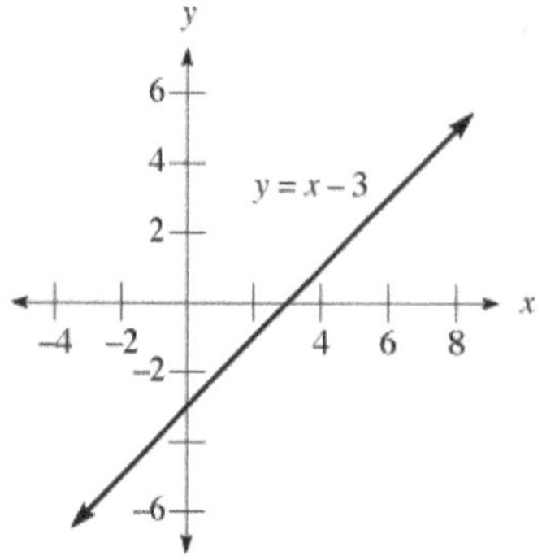

35. a. Solving the equation:

$$\begin{aligned} 4x+12&=-16 \\ 4x&=-28 \\ x&=-7 \end{aligned}$$

b. Substituting $y=0$:

$$\begin{aligned} 4x+12(0)&=-16 \\ 4x&=-16 \\ x&=-4 \end{aligned}$$

c. Substituting $x=0$:

$$\begin{aligned} 4(0)+12y&=-16 \\ 12y&=-16 \\ y&=-\frac{4}{3} \end{aligned}$$

d. Graphing the line:

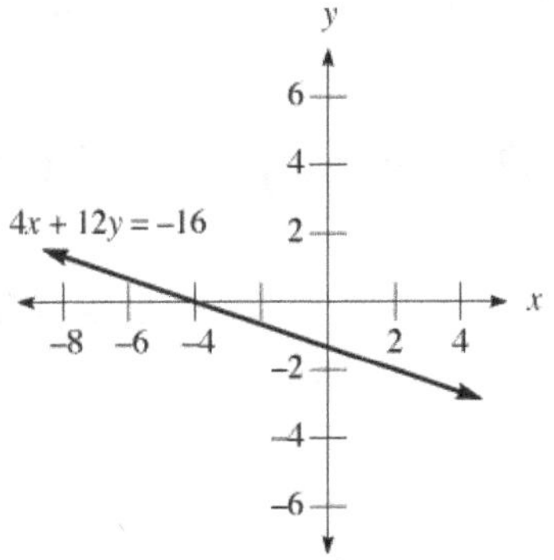

e. Solving for y:

$$\begin{aligned} 4x+12y&=-16 \\ 12y&=-4x-16 \\ y&=-\frac{1}{3}x-\frac{4}{3} \end{aligned}$$

37. a. Yes, (2000, 7500) is a point on the graph.
b. No, (2004, 15000) is not a point on the graph.
c. Yes, (2005, 15000) is a point on the graph.

39. Sketching the line graph:

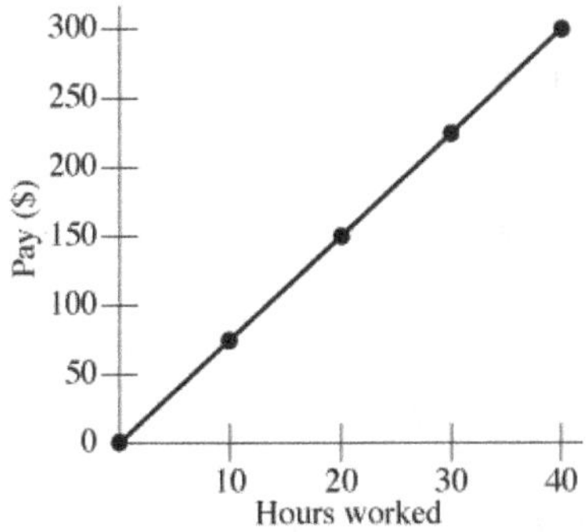

41. Sketching the bar chart:

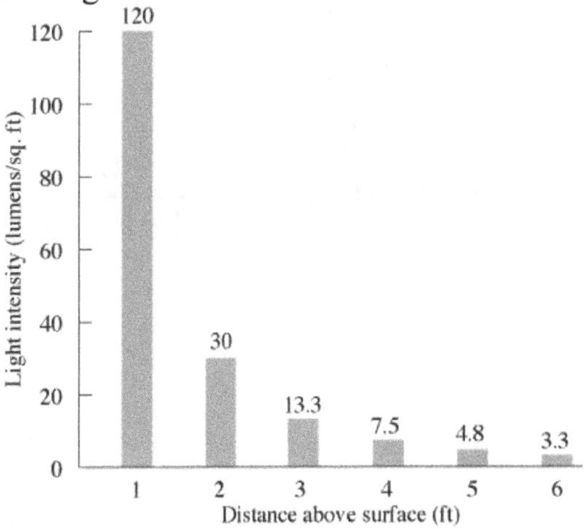

43. a. At 6:30, there are 60 people in line.
b. At 6:45, there are 70 people in line.
c. At 7:30, there are 10 people in line.
d. There are 60 people in line at 6:30 and at 7:00.
e. There are no people in line about 22 minutes after the show starts.

45. Writing as a fraction: $-0.06 = -\frac{6}{100}$

47. Substituting $x = 2$:

$$y = 2(2) - 3$$
$$y = 4 - 3$$
$$y = 1$$

49. Simplifying: $\frac{1-(-3)}{-5-(-2)} = \frac{4}{-3} = -\frac{4}{3}$

51. Simplifying: $\frac{-1-4}{3-3} = \frac{-5}{0}$, which is undefined

53. **a.** The number is $\frac{3}{2}$, since $\frac{2}{3} \bullet \frac{3}{2} = 1$.

b. The number is $-\frac{3}{2}$, since $\frac{2}{3} \bullet \left(-\frac{3}{2}\right) = -1$.

3.2 The Slope of a Line

1. The slope is $\frac{3}{2}$.

3. There is no slope (undefined).

5. The slope is $\frac{2}{3}$.

7. Finding the slope: $m = \frac{4-1}{4-2} = \frac{3}{2}$

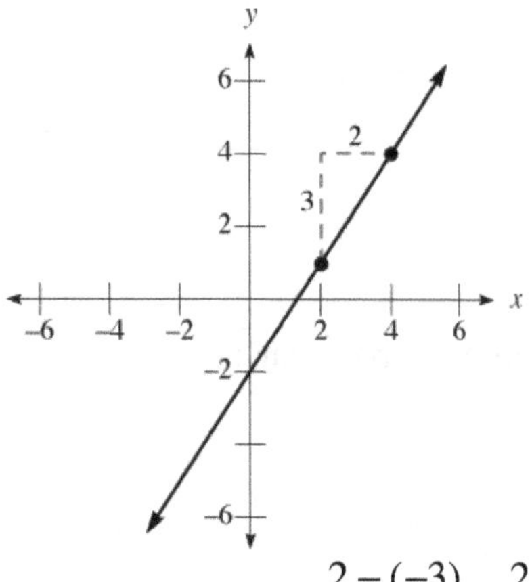

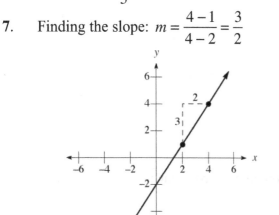

9. Finding the slope: $m = \frac{2-4}{5-1} = \frac{-2}{4} = -\frac{1}{2}$

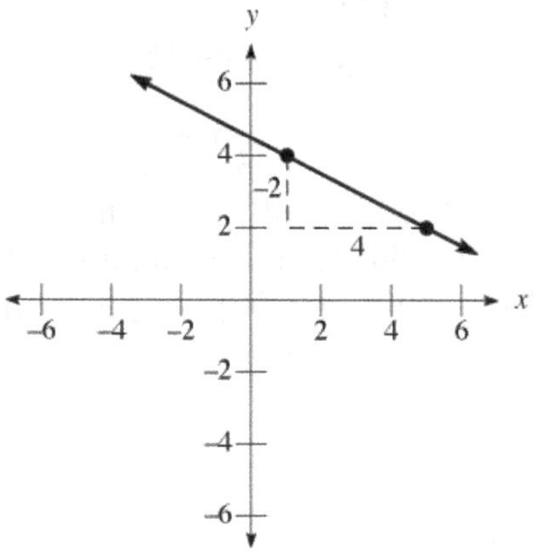

11. Finding the slope: $m = \frac{2-(-3)}{4-1} = \frac{2+3}{3} = \frac{5}{3}$

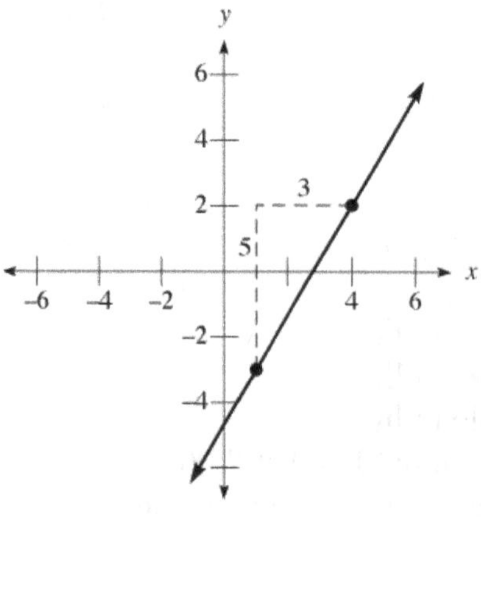

13. Finding the slope: $m = \dfrac{-9-(-4)}{5-2} = \dfrac{-9+4}{3} = -\dfrac{5}{3}$

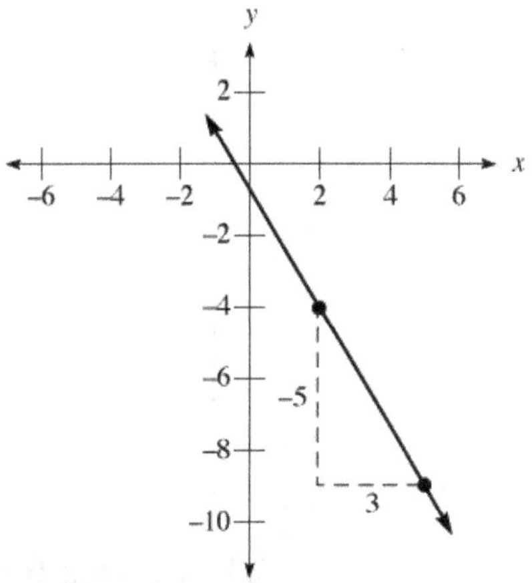

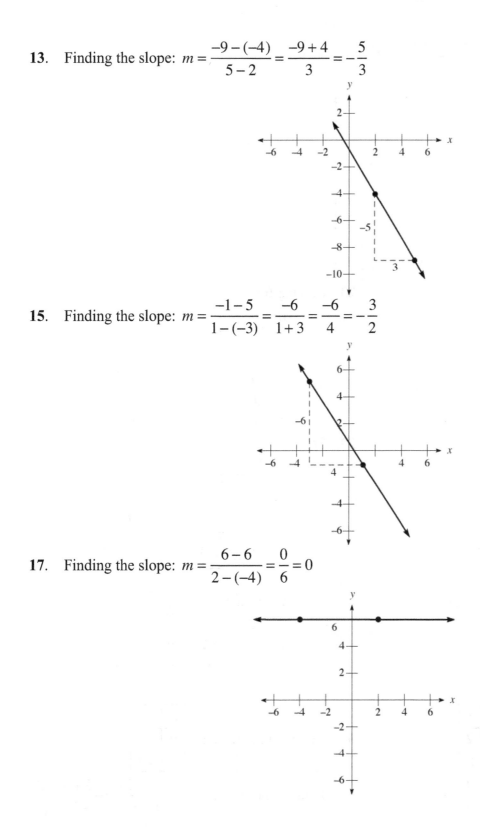

15. Finding the slope: $m = \dfrac{-1-5}{1-(-3)} = \dfrac{-6}{1+3} = \dfrac{-6}{4} = -\dfrac{3}{2}$

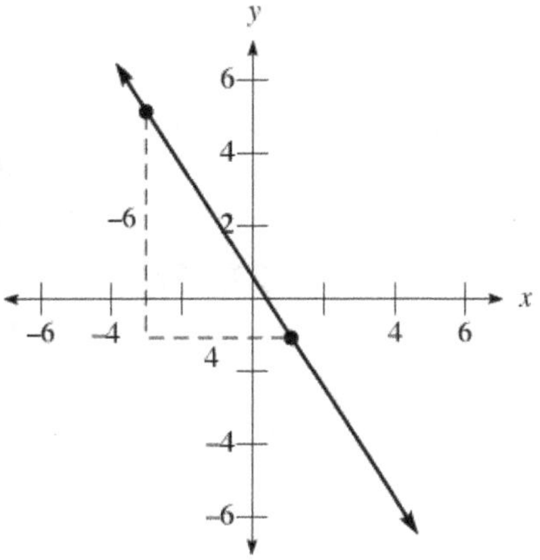

17. Finding the slope: $m = \dfrac{6-6}{2-(-4)} = \dfrac{0}{6} = 0$

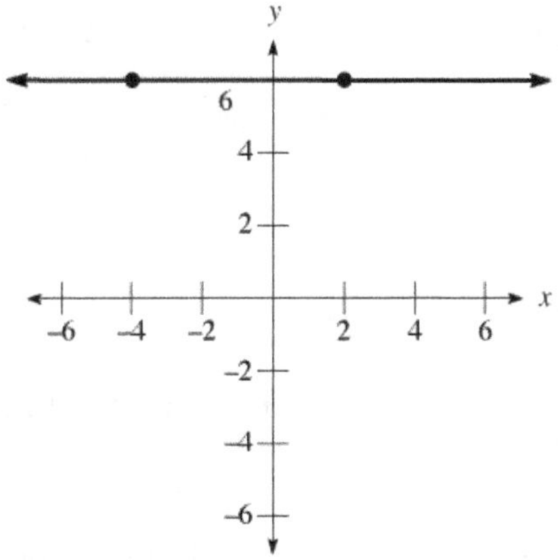

19. Finding the slope: $m=\dfrac{5-(-3)}{a-a}=\dfrac{5+3}{0}$, which is undefined

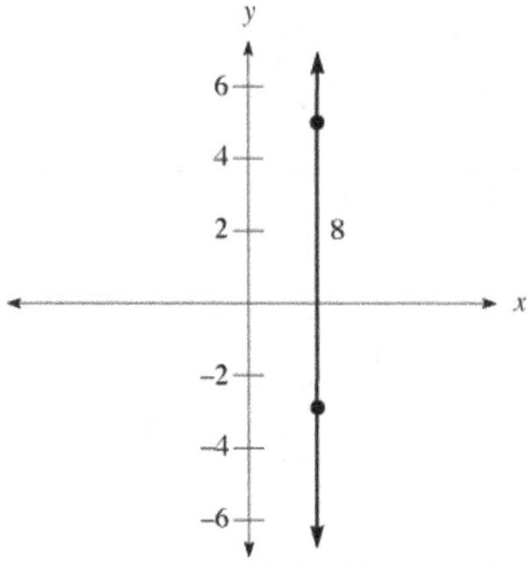

21. Solving for a:

$$\frac{3-6}{a-2}=-1$$
$$-3=-1(a-2)$$
$$-3=-a+2$$
$$-5=-a$$
$$a=5$$

23. Solving for b:

$$\frac{4b-b}{-1-2}=-2$$
$$\frac{3b}{-3}=-2$$
$$3b=6$$
$$b=2$$

25. Solving for x:

$$\frac{x^2-4}{x-2}=5$$
$$\frac{(x+2)(x-2)}{x-2}=5$$
$$x+2=5$$
$$x=3$$

27. Solving for x:

$$\frac{2x^2+1-3}{x-1}=-6$$
$$\frac{2x^2-2}{x-1}=-6$$
$$\frac{2(x^2-1)}{x-1}=-6$$
$$\frac{2(x+1)(x-1)}{x-1}=-6$$
$$2x+2=-6$$
$$2x=-8$$
$$x=-4$$

29. Completing the table:

x	y
0	2
3	0

Finding the slope: $m=\dfrac{2-0}{0-3}=-\dfrac{2}{3}$

31. Completing the table:

x	y
0	–5
3	–3

Finding the slope: $m=\dfrac{-5-(-3)}{0-3}=\dfrac{-5+3}{-3}=\dfrac{2}{3}$

33. Graphing the line:

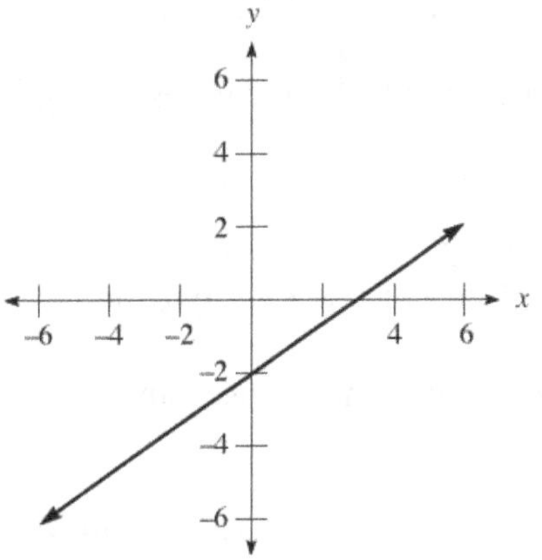

The slope is $m = \frac{2}{3}$.

35. Finding the slope of this line: $m = \frac{1-3}{-8-2} = \frac{-2}{-10} = \frac{1}{5}$.

Since the parallel slope is the same, its slope is $\frac{1}{5}$.

37. Finding the slope of this line: $m = \frac{2-(-6)}{5-5} = \frac{8}{0}$, which is undefined

Since the perpendicular slope is a horizontal line, its slope is 0.

39. Finding the slope of this line: $m = \frac{-5-1}{4-(-2)} = \frac{-6}{6} = -1$.

Since the parallel slope is the same, its slope is -1.

41. Finding the slope of this line: $m = \frac{-3-(-5)}{1-(-2)} = \frac{2}{3}$

Since the perpendicular slope is the negative reciprocal, its slope is $-\frac{3}{2}$.

43. **a.** Since the slopes between each successive pairs of points is 2, this could represent ordered pairs from a line.

b. Since the slopes between each successive pairs of points is not the same, this could not represent ordered pairs from a line.

45. Finding the slope: $m = \frac{105-0}{6-0} = 17.5$ miles/hour

47. Finding the slope: $m = \frac{3600-0}{30-0} = 120$ feet/second

49. **a.** It takes 10 minutes for all the ice to melt. **b.** It takes 20 minutes before the water boils.

c. The slope of A is 20°C per minute. **d.** The slope of C is 10°C per minute.

e. It is changing faster during the first minute, since its slope is greater.

51. Finding the slope: $m = \frac{20,000-8,000}{2006-1997} = \frac{22,000}{9} \approx 1,333$

The number of solar thermal collectors increased at a rate of 1,333 shipments per year.

53. **a.** Computing the slope: $m \approx \dfrac{150-40}{2600-450} = \dfrac{110}{2150} \approx 0.05$

For each additional lumen of output, the incandescent light bulb uses an average of 0.05 watts of energy.

b. Computing the slope: $m \approx \dfrac{40-10}{2600-450} = \dfrac{30}{2150} \approx 0.014$

For each additional lumen of output, the energy efficient bulb uses an average of 0.014 watts of energy.

c. The energy efficient bulb is better, since it uses an average amount of energy which is less per lumen of output.

55. Simplifying: $2\left(-\dfrac{1}{2}\right) = -1$

57. Simplifying: $\dfrac{5-(-3)}{2-6} = \dfrac{8}{-4} = -2$

59. Solving for y:

$$\begin{aligned} \frac{y-b}{x-0} &= m \\ y-b &= mx \\ y &= mx+b \end{aligned}$$

61. Solving for y:

$$\begin{aligned} y-3 &= -2(x+4) \\ y-3 &= -2x-8 \\ y &= -2x-5 \end{aligned}$$

63. Solving for y: $y = -\dfrac{4}{3}(0)+5 = 0+5 = 5$

3.3 The Equation of a Line

1. Using the slope-intercept formula: $y = -4x-3$

3. Using the slope-intercept formula: $y = -\dfrac{2}{3}x$

5. Using the slope-intercept formula: $y = -\dfrac{2}{3}x + \dfrac{1}{4}$

7. **a.** The parallel slope will be the same, which is 3.

b. The perpendicular slope will be the negative reciprocal, which is $-\dfrac{1}{3}$.

9. **a.** First solve for y to find the slope:

$$\begin{aligned} 3y+y &= -2 \\ y &= -3x-2 \end{aligned}$$

The parallel slope will be the same, which is –3.

b. The perpendicular slope will be the negative reciprocal, which is $\dfrac{1}{3}$.

11. **a.** First solve for y to find the slope:

$$\begin{aligned} 2x+5y &= -11 \\ 5y &= -2x-11 \\ y &= -\frac{2}{5}x - \frac{11}{5} \end{aligned}$$

The parallel slope will be the same, which is $-\dfrac{2}{5}$.

b. The perpendicular slope will be the negative reciprocal, which is $\frac{5}{2}$.

13. The slope is 3, the *y*-intercept is –2, and the perpendicular slope is $-\frac{1}{3}$.

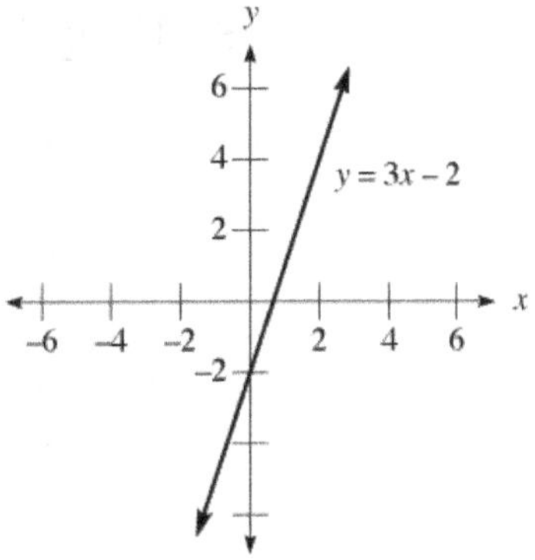

15. The slope is $\frac{2}{3}$, the *y*-intercept is –4, and the perpendicular slope is $-\frac{3}{2}$.

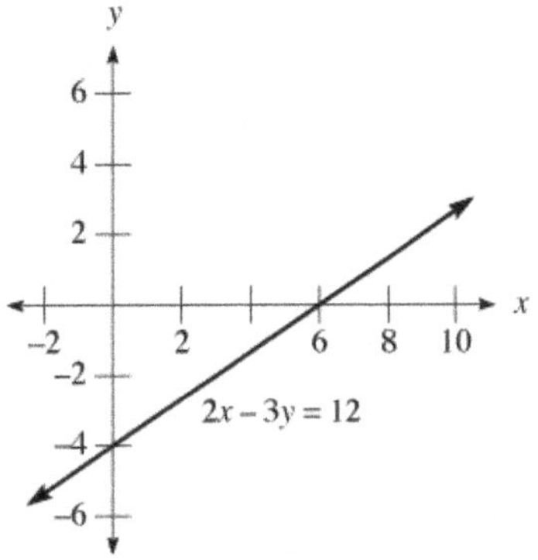

17. The slope is $-\frac{4}{5}$, the *y*-intercept is 4, and the perpendicular slope is $\frac{5}{4}$.

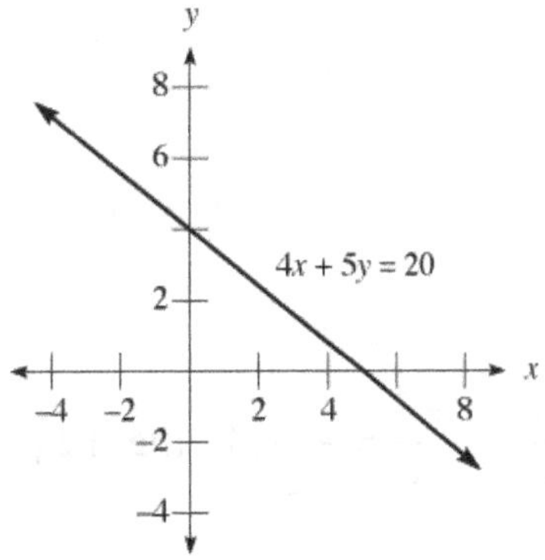

19. The slope is $\frac{1}{2}$ and the *y*-intercept is –4. Using the slope-intercept form, the equation is $y = \frac{1}{2}x - 4$.

21. The slope is $-\frac{2}{3}$ and the y-intercept is 3. Using the slope-intercept form, the equation is $y=-\frac{2}{3}x+3$.

23. Using the point-slope formula:

$$\begin{aligned} y-(-5)&=2(x-(-2)) \\ y+5&=2(x+2) \\ y+5&=2x+4 \\ y&=2x-1 \end{aligned}$$

25. Using the point-slope formula:

$$\begin{aligned} y-1&=-\frac{1}{2}(x-(-4)) \\ y-1&=-\frac{1}{2}(x+4) \\ y-1&=-\frac{1}{2}x-2 \\ y&=-\frac{1}{2}x-1 \end{aligned}$$

27. Using the point-slope formula:

$$\begin{aligned} y-2&=-3\left(x-\left(-\frac{1}{3}\right)\right) \\ y-2&=-3\left(x+\frac{1}{3}\right) \\ y-2&=-3x-1 \\ y&=-3x+1 \end{aligned}$$

29. Using the point-slope formula:

$$\begin{aligned} y-2&=\frac{2}{3}(x-(-4)) \\ y-2&=\frac{2}{3}(x+4) \\ y-2&=\frac{2}{3}x+\frac{8}{3} \\ y&=\frac{2}{3}x+\frac{14}{3} \end{aligned}$$

31. Using the point-slope formula:

$$\begin{aligned} y-(-2)&=-\frac{1}{4}(x-(-5)) \\ y+2&=-\frac{1}{4}(x+5) \\ y+2&=-\frac{1}{4}x-\frac{5}{4} \\ y&=-\frac{1}{4}x-\frac{13}{4} \end{aligned}$$

33. First find the slope: $m=\frac{1-(-2)}{-2-3}=\frac{1+2}{-5}=-\frac{3}{5}$. Using the point-slope formula:

$$\begin{aligned} y-(-2)&=-\frac{3}{5}(x-3) \\ 5(y+2)&=-3(x-3) \\ 5y+10&=-3x+9 \\ 3x+5y&=-1 \end{aligned}$$

35. First find the slope: $m = \dfrac{\frac{1}{3}-\frac{1}{2}}{-4-(-2)} = \dfrac{-\frac{1}{6}}{-4+2} = \dfrac{-\frac{1}{6}}{-2} = \dfrac{1}{12}$. Using the point-slope formula:

$$\begin{aligned} y-\frac{1}{2} &= \frac{1}{12}(x-(-2)) \\ 12\left(y-\frac{1}{2}\right) &= 1(x+2) \\ 12y-6 &= x+2 \\ x-12y &= -8 \end{aligned}$$

37. First find the slope: $m = \dfrac{-1-\left(-\frac{1}{5}\right)}{-\frac{1}{3}-\frac{1}{3}} = \dfrac{-1+\frac{1}{5}}{-\frac{2}{3}} = \dfrac{-\frac{4}{5}}{-\frac{2}{3}} = \dfrac{4}{5} \bullet \dfrac{3}{2} = \dfrac{6}{5}$. Using the point-slope formula:

$$\begin{aligned} y-(-1) &= \frac{6}{5}\left(x-\left(-\frac{1}{3}\right)\right) \\ y+1 &= \frac{6}{5}\left(x+\frac{1}{3}\right) \\ 5(y+1) &= 6\left(x+\frac{1}{3}\right) \\ 5y+5 &= 6x+2 \\ 6x-5y &= 3 \end{aligned}$$

39. Two points on the line are (0,–4) and (2,0). Finding the slope: $m = \dfrac{0-(-4)}{2-0} = \dfrac{4}{2} = 2$

Using the slope-intercept form, the equation is $y = 2x-4$.

41. Two points on the line are (0,4) and (–2,0). Finding the slope: $m = \dfrac{0-4}{-2-0} = \dfrac{-4}{-2} = 2$

Using the slope-intercept form, the equation is $y = 2x+4$.

43. **a.** For the x-intercept, substitute $y = 0$:

$$\begin{aligned} 3x-2(0) &= 10 \\ 3x &= 10 \\ x &= \frac{10}{3} \end{aligned}$$

For the y-intercept, substitute $x = 0$:

$$\begin{aligned} 3(0)-2y &= 10 \\ -2y &= 10 \\ y &= -5 \end{aligned}$$

b. Substituting $y = 1$:

$$\begin{aligned} 3x-2(1) &= 10 \\ 3x-2 &= 10 \\ 3x &= 12 \\ x &= 4 \end{aligned}$$

Another solution is (4,1). Other answers are possible.

c. Solving for y:

$$3x-2y=10$$
$$-2y=-3x+10$$
$$y=\frac{3}{2}x-5$$

d. Substituting $x=2$: $y=\frac{3}{2}(2)-5=3-5=-2$.

No, the point (2,2) is not a solution to the equation.

45. a. Solving for x:

$$-2x+1=-3$$
$$-2x=-4$$
$$x=2$$

b. Substituting $y=0$:

$$-2x+0=-3$$
$$-2x=-3$$
$$x=\frac{3}{2}$$

c. Substituting $x=0$: $y=2(0)-3=-3$

d. Sketching the graph:

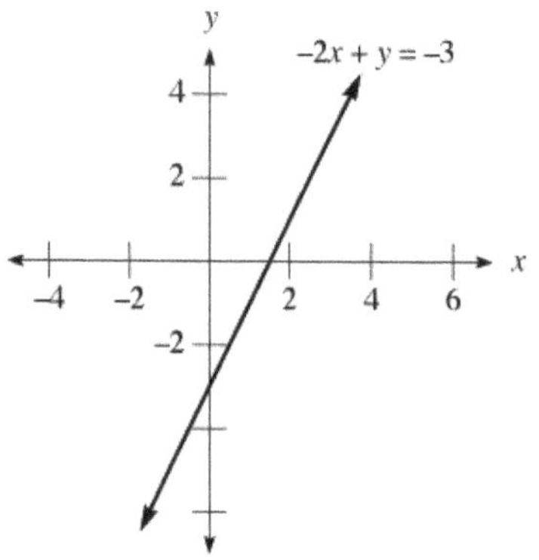

e. Solving for y:

$$-2x+y=-3$$
$$y=2x-3$$

47. a. The slope is $\frac{1}{2}$, the x-intercept is 0, and the y-intercept is 0.

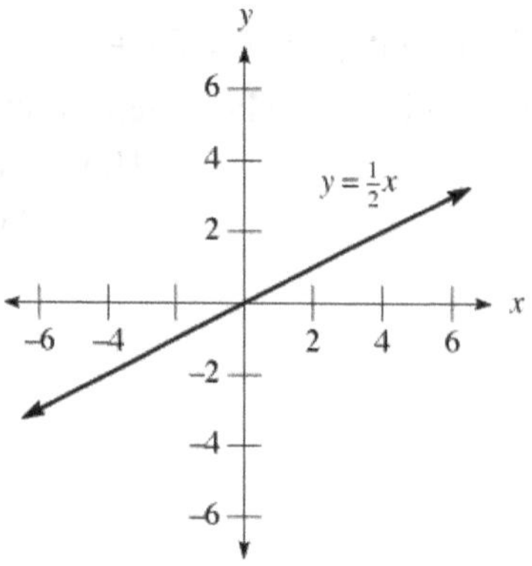

b. There is no slope, the x-intercept is 3, and there is no y-intercept.

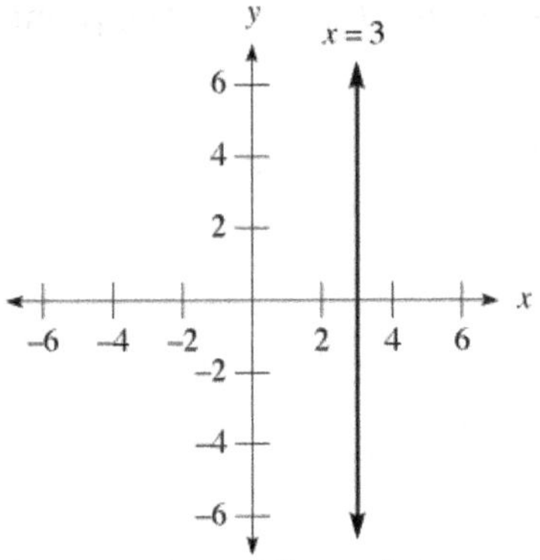

c. The slope is 0, there is no x-intercept, and the y-intercept is –2.

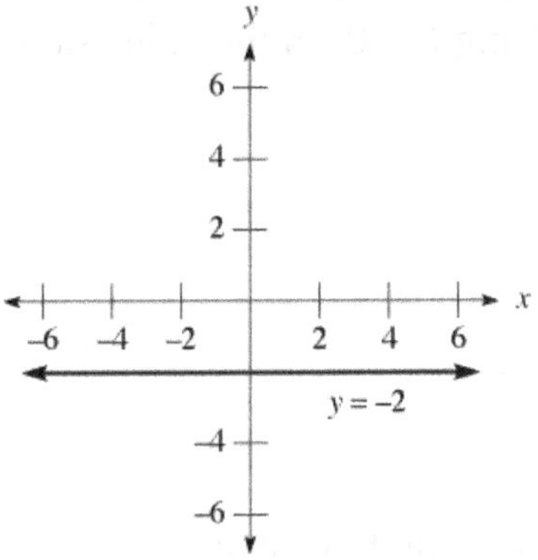

49. First find the slope:

$$\begin{aligned} 3x - y &= 5 \\ -y &= -3x + 5 \\ y &= 3x - 5 \end{aligned}$$

So the slope is 3. Using (–1,4) in the point-slope formula:

$$\begin{aligned} y - 4 &= 3(x - (-1)) \\ y - 4 &= 3(x + 1) \\ y - 4 &= 3x + 3 \\ y &= 3x + 7 \end{aligned}$$

51. First find the slope:

$$\begin{aligned} 2x - 5y &= 10 \\ -5y &= -2x + 10 \\ y &= \frac{2}{5}x - 2 \end{aligned}$$

So the perpendicular slope is $-\frac{5}{2}$. Using (–4,–3) in the point-slope formula:

$$y-(-3)=-\frac{5}{2}(x-(-4))$$
$$y+3=-\frac{5}{2}(x+4)$$
$$y+3=-\frac{5}{2}x-10$$
$$y=-\frac{5}{2}x-13$$

53. The perpendicular slope is $\frac{1}{4}$. Using (–1,0) in the point-slope formula:

$$y-0=\frac{1}{4}(x-(-1))$$
$$y=\frac{1}{4}(x+1)$$
$$y=\frac{1}{4}x+\frac{1}{4}$$

55. Using the points (3,0) and (0,2), first find the slope: $m=\frac{2-0}{0-3}=-\frac{2}{3}$

Using the slope-intercept formula, the equation is: $y=-\frac{2}{3}x+2$

57. **a.** Using the points (0,32) and (25,77), first find the slope: $m=\frac{77-32}{25-0}=\frac{45}{25}=\frac{9}{5}$

Using the slope-intercept formula, the equation is: $F=\frac{9}{5}C+32$

b. Substituting $C = 30$: $F=\frac{9}{5}(30)+32=54+32=86°$

59. **a.** Substituting $n = 10{,}000$: $C=125{,}000+6.5(10{,}000)=\$190{,}000$

b. Finding the average cost: $\frac{\$190{,}000}{10{,}000}=\$19$ per textbook

c. Since each textbook costs \$6.50 in materials, this is the cost to produce the next textbook.

61. Using the points (2000, 65.4) and (2005, 104), first find the slope: $m=\frac{104-65.4}{2005-2000}=\frac{38.6}{5}=7.72$

Using (2005, 104) in the point-slope formula:

$$y-104=7.72(x-2005)$$
$$y-104=7.72x-15{,}478.6$$
$$y\approx 7.7x-15{,}374.6$$

63. Since $0+0\le 4$ and $4+0\le 4$, but $2+3>4$, the points (0,0) and (4,0) are solutions.

65. Since $0\le\frac{1}{2}(0)$ and $0\le\frac{1}{2}(2)$, but $0>\frac{1}{2}(-2)$, the points (0,0) and (2,0) are solutions.

3.4 Linear Inequalities in Two Variables

1. Graphing the solution set:

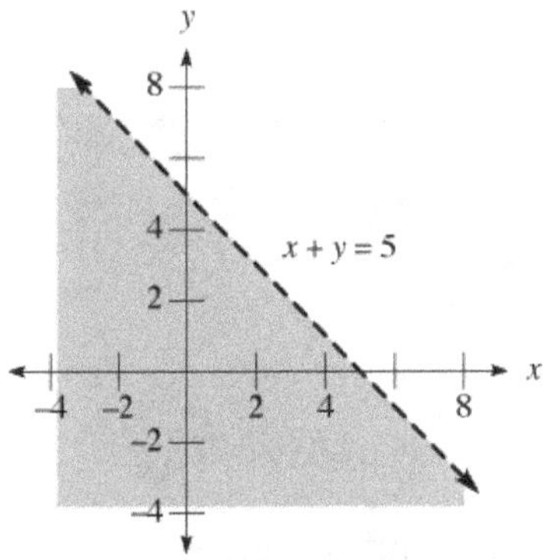

3. Graphing the solution set:

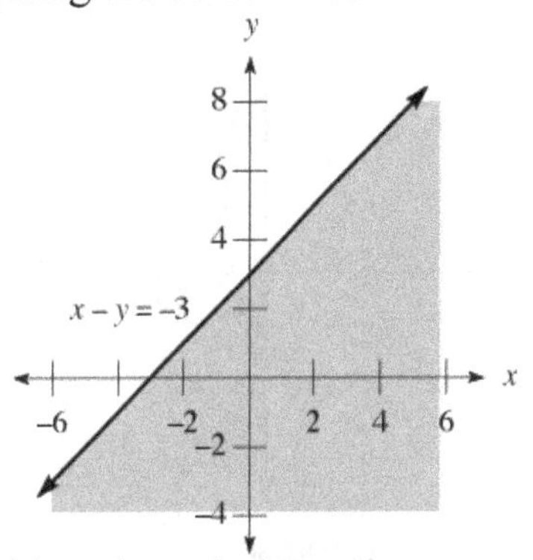

5. Graphing the solution set:

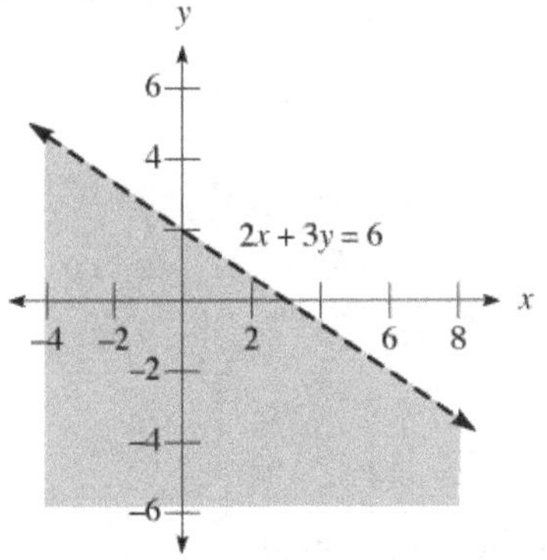

7. Graphing the solution set:

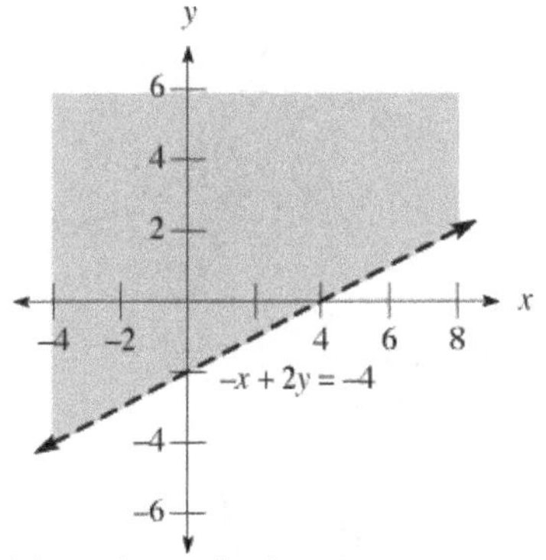

9. Graphing the solution set:

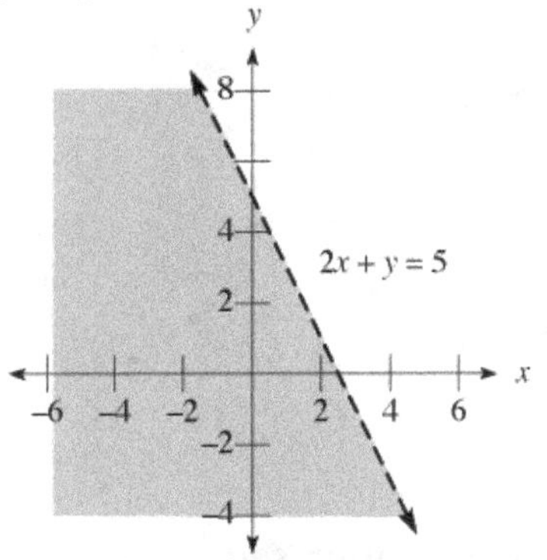

11. Graphing the solution set:

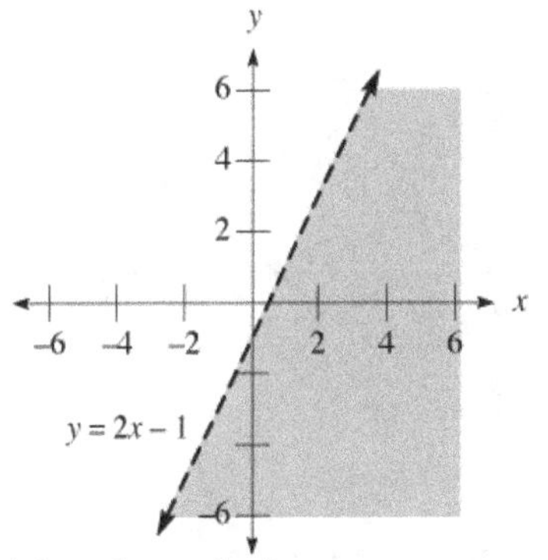

13. Graphing the solution set:

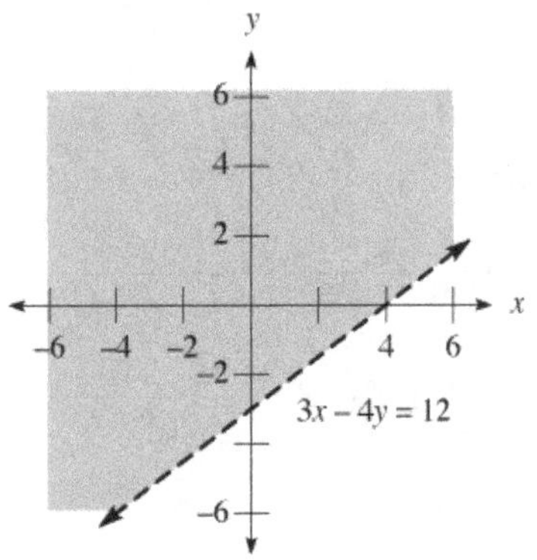

15. Graphing the solution set:

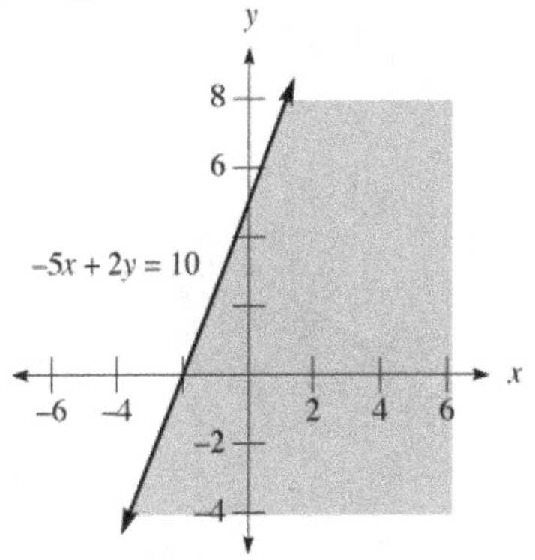

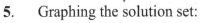

17. The inequality is $x+y>4$, or $y>-x+4$.

19. The inequality is $-x+2y\le 4$ or $y\le\frac{1}{2}x+2$.

21. Graphing the solution set:

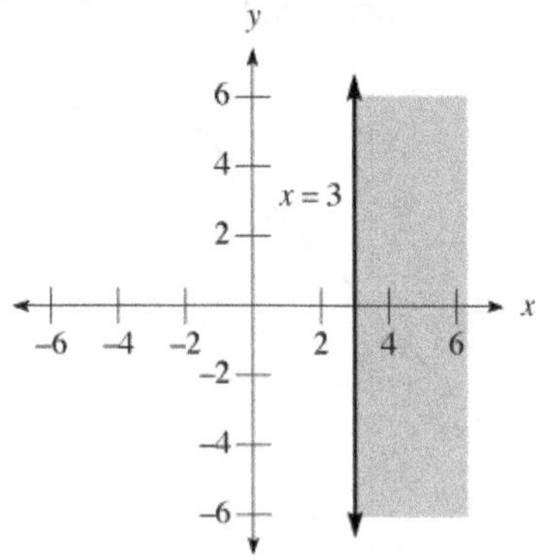

23. Graphing the solution set:

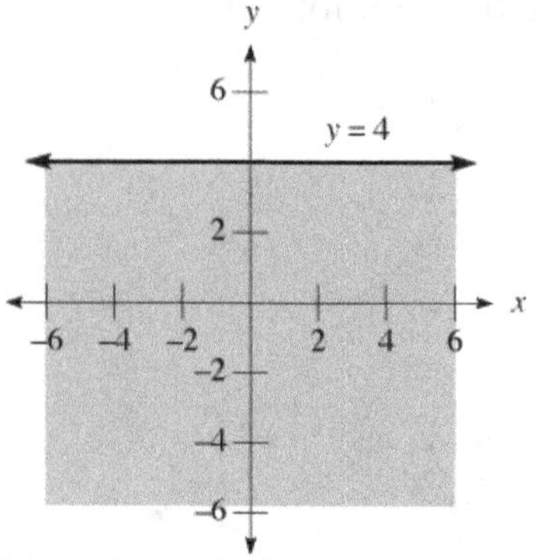

25. Graphing the solution set:

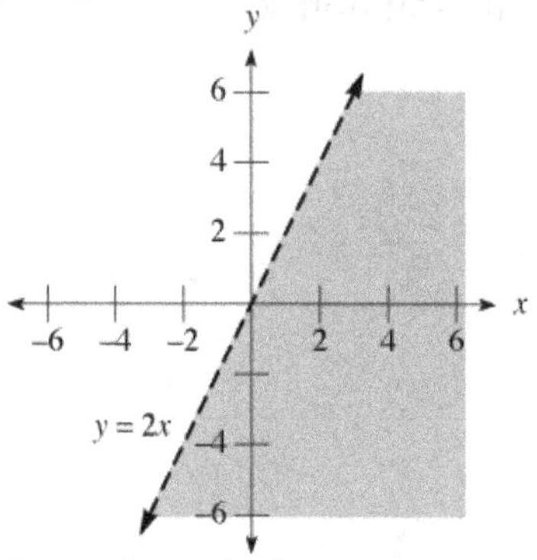

27. Graphing the solution set:

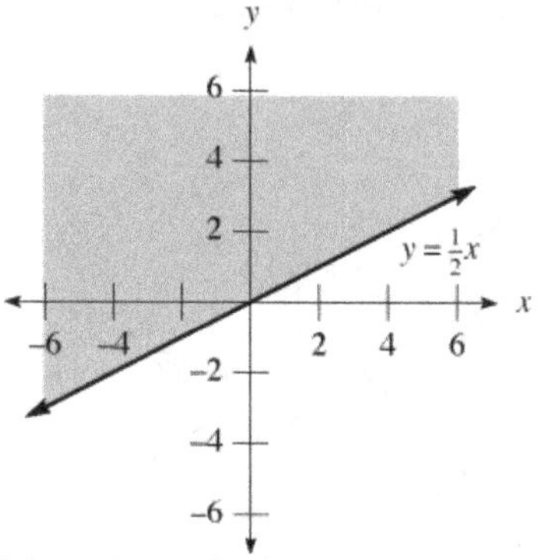

29. Graphing the solution set:

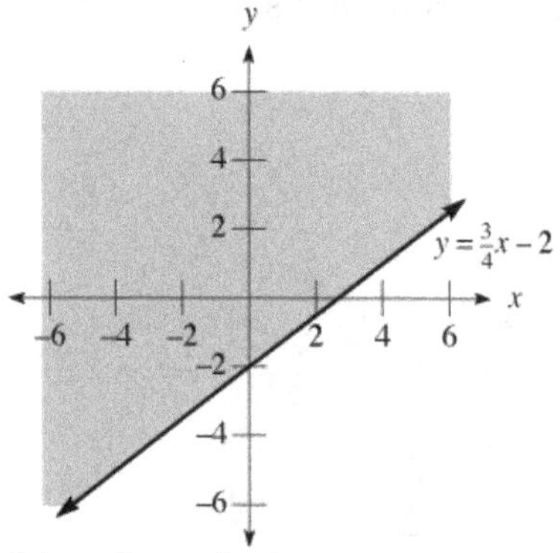

31. Graphing the solution set:

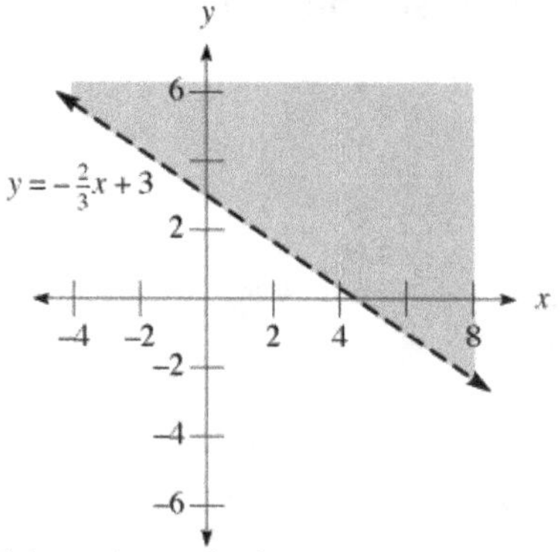

33. Graphing the solution set:

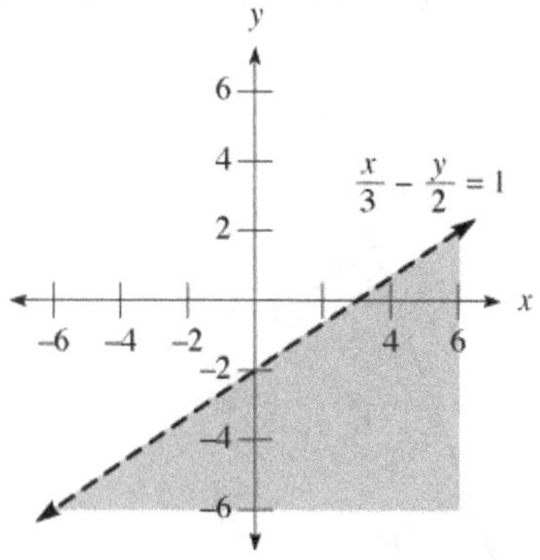

35. Graphing the solution set:

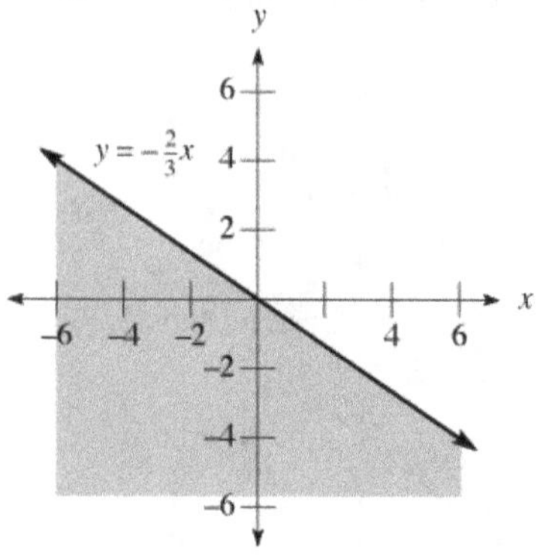

37. Graphing the solution set:

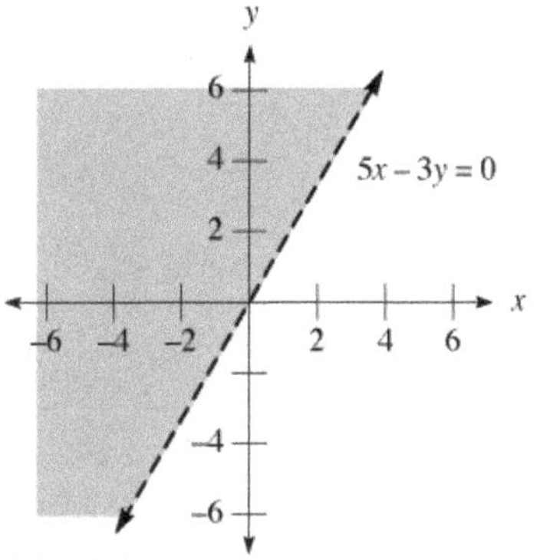

39. Graphing the solution set:

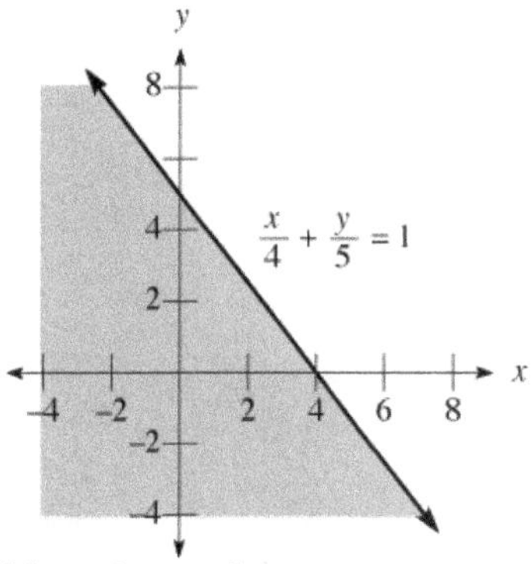

41. Graphing the region:

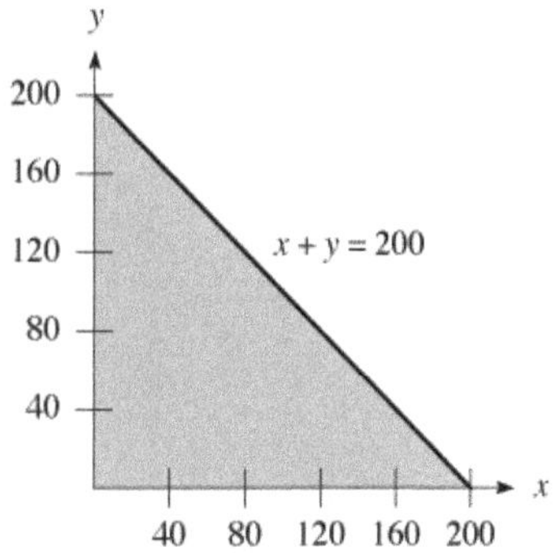

43. Graphing the region:

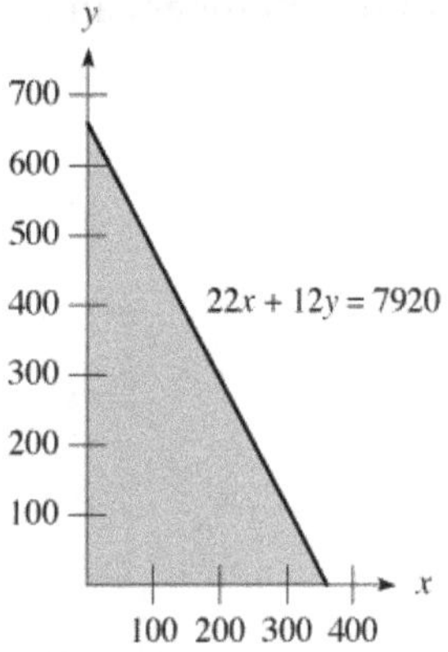

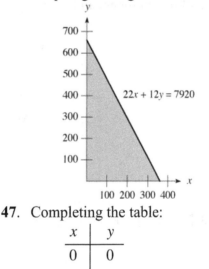

45. Completing the table:

x	y
0	0
10	75
20	150

47. Completing the table:

x	y
0	0
$\frac{1}{2}$	3.75
1	7.5

3.5 Introduction to Functions

1. The domain is $\{1,3,5,7\}$ and the range is $\{2,4,6,8\}$. This is a function.
3. The domain is $\{0,1,2,3\}$ and the range is $\{4,5,6\}$. This is a function.
5. The domain is $\{a,b,c,d\}$ and the range is $\{3,4,5\}$. This is a function.
7. The domain is $\{a\}$ and the range is $\{1,2,3,4\}$. This is not a function.
9. Yes, since it passes the vertical line test.
11. No, since it fails the vertical line test.
13. No, since it fails the vertical line test.
15. Yes, since it passes the vertical line test.
17. Yes, since it passes the vertical line test.
19. The domain is $\{x \mid -5 \le x \le 5\}$ and the range is $\{y \mid 0 \le y \le 5\}$.
21. The domain is $\{x \mid -5 \le x \le 3\}$ and the range is $\{y \mid y = 3\}$.

23. The domain is all real numbers and the range is $\{y \mid y \geq -1\}$. This is a function.

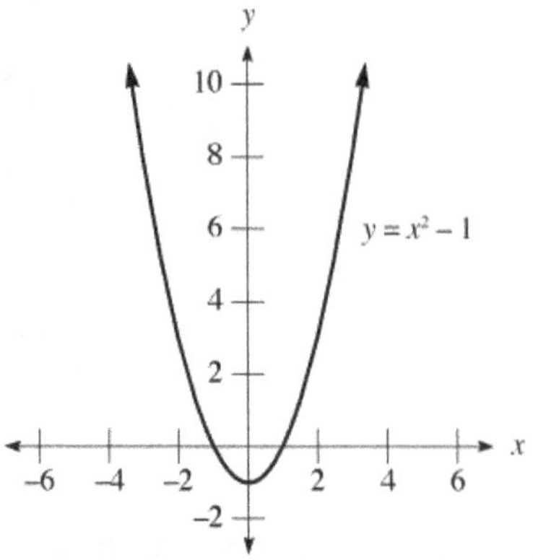

25. The domain is all real numbers and the range is $\{y \mid y \geq 4\}$. This is a function.

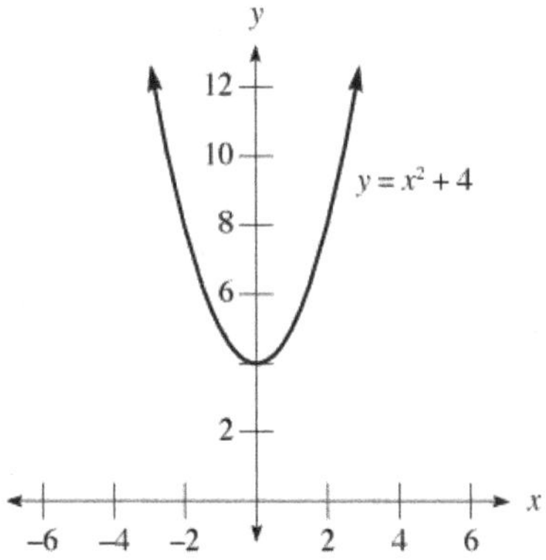

27. The domain is $\{x \mid x \geq -1\}$ and the range is all real numbers. This is not a function.

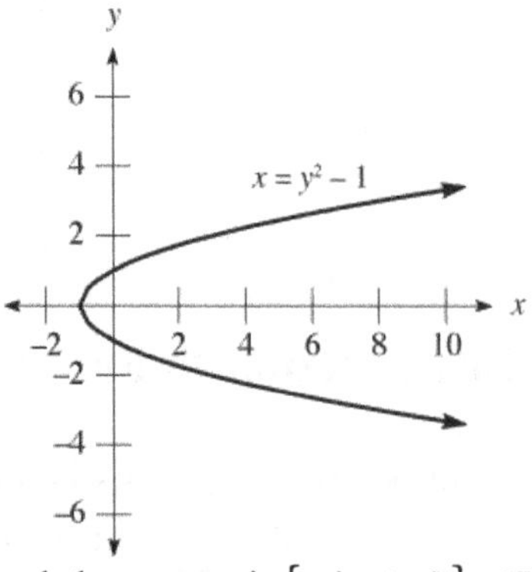

29. The domain is all real numbers and the range is $\{y \mid y \geq 0\}$. This is a function.

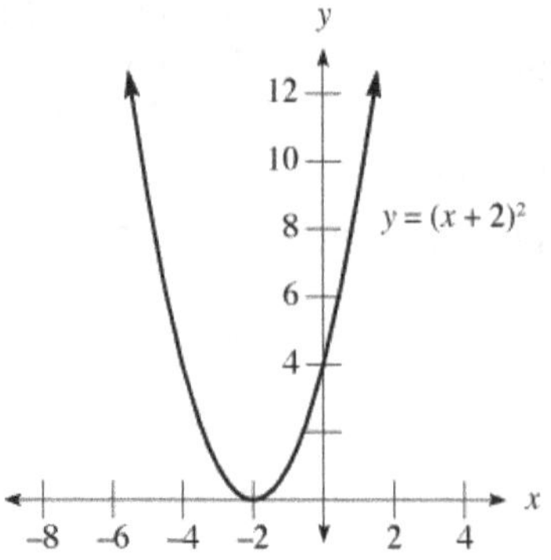

31. The domain is $\{x \mid x \geq 0\}$ and the range is all real numbers. This is not a function.

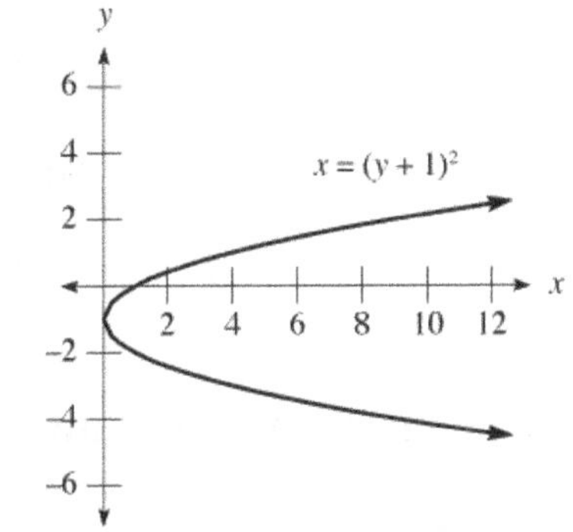

33. **a.** The equation is $y = 8.5x$ for $10 \leq x \leq 40$.

b. Completing the table:

Hours Worked x	Function Rule $y = 8.5x$	Gross Pay (\$) y
10	$y = 8.5(10) = 85$	85
20	$y = 8.5(20) = 170$	170
30	$y = 8.5(30) = 255$	255
40	$y = 8.5(40) = 340$	340

c. Constructing a line graph:

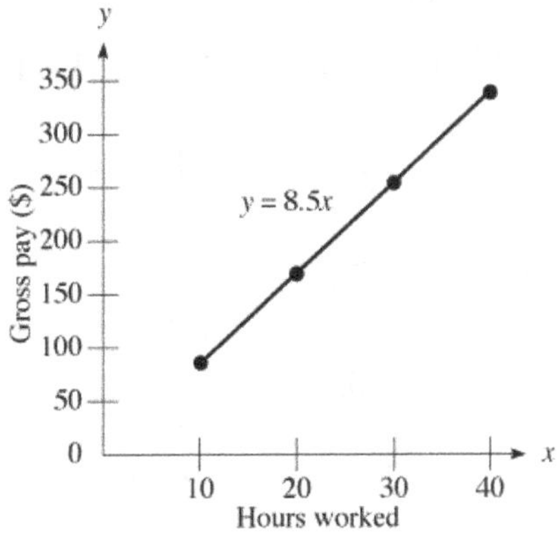

d. The domain is $\{x \mid 10 \leq x \leq 40\}$ and the range is $\{y \mid 85 \leq y \leq 340\}$.

e. The minimum is \$85 and the maximum is \$340.

35. The domain is {2004, 2005, 2006, 2007, 2008, 2009, 2010} and the range is {680, 730, 800, 900, 920, 990, 1030}.

37. **a.** Figure III **b.** Figure I
c. Figure II **d.** Figure IV

39. Simplifying: $4(3.14)(9) \approx 113$

41. Simplifying: $4(-2) - 1 = -8 - 1 = -9$

43. **a.** Substituting $t = 10$: $s = \dfrac{60}{10} = 6$ **b.** Substituting $t = 8$: $s = \dfrac{60}{8} = 7.5$

45. **a.** Substituting $x = 5$: $(5)^2 + 2 = 25 + 2 = 27$

b. Substituting $x = -2$: $(-2)^2 + 2 = 4 + 2 = 6$

47. Substituting $x = 2$: $y = (2)^2 - 3 = 4 - 3 = 1$

49. Substituting $x = 0$: $y = (0)^2 - 3 = 0 - 3 = -3$

51. Solving for y:

$$\begin{aligned}\frac{8}{5}-2y&=4\\-2y&=\frac{12}{5}\\y&=-\frac{6}{5}\end{aligned}$$

53. Substituting $x=0$ and $y=0$:

$$\begin{aligned}0&=a(0-8)^2+70\\0&=64a+70\\-64a&=70\\a&=-\frac{70}{64}=-\frac{35}{32}\end{aligned}$$

3.6 Function Notation

1. Evaluating the function: $f(2)=2(2)-5=4-5=-1$

3. Evaluating the function: $f(-3)=2(-3)-5=-6-5=-11$

5. Evaluating the function: $g(-1)=(-1)^2+3(-1)+4=1-3+4=2$

7. Evaluating the function: $g(-3)=(-3)^2+3(-3)+4=9-9+4=4$

9. Evaluating the function: $g(a)=a^2+3a+4$

11. Evaluating the function: $f(a+6)=2(a+6)-5=2a+12-5=2a+7$

13. Evaluating the function: $f(0)=3(0)^2-4(0)+1=0-0+1=1$

15. Evaluating the function: $g(-4)=2(-4)-1=-8-1=-9$

17. Evaluating the function: $f(-1)=3(-1)^2-4(-1)+1=3+4+1=8$

19. Evaluating the function: $g\left(\frac{1}{2}\right)=2\left(\frac{1}{2}\right)-1=1-1=0$

21. Evaluating the function: $f(a)=3a^2-4a+1$

23. Evaluating the function:

$$f(a+2)=3(a+2)^2-4(a+2)+1=3a^2+12a+12-4a-8+1=3a^2+8a+5$$

25. Evaluating: $f(1)=4$

27. Evaluating: $g\left(\frac{1}{2}\right)=0$

29. Evaluating: $g(-2)=2$

31. Evaluating the function: $f(-4)=(-4)^2-2(-4)=16+8=24$

33. Evaluating the function:

$$f(-2)+g(-1)=\left[(-2)^2-2(-2)\right]+\left[5(-1)-4\right]=(4+4)+(-5-4)=8-9=-1$$

35. Evaluating the function:

$$2f(x)-3g(x)=2(x^2-2x)-3(5x-4)=2x^2-4x-15x+12=2x^2-19x+12$$

37. Evaluating the function:

$$f[g(3)]=f[5(3)-4]=f(15-4)=f(11)=11^2-2(11)=121-22=99$$

39. Evaluating the function: $f\left(\frac{1}{3}\right)=\dfrac{1}{\frac{1}{3}+3}=\dfrac{1}{\frac{10}{3}}=\dfrac{3}{10}$

41. Evaluating the function: $f\left(-\frac{1}{2}\right)=\frac{1}{-\frac{1}{2}+3}=\frac{1}{\frac{5}{2}}=\frac{2}{5}$

43. Evaluating the function: $f(-3)=\frac{1}{-3+3}=\frac{1}{0}$, which is undefined

45. **a.** Evaluating the function: $f(a)-3=a^2-4-3=a^2-7$

b. Evaluating the function: $f(a-3)=(a-3)^2-4=a^2-6a+9-4=a^2-6a+5$

c. Evaluating the function: $f(x)+2=x^2-4+2=x^2-2$

d. Evaluating the function: $f(x+2)=(x+2)^2-4=x^2+4x+4-4=x^2+4x$

e. Evaluating the function: $f(a+b)=(a+b)^2-4=a^2+2ab+b^2-4$

f. Evaluating the function: $f(x+h)=(x+h)^2-4=x^2+2xh+h^2-4$

47. Graphing the function:

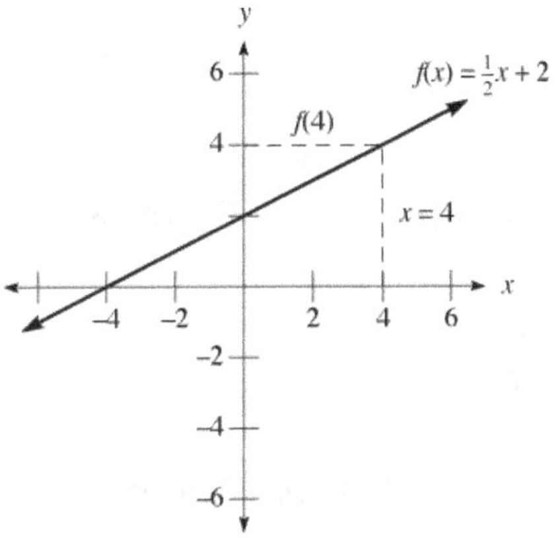

49. Finding where $f(x)=x$:

$$\frac{1}{2}x+2=x$$
$$2=\frac{1}{2}x$$
$$x=4$$

51. Graphing the function:

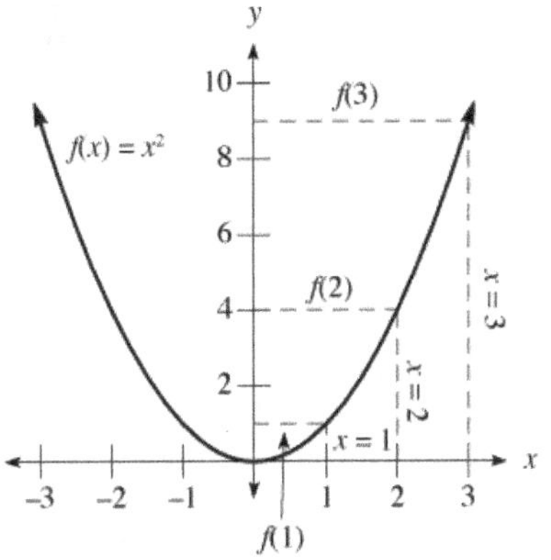

53. Evaluating: $V(3)=150\bullet 2^{3/3}=150\bullet 2=300$. The painting is worth \$300 in 3 years.

Evaluating: $V(6)=150\bullet 2^{6/3}=150\bullet 4=600$. The painting is worth \$600 in 6 years.

55. a. True b. True
c. True d. False
e. True

57. a. Evaluating: $V(3.75) = -3,300(3.75) + 18,000 = \$5,625$
b. Evaluating: $V(5) = -3,300(5) + 18,000 = \$1,500$
c. The domain of this function is $\{t \mid 0 \le t \le 5\}$.
d. Sketching the graph:

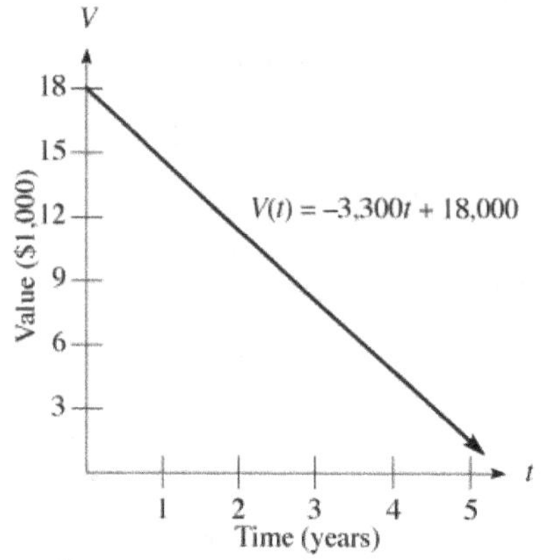

e. The range of this function is $\{V(t) \mid 1,500 \le V(t) \le 18,000\}$.
f. Solving $V(t) = 10,000$:

$$\begin{aligned} -3,300t + 18,000 &= 10,000 \\ -3,300t &= -8,000 \\ t &\approx 2.42 \end{aligned}$$

The copier will be worth $10,000 after approximately 2.42 years.

59. Simplifying: $16(3.5)^2 = 16(12.25) = 196$

61. Simplifying: $\dfrac{180}{45} = 4$

63. Simplifying: $\dfrac{0.0005(200)}{(0.25)^2} = \dfrac{0.1}{0.0625} = 1.6$

65. Solving for *K*:

$$\begin{aligned} 15 &= K(5) \\ K &= 3 \end{aligned}$$

67. Solving for *K*:

$$\begin{aligned} 50 &= \frac{K}{48} \\ K &= 50 \bullet 48 = 2,400 \end{aligned}$$

3.7 Variation

1. The variation equation is $y = Kx$. Substituting $x = 2$ and $y = 10$:

$$10 = K \bullet 2$$
$$K = 5$$

So $y = 5x$. Substituting $x = 6$: $y = 5 \bullet 6 = 30$

3. The variation equation is $r = \frac{K}{s}$. Substituting $s = 4$ and $r = -3$:

$$-3 = \frac{K}{4}$$
$$K = -12$$

So $r = \frac{-12}{s}$. Substituting $s = 2$: $r = \frac{-12}{2} = -6$

5. The variation equation is $d = Kr^2$. Substituting $r = 5$ and $d = 10$:

$$10 = K \bullet 5^2$$
$$10 = 25K$$
$$K = \frac{2}{5}$$

So $d = \frac{2}{5}r^2$. Substituting $r = 10$: $d = \frac{2}{5}(10)^2 = \frac{2}{5} \bullet 100 = 40$

7. The variation equation is $y = \frac{K}{x^2}$. Substituting $x = 3$ and $y = 45$:

$$45 = \frac{K}{3^2}$$
$$45 = \frac{K}{9}$$
$$K = 405$$

So $y = \frac{405}{x^2}$. Substituting $x = 5$: $y = \frac{405}{5^2} = \frac{405}{25} = \frac{81}{5}$

9. The variation equation is $z = Kxy^2$. Substituting $x = 3$, $y = 3$, and $z = 54$:

$$54 = K(3)(3)^2$$
$$54 = 27K$$
$$K = 2$$

So $z = 2xy^2$. Substituting $x = 2$ and $y = 4$: $z = 2(2)(4)^2 = 64$

11. The variation equation is $I = \dfrac{K}{w^3}$. Substituting $w = \dfrac{1}{2}$ and $I = 32$:

$$32 = \frac{K}{\left(\frac{1}{2}\right)^3}$$

$$32 = \frac{K}{1/8}$$

$$K = 4$$

So $I = \dfrac{4}{w^3}$. Substituting $w = \dfrac{1}{3}$: $I = \dfrac{4}{\left(\frac{1}{3}\right)^3} = \dfrac{4}{1/27} = 108$

13. The variation equation is $z = Kyx^2$. Substituting $x = 3, y = 2$, and $z = 72$:

$$72 = K(2)(3)^2$$

$$72 = 18K$$

$$K = 4$$

So $z = 4yx^2$. Substituting $x = 5$ and $y = 3$: $z = 4(3)(5)^2 = 300$

15. The variation equation is $z = Kyx^2$. Substituting $x = 1, y = 5$, and $z = 25$:

$$25 = K(5)(1)^2$$

$$25 = 5K$$

$$K = 5$$

So $z = 5yx^2$. Substituting $z = 160$ and $y = 8$:

$$160 = 5(8)x^2$$

$$160 = 40x^2$$

$$x^2 = 4$$

$$x = \pm 2$$

17. The variation equation is $F = \dfrac{Km}{d^2}$. Substituting $F = 150$, $m = 240$, and $d = 8$:

$$150 = \frac{K(240)}{8^2}$$

$$150 = \frac{240K}{64}$$

$$240K = 9600$$

$$K = 40$$

So $F = \dfrac{40m}{d^2}$. Substituting $m = 360$ and $d = 3$: $F = \dfrac{40(360)}{3^2} = \dfrac{14400}{9} = 1600$

19. The variation equation is $F=\dfrac{Km}{d^2}$. Substituting $F = 24$, $m = 20$, and $d = 5$:

$$\begin{aligned}24&=\frac{K(20)}{5^2}\\24&=\frac{20K}{25}\\20K&=600\\K&=30\end{aligned}$$

So $F=\dfrac{30m}{d^2}$. Substituting $F = 18.75$ and $m = 40$:

$$\begin{aligned}18.75&=\frac{30(40)}{d^2}\\18.75&=\frac{1200}{d^2}\\18.75d^2&=1200\\d^2&=64\\d&=\pm 8\end{aligned}$$

21. Let l represent the length and f represent the force. The variation equation is $l = Kf$. Substituting $f = 5$ and $l = 3$:

$$\begin{aligned}3&=K\bullet 5\\K&=\frac{3}{5}\end{aligned}$$

So $l=\dfrac{3}{5}f$. Substituting $l = 10$:

$$\begin{aligned}10&=\frac{3}{5}f\\50&=3f\\f&=\frac{50}{3}\end{aligned}$$

The force required is $\dfrac{50}{3}$ pounds.

23. **a.** Let T represent the temperature and P represent the pressure. The variation equation is $T = KP$. Substituting $T = 200$ and $P = 50$:

$$\begin{aligned}200&=K\bullet 50\\K&=4\end{aligned}$$

The variation equation is $T = 4P$.

b. Graphing the equation:

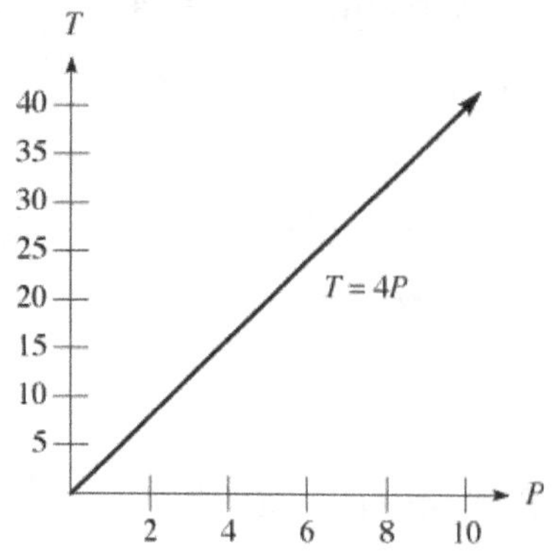

c. Substituting $T = 280$:

$$280 = 4P$$
$$P = 70$$

The pressure is 70 pounds per square inch.

25. Let v represent the volume and p represent the pressure. The variation equation is $v = \frac{K}{p}$.

Substituting $p = 36$ and $v = 25$:

$$25 = \frac{K}{36}$$
$$K = 900$$

The equation is $v = \frac{900}{p}$. Substituting $v = 75$:

$$75 = \frac{900}{p}$$
$$75p = 900$$
$$p = 12$$

The pressure is 12 pounds per square inch.

27. **a.** Let f represent the aperture and d represent the diameter. The variation equation is $f = \frac{K}{d}$.

Substituting $f = 2$ and $d = 40$:

$$2 = \frac{K}{40}$$
$$K = 80$$

The variation equation is $f = \frac{80}{d}$.

b. Graphing the equation:

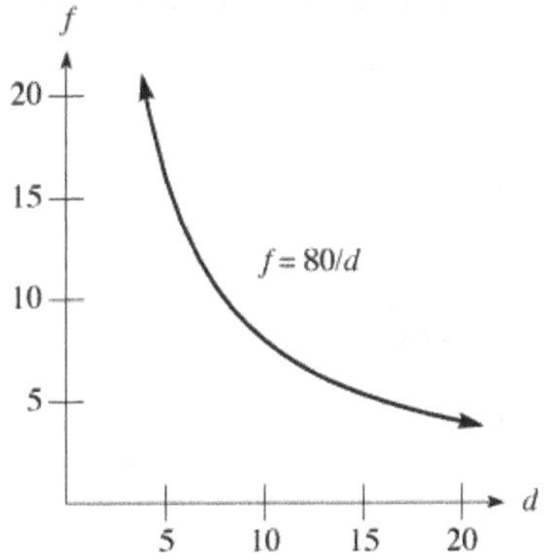

c. Substituting $d = 10$:

$$f = \frac{80}{10}$$

$$f = 8$$

The f-stop is 8.

29. Let A represent the surface area, h represent the height, and r represent the radius. The variation equation is $A = Khr$. Substituting $A = 94$, $r = 3$, and $h = 5$:

$$94 = K(3)(5)$$

$$94 = 15K$$

$$K = \frac{94}{15}$$

The equation is $A = \frac{94}{15}hr$. Substituting $r = 2$ and $h = 8$: $A = \frac{94}{15}(8)(2) = \frac{1504}{15}$.

The surface area is $\frac{1504}{15}$ square inches

31. Let R represent the resistance, l represent the length, and d represent the diameter. The variation equation is $R = \frac{Kl}{d^2}$. Substituting $R = 10$, $l = 100$, and $d = 0.01$:

$$10 = \frac{K(100)}{(0.01)^2}$$

$$0.001 = 100K$$

$$K = 0.00001$$

The equation is $R = \frac{0.00001l}{d^2}$. Substituting $l = 60$ and $d = 0.02$: $R = \frac{0.00001(60)}{(0.02)^2} = 1.5$.

The resistance is 1.5 ohms.

33. **a.** Let P represent the period and L represent the length. The variation equation is $P = K\sqrt{L}$. Substituting $P = 2.1$ and $L = 100$:

$$2.1 = K\sqrt{100}$$
$$2.1 = 10K$$
$$K = 0.21$$

The variation equation is $P = 0.21\sqrt{L}$.

b. Graphing the equation:

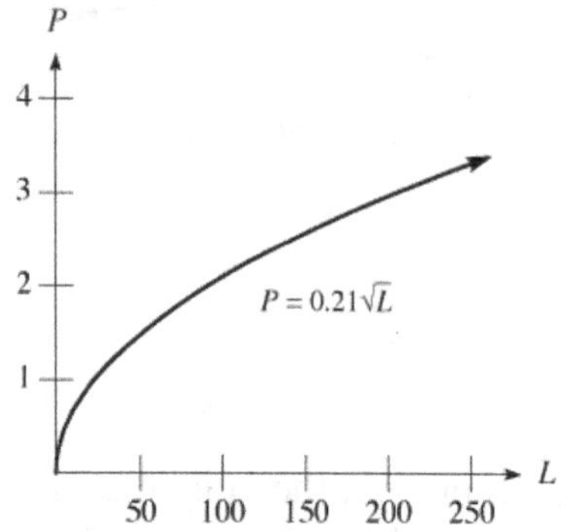

c. Substituting $L = 225$: $P = 0.21\sqrt{225} = 3.15$
The period is 3.15 seconds.

35. Multiplying: $0.6(M-70) = 0.6M - 42$

37. Multiplying:
$$(4x-3)(4x^2-7x+3) = 16x^3 - 28x^2 + 12x - 12x^2 + 21x - 9 = 16x^3 - 40x^2 + 33x - 9$$

39. Simplifying: $(4x-3)+(4x^2-7x+3) = 4x - 3 + 4x^2 - 7x + 3 = 4x^2 - 3x$

41. Simplifying: $(4x^2+3x+2)+(2x^2-5x-6) = 4x^2 + 3x + 2 + 2x^2 - 5x - 6 = 6x^2 - 2x - 4$

43. Simplifying: $4(-1)^2 - 7(-1) = 4(1) - 7(-1) = 4 + 7 = 11$

3.8 Algebra and Composition with Functions

1. Writing the formula: $f + g = f(x) + g(x) = (4x-3) + (2x+5) = 6x + 2$

3. Writing the formula: $g - f = g(x) - f(x) = (2x+5) - (4x-3) = -2x + 8$

5. Writing the formula: $fg = f(x) \bullet g(x) = (4x-3)(2x+5) = 8x^2 + 14x - 15$

7. Writing the formula: $g / f = \dfrac{g(x)}{f(x)} = \dfrac{2x+5}{4x-3}$

9. Writing the formula: $g + f = g(x) + f(x) = (x-2) + (3x-5) = 4x - 7$

11. Writing the formula: $g + h = g(x) + h(x) = (x-2) + (3x^2 - 11x + 10) = 3x^2 - 10x + 8$

13. Writing the formula: $g - f = g(x) - f(x) = (x-2) - (3x-5) = -2x + 3$

15. Writing the formula: $fg = f(x) \bullet g(x) = (3x-5)(x-2) = 3x^2 - 11x + 10$

17. Writing the formula:

$$\begin{aligned} fh &= f(x) \bullet h(x) \\ &= (3x-5)(3x^2-11x+10) \\ &= 9x^3-33x^2+30x-15x^2+55x-50 \\ &= 9x^3-48x^2+85x-50 \end{aligned}$$

19. Writing the formula: $h/f = \dfrac{h(x)}{f(x)} = \dfrac{3x^2-11x+10}{3x-5} = \dfrac{(3x-5)(x-2)}{3x-5} = x-2$

Note: We will cover reducing rational expressions more extensively in chapter 5.

21. Writing the formula: $f/h = \dfrac{f(x)}{h(x)} = \dfrac{3x-5}{3x^2-11x+10} = \dfrac{3x-5}{(3x-5)(x-2)} = \dfrac{1}{x-2}$

Note: We will cover reducing rational expressions more extensively in chapter 5.

23. Writing the formula:

$$f+g+h = f(x)+g(x)+h(x) = (3x-5)+(x-2)+(3x^2-11x+10) = 3x^2-7x+3$$

25. Writing the formula:

$$\begin{aligned} h+fg &= h(x)+f(x)g(x) \\ &= (3x^2-11x+10)+(3x-5)(x-2) \\ &= 3x^2-11x+10+3x^2-11x+10 \\ &= 6x^2-22x+20 \end{aligned}$$

27. Evaluating: $(f+g)(2) = f(2)+g(2) = (2 \bullet 2+1)+(4 \bullet 2+2) = 5+10 = 15$

29. Evaluating: $(fg)(3) = f(3) \bullet g(3) = (2 \bullet 3+1)(4 \bullet 3+2) = 7 \bullet 14 = 98$

31. Evaluating: $(h/g)(1) = \dfrac{h(1)}{g(1)} = \dfrac{4(1)^2+4(1)+1}{4(1)+2} = \dfrac{9}{6} = \dfrac{3}{2}$

33. Evaluating: $(fh)(0) = f(0) \bullet h(0) = (2(0)+1)(4(0)^2+4(0)+1) = (1)(1) = 1$

35. Evaluating:

$$\begin{aligned} (f+g+h)(2) &= f(2)+g(2)+h(2) \\ &= (2(2)+1)+(4(2)+2)+(4(2)^2+4(2)+1) \\ &= 5+10+25 \\ &= 40 \end{aligned}$$

37. Evaluating:

$$\begin{aligned} (h+fg)(3) &= h(3)+f(3) \bullet g(3) \\ &= (4(3)^2+4(3)+1)+(2(3)+1) \bullet (4(3)+2) \\ &= 49+7 \bullet 14 \\ &= 49+98 \\ &= 147 \end{aligned}$$

39. **a.** Evaluating: $(f\circ g)(5)=f(g(5))=f(5+4)=f(9)=9^2=81$

b. Evaluating: $(g\circ f)(5)=g(f(5))=g(5^2)=g(25)=25+4=29$

c. Evaluating: $(f\circ g)(x)=f(g(x))=f(x+4)=(x+4)^2$

d. Evaluating: $(g\circ f)(x)=g(f(x))=g(x^2)=x^2+4$

41. **a.** Evaluating: $(f\circ g)(0)=f(g(0))=f(4\bullet 0-1)=f(-1)=(-1)^2+3(-1)=1-3=-2$

b. Evaluating: $(g\circ f)(0)=g(f(0))=g(0^2+3\bullet 0)=g(0)=4(0)-1=-1$

c. Evaluating:

$$\begin{aligned}(f\circ g)(x)&=f(g(x))\\&=f(4x-1)\\&=(4x-1)^2+3(4x-1)\\&=16x^2-8x+1+12x-3\\&=16x^2+4x-2\end{aligned}$$

d. Evaluating: $(g\circ f)(x)=g(f(x))=g(x^2+3x)=4(x^2+3x)-1=4x^2+12x-1$

43. Evaluating each composition:

$$(f\circ g)(x)=f(g(x))=f\left(\frac{x+4}{5}\right)=5\left(\frac{x+4}{5}\right)-4=x+4-4=x$$

$$(g\circ f)(x)=g(f(x))=g(5x-4)=\frac{5x-4+4}{5}=\frac{5x}{5}=x$$

Thus $(f\circ g)(x)=(g\circ f)(x)=x$.

45. **a.** Finding the revenue: $R(x)=x(11.5-0.05x)=11.5x-0.05x^2$

b. Finding the cost: $C(x)=2x+200$

c. Finding the profit:

$$P(x)=R(x)-C(x)=(11.5x-0.05x^2)-(2x+200)=-0.05x^2+9.5x-200$$

d. Finding the average cost: $\bar{C}(x)=\dfrac{C(x)}{x}=\dfrac{2x+200}{x}=2+\dfrac{200}{x}$

47. **a.** The function is $M(x)=220-x$.

b. Evaluating: $M(24)=220-24=196$ beats per minute

c. The training heart rate function is: $T(M)=62+0.6(M-62)=0.6M+24.8$

Finding the composition: $T(M(x))=T(220-x)=0.6(220-x)+24.8=156.8-0.6x$

Evaluating: $T(M(24))=156.8-0.6(24)\approx 142$ beats per minute

d. Evaluating: $T(M(36))=156.8-0.6(36)\approx 135$ beats per minute

e. Evaluating: $T(M(48))=156.8-0.6(48)\approx 128$ beats per minute

49. Solving the equation:

$$x - 5 = 7$$
$$x = 12$$

51. Solving the equation:

$$5 - \frac{4}{7}a = -11$$
$$7\left(5 - \frac{4}{7}a\right) = 7(-11)$$
$$35 - 4a = -77$$
$$-4a = -112$$
$$a = 28$$

53. Solving the equation:

$$5(x-1) - 2(2x+3) = 5x - 4$$
$$5x - 5 - 4x - 6 = 5x - 4$$
$$x - 11 = 5x - 4$$
$$-4x = 7$$
$$x = -\frac{7}{4}$$

55. Solving for w:

$$P = 2l + 2w$$
$$P - 2l = 2w$$
$$w = \frac{P - 2l}{2}$$

57. Solving the inequality:

$$-5t \le 30$$
$$t \ge -6$$

The solution set is $[-6, \infty)$. Graphing:

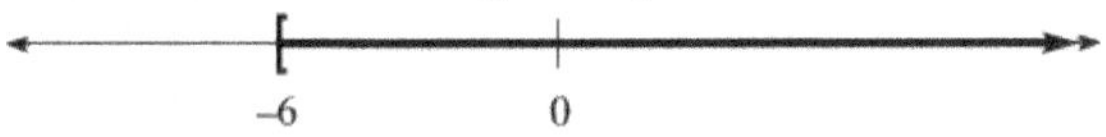

59. Solving the inequality:

$$1.6x - 2 < 0.8x + 2.8$$
$$0.8x - 2 < 2.8$$
$$0.8x < 4.8$$
$$x < 6$$

The solution set is $(-\infty, 6)$. Graphing:

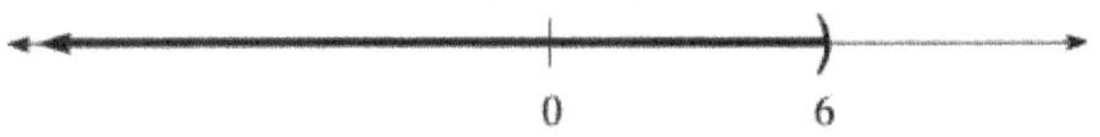

61. Solving the equation:

$$\left|\frac{1}{4}x - 1\right| = \frac{1}{2}$$
$$\frac{1}{4}x - 1 = -\frac{1}{2}, \frac{1}{2}$$
$$\frac{1}{4}x = \frac{1}{2}, \frac{3}{2}$$
$$x = 2, 6$$

63. Solving the equation:

$$|3 - 2x| + 5 = 2$$
$$|3 - 2x| = -3$$

Since this statement is false, there is no solution, or $\varnothing$.

Chapter 3 Test

See www.mathtv.com for video solutions to all problems in this chapter test.

Chapter 4
Systems of Equations

4.1 Systems of Linear Equations in Two Variables

1. The intersection point is (4,3).

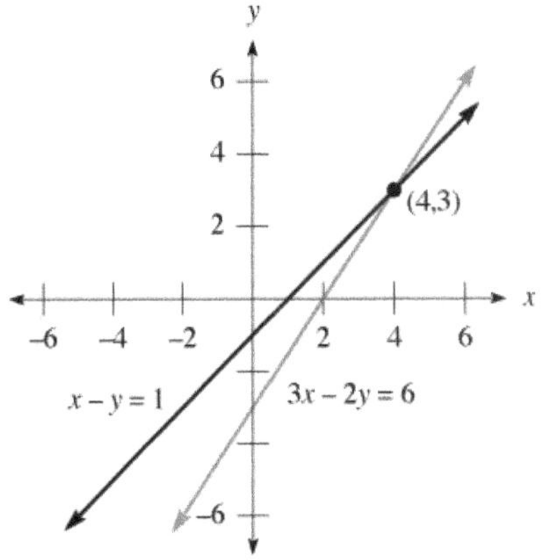

3. The intersection point is (–5,–6).

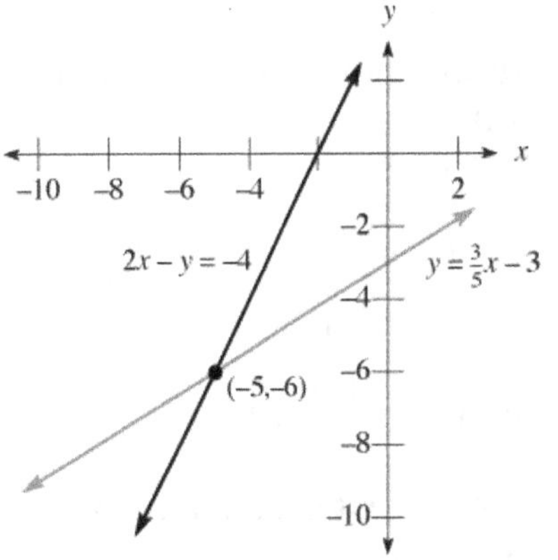

5. The intersection point is (4,2).

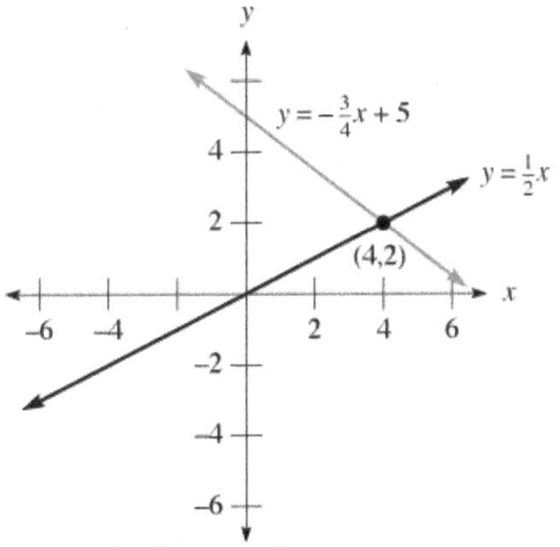

7. The lines are parallel. There is no solution to the system.

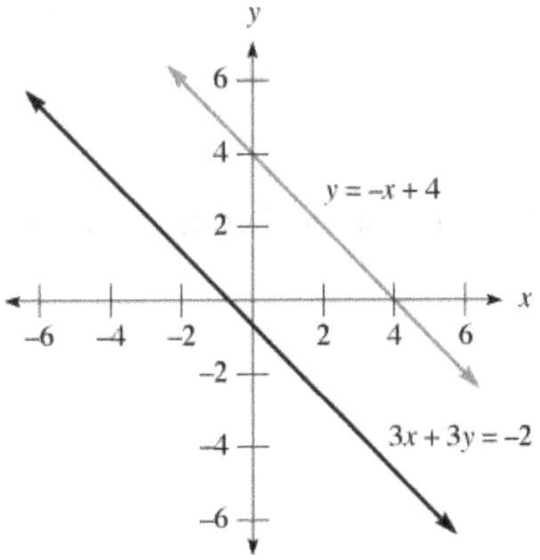

9. Solving the two equations:

$$3x + y = 5$$
$$3x - y = 3$$

Adding yields:

$$6x = 8$$
$$x = \frac{4}{3}$$

Substituting into the first equation:

$$3\left(\frac{4}{3}\right) + y = 5$$
$$4 + y = 5$$
$$y = 1$$

The solution is $\left(\frac{4}{3}, 1\right)$.

11. Multiply the first equation by 3:

$$3x + 6y = 0$$
$$2x - 6y = 5$$

Adding yields:

$$5x = 5$$
$$x = 1$$

The solution is $\left(1, -\frac{1}{2}\right)$.

13. Multiply the first equation by –2:

$$-4x + 10y = -32$$
$$4x - 3y = 11$$

Adding yields:

$$7y = -21$$
$$y = -3$$

The solution is $\left(\frac{1}{2}, -3\right)$.

15. Multiply the first equation by 3 and the second equation by –2:

$$18x + 9y = -3$$
$$-18x - 10y = -2$$

Adding yields:

$$-y = -5$$
$$y = 5$$

The solution is $\left(-\frac{8}{3}, 5\right)$.

17. Multiply the first equation by 2 and the second equation by 3:

$$8x+6y=28$$
$$27x-6y=42$$

Adding yields:

$$35x=70$$
$$x=2$$

The solution is (2,2).

19. Multiply the first equation by 2:

$$4x-10y=6$$
$$-4x+10y=3$$

Adding yields $0=9$, which is false. There is no solution ($\varnothing$).

21. To clear each equation of fractions, multiply the first equation by 6 and the second equation by 20:

$$3x+2y=78$$
$$8x+5y=200$$

Multiply the first equation by 5 and the second equation by –2:

$$15x+10y=390$$
$$-16x-10y=-400$$

Adding yields:

$$-x=-10$$
$$x=10$$

The solution is (10,24).

23. To clear each equation of fractions, multiply the first equation by 15 and the second equation by 6:

$$10x+6y=-60$$
$$2x-3y=-2$$

Multiply the second equation by 2:

$$10x+6y=-60$$
$$4x-6y=-4$$

Adding yields:

$$14x=-64$$
$$x=-\frac{32}{7}$$

Substituting into the second equation:

$$2\left(-\frac{32}{7}\right)-3y=-2$$
$$-\frac{64}{7}-3y=-2$$
$$-3y=\frac{50}{7}$$
$$y=-\frac{50}{21}$$

The solution is $\left(-\frac{32}{7},-\frac{50}{21}\right)$.

25. Substituting into the first equation:

$$\begin{aligned} 7(2y+9)-y&=24\\ 14y+63-y&=24\\ 13y&=-39\\ y&=-3 \end{aligned}$$

The solution is (3,–3).

27. Substituting into the first equation:

$$\begin{aligned} 6x-\left(-\frac{3}{4}x-1\right)&=10\\ 6x+\frac{3}{4}x+1&=10\\ \frac{27}{4}x&=9\\ 27x&=36\\ x&=\frac{4}{3} \end{aligned}$$

The solution is $\left(\frac{4}{3},-2\right)$.

29. Substituting into the first equation:

$$\begin{aligned} 4x-4&=3x-2\\ x-4&=-2\\ x&=2 \end{aligned}$$

The solution is (2,4).

31. Solving the first equation for y yields $y=2x-5$. Substituting into the second equation:

$$\begin{aligned} 4x-2(2x-5)&=10\\ 4x-4x+10&=10\\ 10&=10 \end{aligned}$$

Since this statement is true, the two lines coincide. The solution is $\{(x,y)\mid 2x-y=5\}$.

33. Substituting into the first equation:

$$\begin{aligned} \frac{1}{3}\left(\frac{3}{2}y\right)-\frac{1}{2}y&=0\\ \frac{1}{2}y-\frac{1}{2}y&=0\\ 0&=0 \end{aligned}$$

Since this statement is true, the two lines coincide. The solution is $\left\{(x,y)\mid x=\frac{3}{2}y\right\}$.

35. Multiply the first equation by 2 and the second equation by 7:

$$\begin{aligned} 8x-14y&=6\\ 35x+14y&=-21 \end{aligned}$$

Adding yields:

$$\begin{aligned} 43x&=-15\\ x&=-\frac{15}{43} \end{aligned}$$

Substituting into the original second equation:

$$5\left(-\frac{15}{43}\right)+2y=-3$$

$$-\frac{75}{43}+2y=-3$$

$$2y=-\frac{54}{43}$$

$$y=-\frac{27}{43}$$

The solution is $\left(-\frac{15}{43},-\frac{27}{43}\right)$.

37. Multiply the first equation by 3 and the second equation by 8:

$$27x-24y=12$$

$$16x+24y=48$$

Adding yields:

$$43x=60$$

$$x=\frac{60}{43}$$

Substituting into the original second equation:

$$2\left(\frac{60}{43}\right)+3y=6$$

$$\frac{120}{43}+3y=6$$

$$3y=\frac{138}{43}$$

$$y=\frac{46}{43}$$

The solution is $\left(\frac{60}{43},\frac{46}{43}\right)$.

39. Multiply the first equation by 2 and the second equation by 5:

$$6x-10y=4$$

$$35x+10y=5$$

Adding yields:

$$41x=9$$

$$x=\frac{9}{41}$$

Substituting into the original second equation:

$$7\left(\frac{9}{41}\right)+2y=1$$

$$\frac{63}{41}+2y=1$$

$$2y=-\frac{22}{41}$$

$$y=-\frac{11}{41}$$

The solution is $\left(\frac{9}{41},-\frac{11}{41}\right)$.

41. Multiply the second equation by 3:

$$x-3y=7$$

$$6x+3y=-18$$

Adding yields:

$$7x=-11$$

$$x=-\frac{11}{7}$$

Substituting into the original second equation:

$$2\left(-\frac{11}{7}\right)+y=-6$$

$$-\frac{22}{7}+y=-6$$

$$y=-\frac{20}{7}$$

The solution is $\left(-\frac{11}{7},-\frac{20}{7}\right)$.

43. Substituting into the first equation:

$$-\frac{1}{3}x+2=\frac{1}{2}x+\frac{1}{3}$$

$$6\left(-\frac{1}{3}x+2\right)=6\left(\frac{1}{2}x+\frac{1}{3}\right)$$

$$-2x+12=3x+2$$

$$-5x=-10$$

$$x=2$$

Substituting into the first equation: $y=\frac{1}{2}(2)+\frac{1}{3}=1+\frac{1}{3}=\frac{4}{3}$. The solution is $\left(2,\frac{4}{3}\right)$.

45. Substituting into the first equation:

$$\begin{aligned} 3\left(\frac{2}{3}y-4\right)-4y&=12\\ 2y-12-4y&=12\\ -2y-12&=12\\ -2y&=24\\ y&=-12 \end{aligned}$$

Substituting into the second equation: $x=\frac{2}{3}(-12)-4=-8-4=-12$. The solution is (–12,–12).

47. Multiply the first equation by 2:

$$\begin{aligned} 8x-6y&=-14\\ -8x+6y&=-11\\ 0&=-25 \end{aligned}$$

Since this statement is false, there is no solution ($\varnothing$).

49. Multiply the first equation by –20:

$$\begin{aligned} -60y-20z&=-340\\ 5y+20z&=65 \end{aligned}$$

Adding yields:

$$\begin{aligned} -55y&=-275\\ y&=5 \end{aligned}$$

Substituting into the first equation:

$$\begin{aligned} 3(5)+z&=17\\ 15+z&=17\\ z&=2 \end{aligned}$$

The solution is $y = 5$, $z = 2$.

51. Substitute into the first equation:

$$\begin{aligned} \frac{3}{4}x-\frac{1}{3}\left(\frac{1}{4}x\right)&=1\\ \frac{3}{4}x-\frac{1}{12}x&=1\\ \frac{2}{3}x&=1\\ x&=\frac{3}{2} \end{aligned}$$

Substituting into the second equation: $y=\frac{1}{4}\left(\frac{3}{2}\right)=\frac{3}{8}$. The solution is $\left(\frac{3}{2},\frac{3}{8}\right)$.

53. To clear each equation of fractions, multiply the first equation by 12 and the second equation by 12:

$$3x - 6y = 4$$
$$4x - 3y = -8$$

Multiply the second equation by –2:

$$3x - 6y = 4$$
$$-8x + 6y = 16$$

Adding yields:

$$-5x = 20$$
$$x = -4$$

Substituting into the first equation:

$$3(-4) - 6y = 4$$
$$-12 - 6y = 4$$
$$-6y = 16$$
$$y = -\frac{8}{3}$$

The solution is $\left(-4, -\frac{8}{3}\right)$.

55. **a.** Simplifying: $(3x - 4y) - 3(x - y) = 3x - 4y - 3x + 3y = -y$

b. Substituting $x = 0$:

$$3(0) - 4y = 8$$
$$-4y = 8$$
$$y = -2$$

c. From part **b**, the y-intercept is (0,–2).

d. Graphing the line:

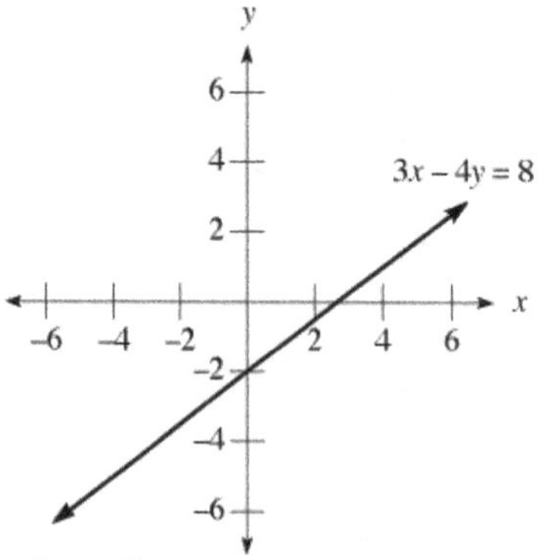

e. Multiply the second equation by –3:

$$3x - 4y = 8$$
$$-3x + 3y = -6$$

Adding yields:

$$-y = 2$$
$$y = -2$$

Substituting into the first equation:

$$3x-4(-2)=8$$
$$3x+8=8$$
$$3x=0$$
$$x=0$$

The lines intersect at the point (0,–2).

57. Multiply the second equation by 100:

$$x+y=10000$$
$$6x+5y=56000$$

Multiply the first equation by –5:

$$-5x-5y=-50000$$
$$6x+5y=56000$$

Adding yields $x = 6000$. The solution is (6000,4000).

59. Substituting $x=4-y$ into the second equation:

$$(4-y)-2y=4$$
$$4-3y=4$$
$$-3y=0$$
$$y=0$$

The solution is (4,0).

61. Simplifying: $2-2(6)=2-12=-10$

63. Simplifying: $(x+3y)-1(x-2z)=x+3y-x+2z=3y+2z$

65. Solving the equation:

$$-9y=-9$$
$$y=1$$

67. Solving the equation:

$$3(1)+2z=9$$
$$3+2z=9$$
$$2z=6$$
$$z=3$$

69. Applying the distributive property: $2(5x-z)=10x-2z$

71. Applying the distributive property: $3(3x+y-2z)=9x+3y-6z$

4.2 Systems of Linear Equations in Three Variables

1. Adding the first two equations and the first and third equations results in the system:

$$2x+3z=5$$
$$2x-2z=0$$

Solving the second equation yields $x = z$, now substituting:

$$2z+3z=5$$
$$5z=5$$
$$z=1$$

So $x = 1$, now substituting into the original first equation:

$$1+y+1=4$$
$$y+2=4$$
$$y=2$$

The solution is (1,2,1).

3. Adding the first two equations and the first and third equations results in the system:

$$2x + 3z = 13$$
$$3x - 3z = -3$$

Adding yields:

$$5x = 10$$
$$x = 2$$

Substituting to find z:

$$2(2) + 3z = 13$$
$$4 + 3z = 13$$
$$3z = 9$$
$$z = 3$$

Substituting into the original first equation:

$$2 + y + 3 = 6$$
$$y + 5 = 6$$
$$y = 1$$

The solution is (2,1,3).

5. Adding the second and third equations:

$$5x + z = 11$$

Multiplying the second equation by 2:

$$x + 2y + z = 3$$
$$4x - 2y + 4z = 12$$

Adding yields:

$$5x + 5z = 15$$
$$x + z = 3$$

So the system becomes:

$$5x + z = 11$$
$$x + z = 3$$

Multiply the second equation by –1:

$$5x + z = 11$$
$$-x - z = -3$$

Adding yields:

$$4x = 8$$
$$x = 2$$

Substituting to find z:

$$5(2) + z = 11$$
$$z + 10 = 11$$
$$z = 1$$

Substituting into the original first equation:

$$2+2y+1=3$$
$$2y+3=3$$
$$2y=0$$
$$y=0$$

The solution is (2,0,1).

7. Multiply the second equation by –1 and add it to the first equation:

$$2x+3y-2z=4$$
$$-x-3y+3z=-4$$

Adding results in the equation $x+z=0$. Multiply the second equation by 2 and add it to the third equation:

$$2x+6y-6z=8$$
$$3x-6y+z=-3$$

Adding results in the equation:

$$5x-5z=5$$
$$x-z=1$$

So the system becomes:

$$x-z=1$$
$$x+z=0$$

Adding yields:

$$2x=1$$
$$x=\frac{1}{2}$$

Substituting to find z:

$$\frac{1}{2}+z=0$$
$$z=-\frac{1}{2}$$

Substituting into the original first equation:

$$2\left(\frac{1}{2}\right)+3y-2\left(-\frac{1}{2}\right)=4$$
$$1+3y+1=4$$
$$3y+2=4$$
$$3y=2$$
$$y=\frac{2}{3}$$

The solution is $\left(\frac{1}{2},\frac{2}{3},-\frac{1}{2}\right)$.

9. Multiply the first equation by 2 and add it to the second equation:

$$-2x+8y-6z=4$$
$$2x-8y+6z=1$$

Adding yields 0 = 5, which is false. There is no solution (inconsistent system).

11. To clear the system of fractions, multiply the first equation by 2 and the second equation by 3:

$$x-2y+2z=0$$
$$6x+y+3z=6$$
$$x+y+z=-4$$

Multiply the third equation by 2 and add it to the first equation:

$$x-2y+2z=0$$
$$2x+2y+2z=-8$$

Adding yields the equation $3x+4z=-8$. Multiply the third equation by –1 and add it to the second equation:

$$6x+y+3z=6$$
$$-x-y-z=4$$

Adding yields the equation $5x+2z=10$. So the system becomes:

$$3x+4z=-8$$
$$5x+2z=10$$

Multiply the second equation by –2:

$$3x+4z=-8$$
$$-10x-4z=-20$$

Adding yields:

$$-7x=-28$$
$$x=4$$

Substituting to find z:

$$3(4)+4z=-8$$
$$12+4z=-8$$
$$4z=-20$$
$$z=-5$$

Substituting into the original third equation:

$$4+y-5=-4$$
$$y-1=-4$$
$$y=-3$$

The solution is (4,–3,–5).

13. Multiply the first equation by –2 and add it to the third equation:

$$-4x+2y+6z=-2$$
$$4x-2y-6z=2$$

Adding yields 0 = 0, which is true. Since there are now less equations than unknowns, there is no unique solution (dependent system).

15. Multiply the second equation by 3 and add it to the first equation:

$$2x - y + 3z = 4$$
$$3x + 6y - 3z = -9$$

Adding yields the equation $5x + 5y = -5$, or $x + y = -1$.

Multiply the second equation by 2 and add it to the third equation:

$$2x + 4y - 2z = -6$$
$$4x + 3y + 2z = -5$$

Adding yields the equation $6x + 7y = -11$. So the system becomes:

$$6x + 7y = -11$$
$$x + y = -1$$

Multiply the second equation by –6:

$$6x + 7y = -11$$
$$-6x - 6y = 6$$

Adding yields $y = -5$. Substituting to find x:

$$6x + 7(-5) = -11$$
$$6x - 35 = -11$$
$$6x = 24$$
$$x = 4$$

Substituting into the original first equation:

$$2(4) - (-5) + 3z = 4$$
$$13 + 3z = 4$$
$$3z = -9$$
$$z = -3$$

The solution is (4,–5,–3).

17. Adding the second and third equations results in the equation $x + y = 9$. Since this is the same as the first equation, there are less equations than unknowns. There is no unique solution (dependent system).

19. Adding the second and third equations results in the equation $4x + y = 3$. So the system becomes:

$$4x + y = 3$$
$$2x + y = 2$$

Multiplying the second equation by –1:

$$4x + y = 3$$
$$-2x - y = -2$$

Adding yields:

$$2x = 1$$
$$x = \frac{1}{2}$$

Substituting to find y:

$$2\left(\frac{1}{2}\right)+y=2$$
$$1+y=2$$
$$y=1$$

Substituting into the original second equation:

$$1+z=3$$
$$z=2$$

The solution is $\left(\frac{1}{2},1,2\right)$.

21. Multiply the third equation by 2 and adding it to the second equation:

$$6y-4z=1$$
$$2x+4z=2$$

Adding yields the equation $2x+6y=3$. So the system becomes:

$$2x-3y=0$$
$$2x+6y=3$$

Multiply the first equation by 2:

$$4x-6y=0$$
$$2x+6y=3$$

Adding yields:

$$6x=3$$
$$x=\frac{1}{2}$$

Substituting to find y:

$$2\left(\frac{1}{2}\right)+6y=3$$
$$1+6y=3$$
$$6y=2$$
$$y=\frac{1}{3}$$

Substituting into the original third equation to find z:

$$\frac{1}{2}+2z=1$$
$$2z=\frac{1}{2}$$
$$z=\frac{1}{4}$$

The solution is $\left(\frac{1}{2},\frac{1}{3},\frac{1}{4}\right)$.

23. Multiply the first equation by –2 and add it to the second equation:

$$-2x - 2y + 2z = -4$$
$$2x + y + 3z = 4$$

Adding yields $-y + 5z = 0$. Multiply the first equation by –1 and add it to the third equation:

$$-x - y + z = -2$$
$$x - 2y + 2z = 6$$

Adding yields $-3y + 3z = 4$. So the system becomes:

$$-y + 5z = 0$$
$$-3y + 3z = 4$$

Multiply the first equation by –3:

$$3y - 15z = 0$$
$$-3y + 3z = 4$$

Adding yields:

$$-12z = 4$$
$$z = -\frac{1}{3}$$

Substituting to find y:

$$-3y + 3\left(-\frac{1}{3}\right) = 4$$
$$-3y - 1 = 4$$
$$-3y = 5$$
$$y = -\frac{5}{3}$$

Substituting into the original first equation:

$$x - \frac{5}{3} + \frac{1}{3} = 2$$
$$x - \frac{4}{3} = 2$$
$$x = \frac{10}{3}$$

The solution is $\left(\frac{10}{3}, -\frac{5}{3}, -\frac{1}{3}\right)$.

25. Multiply the third equation by –1 and add it to the first equation:

$$2x+3y=-\frac{1}{2}$$

$$-3y-2z=\frac{3}{4}$$

Adding yields the equation $2x-2z=\frac{1}{4}$. So the system becomes:

$$2x-2z=\frac{1}{4}$$

$$4x+8z=2$$

Multiply the first equation by 4:

$$8x-8z=1$$

$$4x+8z=2$$

Adding yields:

$$12x=3$$

$$x=\frac{1}{4}$$

Substituting to find z:

$$4\left(\frac{1}{4}\right)+8z=2$$

$$1+8z=2$$

$$8z=1$$

$$z=\frac{1}{8}$$

Substituting to find y:

$$2\left(\frac{1}{4}\right)+3y=-\frac{1}{2}$$

$$\frac{1}{2}+3y=-\frac{1}{2}$$

$$3y=-1$$

$$y=-\frac{1}{3}$$

The solution is $\left(\frac{1}{4},-\frac{1}{3},\frac{1}{8}\right)$.

27. To clear each equation of fractions, multiply the first equation by 6, the second equation by 4, and the third equation by 12:

$$2x+3y-z=24$$
$$x-3y+2z=6$$
$$6x-8y-3z=-64$$

Multiply the first equation by 2 and add it to the second equation:

$$4x+6y-2z=48$$
$$x-3y+2z=6$$

Adding yields the equation $5x+3y=54$. Multiply the first equation by –3 and add it to the third equation:

$$-6x-9y+3z=-72$$
$$6x-8y-3z=-64$$

Adding yields:

$$-17y=-136$$
$$y=8$$

Substituting to find x:

$$5x+3(8)=54$$
$$5x+24=54$$
$$5x=30$$
$$x=6$$

Substituting to find z:

$$6-3(8)+2z=6$$
$$-18+2z=6$$
$$2z=24$$
$$z=12$$

The solution is (6,8,12).

29. To clear each equation of fractions, multiply the first equation by 6, the second equation by 6, and the third equation by 12:

$$6x-3y-2z=-8$$
$$2x-3z=30$$
$$-3x+8y-12z=-9$$

Multiply the first equation by 8 and the third equation by 3:

$$48x-24y-16z=-64$$
$$-9x+24y-36z=-27$$

Adding yields the equation $39x-52z=-91$, or $3x-4z=-7$. So the system becomes:

$$2x-3z=30$$
$$3x-4z=-7$$

Multiply the first equation by 3 and the second equation by –2:

$$6x-9z=90$$
$$-6y+8z=14$$

Adding yields:

$$-z = 104$$
$$z = -104$$

Substituting to find x:

$$2x - 3(-104) = 30$$
$$2x + 312 = 30$$
$$2x = -282$$
$$x = -141$$

Substituting to find y:

$$6(-141) - 3y - 2(-104) = -8$$
$$-846 - 3y + 208 = -8$$
$$-3y - 638 = -8$$
$$-3y = 630$$
$$y = -210$$

The solution is (–141,–210,–104).

31. Divide the second equation by 5 and the third equation by 10 to produce the system:

$$x - y - z = 0$$
$$x + 4y = 16$$
$$2y - z = 5$$

Multiply the third equation by –1 and add it to the first equation:

$$x - y - z = 0$$
$$-2y + z = -5$$

Adding yields the equation $x - 3y = -5$. So the system becomes:

$$x + 4y = 16$$
$$x - 3y = -5$$

Multiply the second equation by –1:

$$x + 4y = 16$$
$$-x + 3y = 5$$

Adding yields:

$$7y = 21$$
$$y = 3$$

Substituting to find x:

$$x + 12 = 16$$
$$x = 4$$

Substituting to find z:

$$6 - z = 5$$
$$z = 1$$

The currents are 4 amps, 3 amps, and 1 amp.

33. Translating into symbols: $3x + 2$

35. Simplifying: $25 - \frac{385}{9} = \frac{225}{9} - \frac{385}{9} = -\frac{160}{9}$

37. Simplifying: $0.08(4{,}000) = 320$

39. Applying the distributive property: $10(0.2x + 0.5y) = 2x + 5y$

41. Solving the equation:

$$\begin{aligned} x + (3x + 2) &= 26 \\ 4x + 2 &= 26 \\ 4x &= 24 \\ x &= 6 \end{aligned}$$

43. Adding the two equations:

$$\begin{aligned} -9y &= 9 \\ y &= -1 \end{aligned}$$

Substituting into the second equation:

$$\begin{aligned} -7(-1) + 4z &= 27 \\ 7 + 4z &= 27 \\ 4z &= 20 \\ z &= 5 \end{aligned}$$

The solution is (–1,5).

4.3 Applications

1. Let x and y represent the two numbers. The system of equations is:

$$\begin{aligned} y &= 2x + 3 \\ x + y &= 18 \end{aligned}$$

Substituting into the second equation:

$$\begin{aligned} x + 2x + 3 &= 18 \\ 3x &= 15 \\ x &= 5 \\ y &= 2(5) + 3 = 13 \end{aligned}$$

The two numbers are 5 and 13.

3. Let x and y represent the two numbers. The system of equations is:

$$\begin{aligned} y - x &= 6 \\ 2x &= 4 + y \end{aligned}$$

The second equation is $y = 2x - 4$. Substituting into the first equation:

$$\begin{aligned} 2x - 4 - x &= 6 \\ x &= 10 \\ y &= 2(10) - 4 = 16 \end{aligned}$$

The two numbers are 10 and 16.

5. Let x, y, and z represent the three numbers. The system of equations is:

$$x + y + z = 8$$
$$2x = z - 2$$
$$x + z = 5$$

The third equation is $z = 5 - x$. Substituting into the second equation:

$$2x = 5 - x - 2$$
$$3x = 3$$
$$x = 1$$
$$z = 5 - 1 = 4$$

Substituting into the first equation:

$$1 + y + 4 = 8$$
$$y = 3$$

The three numbers are 1, 3, and 4.

7. Let a represent the number of adult tickets and c represent the number of children's tickets. The system of equations is:

$$a + c = 925$$
$$2a + c = 1150$$

Multiply the first equation by –1:

$$-a - c = -925$$
$$2a + c = 1150$$

Adding yields:

$$a = 225$$
$$c = 700$$

There were 225 adult tickets and 700 children's tickets sold.

9. Let x represent the amount invested at 6% and y represent the amount invested at 7%. The system of equations is:

$$x + y = 20000$$
$$0.06x + 0.07y = 1280$$

Multiplying the first equation by –0.06:

$$-0.06x - 0.06y = -1200$$
$$0.06x + 0.07y = 1280$$

Adding yields:

$$0.01y = 80$$
$$y = 8000$$
$$x = 12000$$

Mr. Jones invested $12,000 at 6% and $8,000 at 7%.

11. Let x represent the amount invested at 6% and $2x$ represent the amount invested at 7.5%.
The equation is:

$$0.075(2x)+0.06(x)=840$$
$$0.21x=840$$
$$x=4000$$
$$2x=8000$$

Susan invested \$4,000 at 6% and \$8,000 at 7.5%.

13. Let x, y and z represent the amounts invested in the three accounts. The system of equations is:

$$x+y+z=2200$$
$$z=3x$$
$$0.06x+0.08y+0.09z=178$$

Substituting into the first equation:

$$x+y+3x=2200$$
$$4x+y=2200$$

Substituting into the third equation:

$$0.06x+0.08y+0.09(3x)=178$$
$$0.33x+0.08y=178$$

The system of equations becomes:

$$4x+y=2200$$
$$0.33x+0.08y=178$$

Multiply the first equation by –0.08:

$$-0.32x-0.08y=-176$$
$$0.33x+0.08y=178$$

Adding yields:

$$0.01x=2$$
$$x=200$$
$$z=3(200)=600$$
$$y=2200-4(200)=1400$$

He invested \$200 at 6%, \$1,400 at 8%, and \$600 at 9%.

15. Let x represent the amount of 20% alcohol and y represent the amount of 50% alcohol.
The system of equations is:

$$x+y=9$$
$$0.20x+0.50y=0.30(9)$$

Multiplying the first equation by –0.2:

$$-0.20x-0.20y=-1.8$$
$$0.20x+0.50y=2.7$$

Adding yields:

$$0.30y=0.9$$
$$y=3$$
$$x=6$$

The mixture contains 3 gallons of 50% alcohol and 6 gallons of 20% alcohol.

17. Let x represent the amount of 20% disinfectant and y represent the amount of 14% disinfectant. The system of equations is:

$$x+y=15$$
$$0.20x+0.14y=0.16(15)$$

Multiplying the first equation by –0.14:

$$-0.14x-0.14y=-2.1$$
$$0.20x+0.14y=2.4$$

Adding yields:

$$0.06x=0.3$$
$$x=5$$
$$y=10$$

The mixture contains 5 gallons of 20% disinfectant and 10 gallons of 14% disinfectant.

19. Let x represent the amount of nuts and y represent the amount of oats. The system of equations is:

$$x+y=25$$
$$1.55x+1.35y=1.45(25)$$

Multiplying the first equation by –1.35:

$$-1.35x-1.35y=-33.75$$
$$1.55x+1.35y=36.25$$

Adding yields:

$$0.20x=2.5$$
$$x=12.5$$
$$y=12.5$$

The mixture contains 12.5 pounds of oats and 12.5 pounds of nuts.

21. Let b represent the rate of the boat and c represent the rate of the current. The system of equations is:

$$2(b+c)=24$$
$$3(b-c)=18$$

The system of equations simplifies to:

$$b+c=12$$
$$b-c=6$$

Adding yields:

$$2b=18$$
$$b=9$$
$$c=3$$

The rate of the boat is 9 mph and the rate of the current is 3 mph.

23. Let a represent the rate of the airplane and w represent the rate of the wind.
The system of equations is:

$$2(a+w)=600$$
$$\tfrac{5}{2}(a-w)=600$$

The system of equations simplifies to:

$$a+w=300$$
$$a-w=240$$

Adding yields:

$$2a=540$$
$$a=270$$
$$w=30$$

The rate of the airplane is 270 mph and the rate of the wind is 30 mph.

25. Let n represent the number of nickels and d represent the number of dimes.
The system of equations is:

$$n+d=20$$
$$0.05n+0.10d=1.40$$

Multiplying the first equation by –0.05:

$$-0.05n-0.05d=-1$$
$$0.05n+0.10d=1.40$$

Adding yields:

$$0.05d=0.40$$
$$d=8$$
$$n=12$$

Bob has 12 nickels and 8 dimes.

27. Let n, d, and q represent the number of nickels, dimes, and quarters. The system of equations is:

$$n+d+q=9$$
$$0.05n+0.10d+0.25q=1.20$$
$$d=n$$

Substituting into the first equation:

$$n+n+q=9$$
$$2n+q=9$$

Substituting into the second equation:

$$0.05n+0.10n+0.25q=1.20$$
$$0.15n+0.25q=1.20$$

The system of equations becomes:

$$2n+q=9$$
$$0.15n+0.25q=1.20$$

Multiplying the first equation by –0.25:

$$-0.50n-0.25q=-2.25$$
$$0.15n+0.25q=1.20$$

Adding yields:

$$-0.35n = -1.05$$
$$n = 3$$
$$d = 3$$
$$q = 9 - 2(3) = 3$$

The collection contains 3 nickels, 3 dimes, and 3 quarters.

29. Let *n*, *d*, and *q* represent the number of nickels, dimes, and quarters. The system of equations is:

$$n + d + q = 140$$
$$0.05n + 0.10d + 0.25q = 10.00$$
$$d = 2q$$

Substituting into the first equation:

$$n + 2q + q = 140$$
$$n + 3q = 140$$

Substituting into the second equation:

$$0.05n + 0.10(2q) + 0.25q = 10.00$$
$$0.05n + 0.45q = 10.00$$

The system of equations becomes:

$$n + 3q = 140$$
$$0.05n + 0.45q = 10.00$$

Multiplying the first equation by –0.05:

$$-0.05n - 0.15q = -7$$
$$0.05n + 0.45q = 10$$

Adding yields:

$$0.30q = 3$$
$$q = 10$$
$$d = 2(10) = 20$$
$$n = 140 - 3(10) = 110$$

There are 110 nickels in the collection.

31. Let $x = mp + b$ represent the relationship. Using the points (2,300) and (1.5,400) results in the system:

$$300 = 2m + b$$
$$400 = 1.5m + b$$

Multiplying the second equation by –1:

$$300 = 2m + b$$
$$-400 = -1.5m - b$$

Adding yields:

$$-100 = 0.5m$$
$$m = -200$$
$$b = 300 - 2(-200) = 700$$

The equation is $x = -200p + 700$. Substituting $p = 3$: $x = -200(3) + 700 = 100$ items

33. The system of equations is:

$$a+b+c=128$$
$$9a+3b+c=128$$
$$25a+5b+c=0$$

Multiply the first equation by –1 and add it to the second equation:

$$-a-b-c=-128$$
$$9a+3b+c=128$$

Adding yields:

$$8a+2b=0$$
$$4a+b=0$$

Multiply the first equation by –1 and add it to the third equation:

$$-a-b-c=-128$$
$$25a+5b+c=0$$

Adding yields:

$$24a+4b=-128$$
$$6a+b=-32$$

The system simplifies to:

$$4a+b=0$$
$$6a+b=-32$$

Multiplying the first equation by –1:

$$-4a-b=0$$
$$6a+b=-32$$

Adding yields:

$$2a=-32$$
$$a=-16$$

Substituting to find b:

$$4(-16)+b=0$$
$$b=64$$

Substituting to find c:

$$-16+64+c=128$$
$$c=80$$

The equation for the height is $h=-16t^2+64t+80$.

35. No, the graph does not include the boundary line.

37. Substituting $x=4-y$ into the second equation:

$$(4-y)-2y=4$$
$$4-3y=4$$
$$-3y=0$$
$$y=0$$

The solution is (4,0).

39. Solving the inequality:

$$20x+9,300>18,000$$
$$20x>8,700$$
$$x>435$$

4.4 Systems of Linear Inequalities

1. Graphing the solution set:

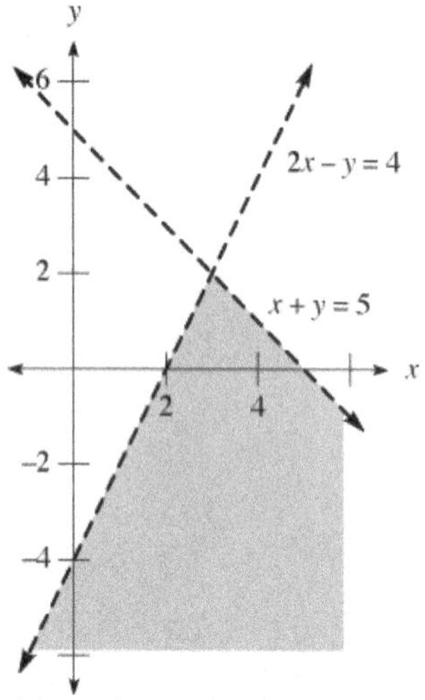

3. Graphing the solution set:

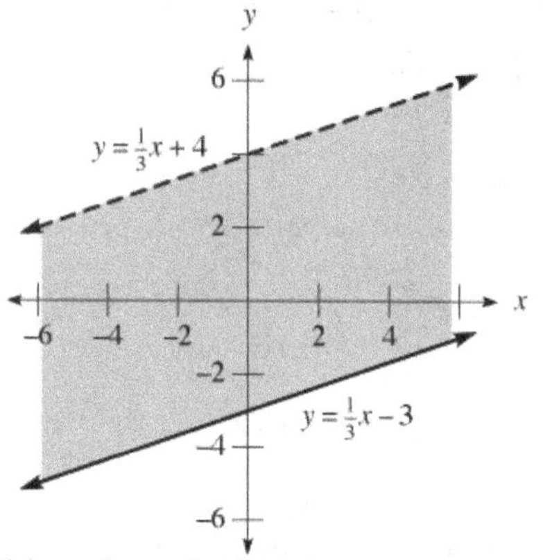

5. Graphing the solution set:

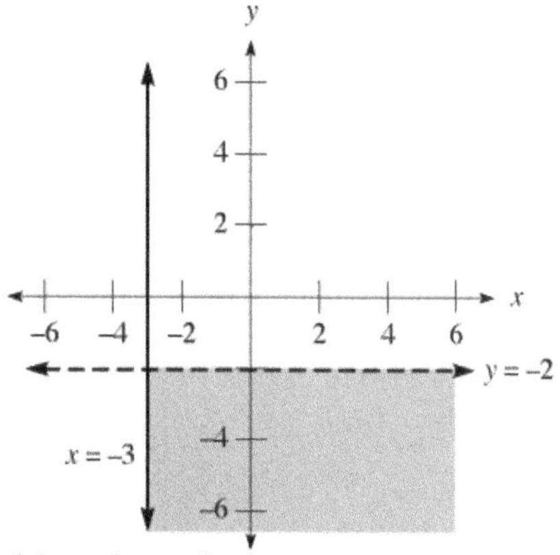

7. Graphing the solution set:

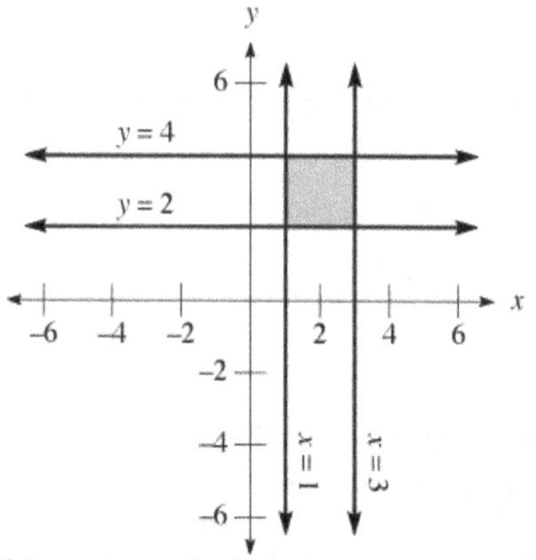

9. Graphing the solution set:

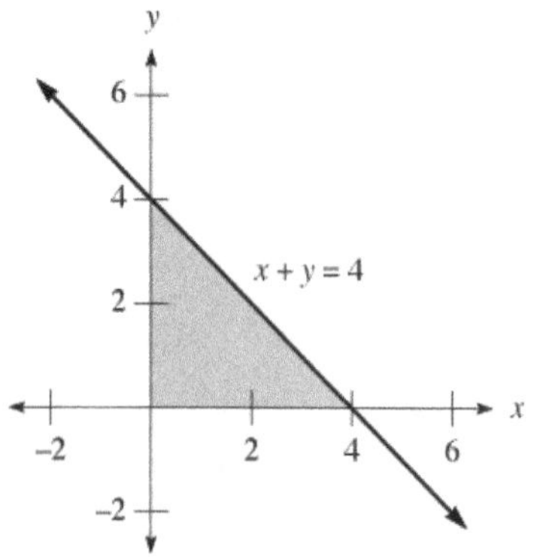

11. Graphing the solution set:

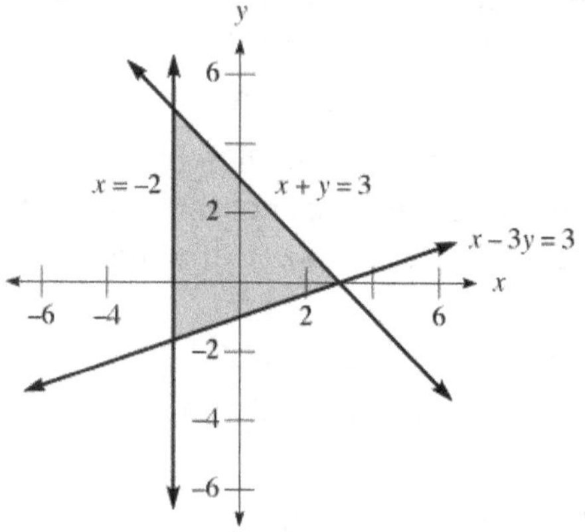

13. Graphing the solution set:

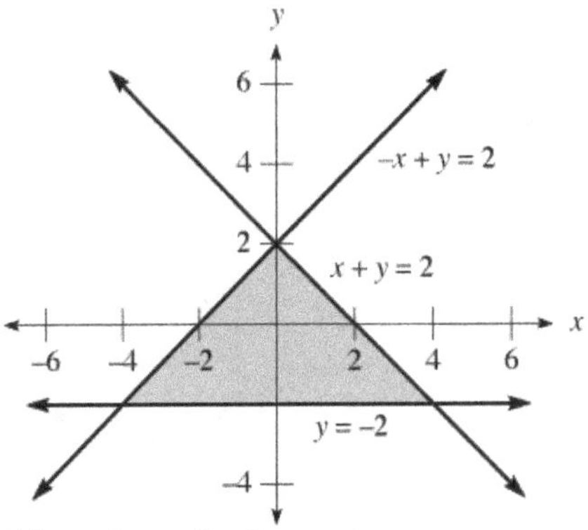

15. Graphing the solution set:

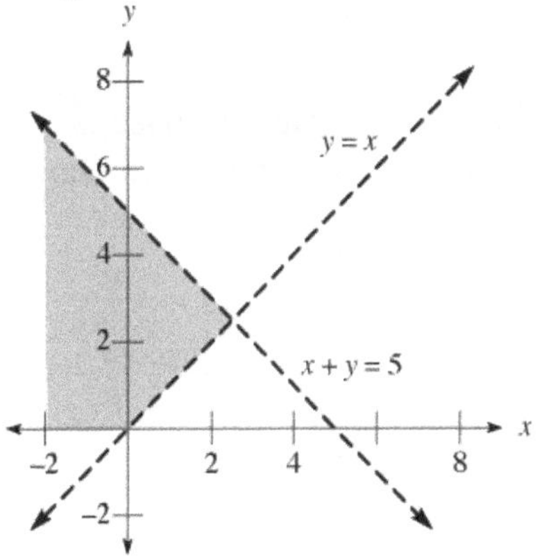

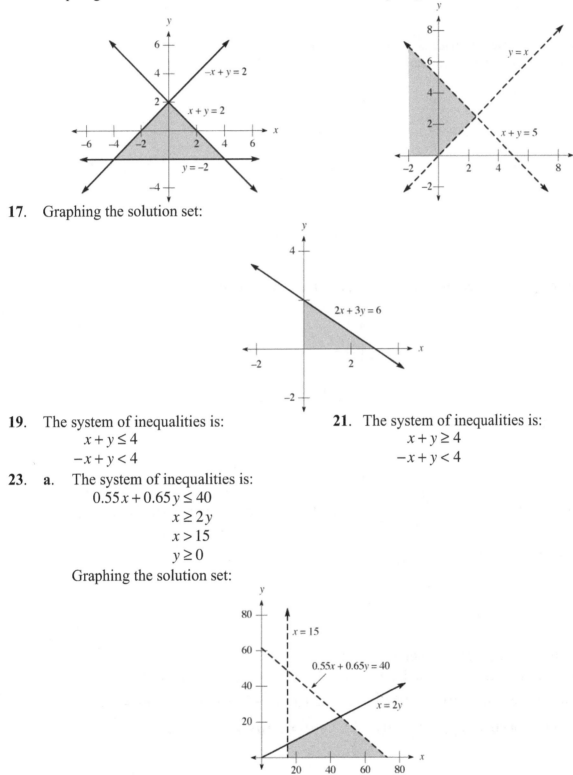

17. Graphing the solution set:

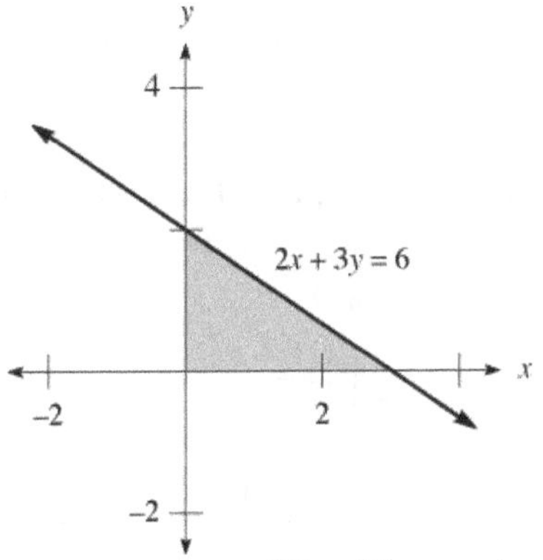

19. The system of inequalities is:

$$x + y \le 4$$
$$-x + y < 4$$

21. The system of inequalities is:

$$x + y \ge 4$$
$$-x + y < 4$$

23. **a.** The system of inequalities is:

$$0.55x + 0.65y \le 40$$
$$x \ge 2y$$
$$x > 15$$
$$y \ge 0$$

Graphing the solution set:

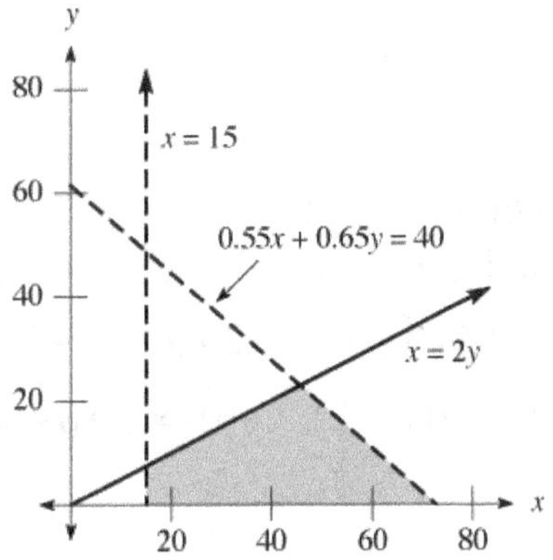

b. Substitute $x = 20$:

$$2y \le 20$$
$$y \le 10$$

The most he can purchase is 10 65-cent stamps.

25. The x-intercept is 3, the y-intercept is 6, and the slope is –2.

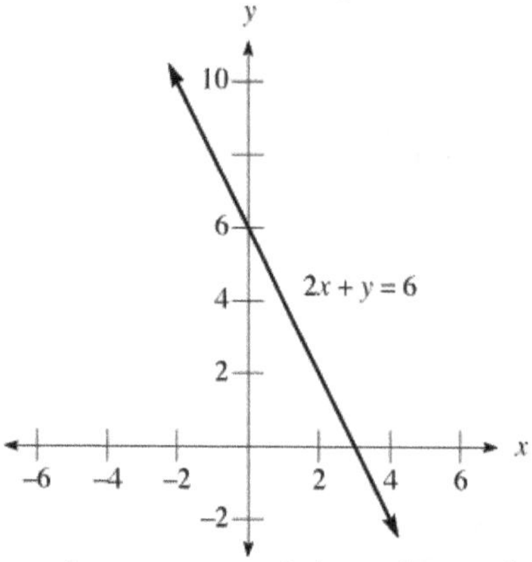

27. The x-intercept is –2, there is no y-intercept, and there is no slope.

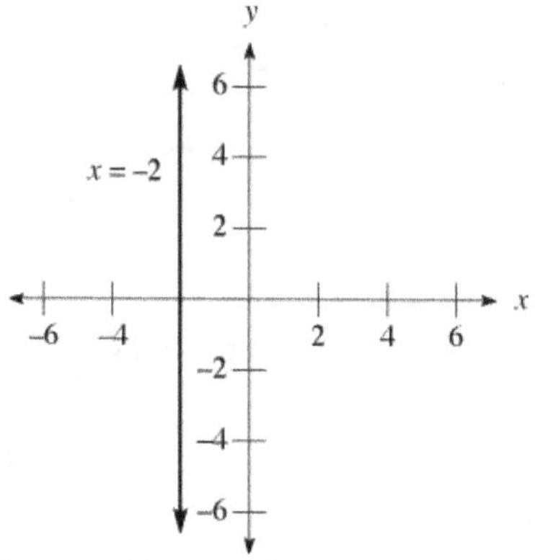

29. First find the slope: $m = \dfrac{-1-2}{4-(-3)} = \dfrac{-3}{4+3} = -\dfrac{3}{7}$. Using the point-slope formula:

$$y-2 = -\frac{3}{7}(x+3)$$
$$y-2 = -\frac{3}{7}x - \frac{9}{7}$$
$$y = -\frac{3}{7}x + \frac{5}{7}$$

31. Since the line is vertical, its equation is $x = 4$.

33. The domain is all real numbers and the range is $\{y \mid y \ge -9\}$. This is a function.

35. Evaluating the function: $h(0)+g(0) = [3 \bullet 0^2 - 2 \bullet 0 - 8] + [3 \bullet 0 + 4] = -8 + 4 = -4$

37. Evaluating the function: $g[f(2)] = g(2-2) = g(0) = 3 \bullet 0 + 4 = 4$

39. The variation equation is $z = Kxy^3$. Substituting $x = 5$, $y = 2$, and $z = 15$:

$$15 = K(5)(2)^3$$
$$15 = 40K$$
$$K = \frac{3}{8}$$

The equation is $z = \frac{3}{8}xy^3$. Substituting $x = 2$ and $y = 3$: $z = \frac{3}{8}(2)(3)^3 = \frac{3}{8} \bullet 54 = \frac{81}{4}$

Chapter 4 Test

See www.mathtv.com for video solutions to all problems in this chapter test.

Chapter 5
Rational Expressions and Rational Functions

5.1 Basic Properties and Reducing to Lowest Terms

1. Finding each function value:

$$g(0)=\frac{0+3}{0-1}=\frac{3}{-1}=-3 \qquad g(-3)=\frac{-3+3}{-3-1}=\frac{0}{-4}=0$$

$$g(3)=\frac{3+3}{3-1}=\frac{6}{2}=3 \qquad g(-1)=\frac{-1+3}{-1-1}=\frac{2}{-2}=-1$$

$$g(1)=\frac{1+3}{1-1}=\frac{4}{0}, \text{ which is undefined}$$

3. Finding each function value:

$$h(0)=\frac{0-3}{0+1}=\frac{-3}{1}=-3 \qquad h(-3)=\frac{-3-3}{-3+1}=\frac{-6}{-2}=3$$

$$h(3)=\frac{3-3}{3+1}=\frac{0}{4}=0 \qquad h(-1)=\frac{-1-3}{-1+1}=\frac{-4}{0}, \text{ which is undefined}$$

$$h(1)=\frac{1-3}{1+1}=\frac{-2}{2}=-1$$

5. The domain is $\{x \mid x \neq 1\}$.

7. The domain is $\{x \mid x \neq 2\}$.

9. Setting the denominator equal to 0:

$$t^2-16=0$$
$$(t+4)(t-4)=0$$
$$t=-4,4$$

The domain is $\{t \mid t \neq -4, t \neq 4\}$.

11. Reducing to lowest terms: $\dfrac{x^2-16}{6x+24}=\dfrac{(x+4)(x-4)}{6(x+4)}=\dfrac{x-4}{6}$

13. Reducing to lowest terms: $\dfrac{a^4-81}{a-3}=\dfrac{(a^2+9)(a^2-9)}{a-3}=\dfrac{(a^2+9)(a+3)(a-3)}{a-3}=(a^2+9)(a+3)$

15. Reducing to lowest terms: $\dfrac{20y^2-45}{10y^2-5y-15}=\dfrac{5(4y^2-9)}{5(2y^2-y-3)}=\dfrac{5(2y+3)(2y-3)}{5(2y-3)(y+1)}=\dfrac{2y+3}{y+1}$

17. Reducing to lowest terms: $\dfrac{12y-2xy-2x^2y}{6y-4xy-2x^2y}=\dfrac{-2y(x^2+x-6)}{-2y(x^2+2x-3)}=\dfrac{-2y(x+3)(x-2)}{-2y(x+3)(x-1)}=\dfrac{x-2}{x-1}$

19. Reducing to lowest terms: $\dfrac{(x-3)^2(x+2)}{(x+2)^2(x-3)}=\dfrac{x-3}{x+2}$

21. Reducing to lowest terms: $\dfrac{x^3+1}{x^2-1}=\dfrac{(x+1)(x^2-x+1)}{(x+1)(x-1)}=\dfrac{x^2-x+1}{x-1}$

23. Reducing to lowest terms: $\dfrac{4am-4an}{3n-3m}=\dfrac{4a(m-n)}{3(n-m)}=\dfrac{-4a(n-m)}{3(n-m)}=-\dfrac{4a}{3}$

25. Reducing to lowest terms: $\dfrac{ab-a+b-1}{ab+a+b+1}=\dfrac{a(b-1)+1(b-1)}{a(b+1)+1(b+1)}=\dfrac{(b-1)(a+1)}{(b+1)(a+1)}=\dfrac{b-1}{b+1}$

27. Reducing to lowest terms: $\dfrac{21x^2-23x+6}{21x^2+x-10}=\dfrac{(7x-3)(3x-2)}{(7x+5)(3x-2)}=\dfrac{7x-3}{7x+5}$

29. Reducing to lowest terms: $\dfrac{8x^2-6x-9}{8x^2-18x+9}=\dfrac{(4x+3)(2x-3)}{(4x-3)(2x-3)}=\dfrac{4x+3}{4x-3}$

31. Reducing to lowest terms: $\dfrac{4x^2+29x+45}{8x^2-10x-63}=\dfrac{(x+5)(4x+9)}{(2x-7)(4x+9)}=\dfrac{x+5}{2x-7}$

33. Reducing to lowest terms: $\dfrac{a^3+b^3}{a^2-b^2}=\dfrac{(a+b)(a^2-ab+b^2)}{(a+b)(a-b)}=\dfrac{a^2-ab+b^2}{a-b}$

35. Reducing to lowest terms: $\dfrac{8x^4-8x}{4x^4+4x^3+4x^2}=\dfrac{8x(x^3-1)}{4x^2(x^2+x+1)}=\dfrac{8x(x-1)(x^2+x+1)}{4x^2(x^2+x+1)}=\dfrac{2(x-1)}{x}$

37. Reducing to lowest terms: $\dfrac{ax+2x+3a+6}{ay+2y-4a-8}=\dfrac{x(a+2)+3(a+2)}{y(a+2)-4(a+2)}=\dfrac{(a+2)(x+3)}{(a+2)(y-4)}=\dfrac{x+3}{y-4}$

39. Reducing to lowest terms:

$$\dfrac{x^3+3x^2-4x-12}{x^2+x-6}=\dfrac{x^2(x+3)-4(x+3)}{(x+3)(x-2)}=\dfrac{(x+3)(x^2-4)}{(x+3)(x-2)}=\dfrac{(x+3)(x+2)(x-2)}{(x+3)(x-2)}=x+2$$

41. Reducing to lowest terms: $\dfrac{x^3-8}{x^2-4}=\dfrac{(x-2)(x^2+2x+4)}{(x-2)(x+2)}=\dfrac{x^2+2x+4}{x+2}$

43. Reducing to lowest terms: $\dfrac{8x^3-27}{4x^2-9}=\dfrac{(2x-3)(4x^2+6x+9)}{(2x-3)(2x+3)}=\dfrac{4x^2+6x+9}{2x+3}$

45. Reducing to lowest terms: $\dfrac{x-4}{4-x}=\dfrac{x-4}{-1(x-4)}=-1$

47. Reducing to lowest terms: $\dfrac{y^2-36}{6-y}=\dfrac{(y+6)(y-6)}{-1(y-6)}=-(y+6)$

49. Reducing to lowest terms: $\dfrac{1-9a^2}{9a^2-6a+1}=\dfrac{-1(9a^2-1)}{(3a-1)^2}=\dfrac{-1(3a+1)(3a-1)}{(3a-1)^2}=-\dfrac{3a+1}{3a-1}$

51. Simplifying: $\dfrac{(3x-5)-(3a-5)}{x-a}=\dfrac{3x-5-3a+5}{x-a}=\dfrac{3x-3a}{x-a}=\dfrac{3(x-a)}{x-a}=3$

53. Simplifying: $\dfrac{(x^2-4)-(a^2-4)}{x-a}=\dfrac{x^2-4-a^2+4}{x-a}=\dfrac{x^2-a^2}{x-a}=\dfrac{(x+a)(x-a)}{x-a}=x+a$

55. **a.** Evaluating the formula: $\dfrac{f(x)-f(a)}{x-a}=\dfrac{4x-4a}{x-a}=\dfrac{4(x-a)}{x-a}=4$

b. Evaluating the formula: $\dfrac{f(x+h)-f(x)}{h}=\dfrac{4(x+h)-4x}{h}=\dfrac{4x+4h-4x}{h}=\dfrac{4h}{h}=4$

57. **a.** Evaluating the formula: $\dfrac{f(x)-f(a)}{x-a}=\dfrac{(5x+3)-(5a+3)}{x-a}=\dfrac{5x-5a}{x-a}=\dfrac{5(x-a)}{x-a}=5$

b. Evaluating the formula:

$$\frac{f(x+h)-f(x)}{h}=\frac{5(x+h)+3-(5x+3)}{h}=\frac{5x+5h+3-5x-3}{h}=\frac{5h}{h}=5$$

59. **a.** Evaluating the formula: $\dfrac{f(x)-f(a)}{x-a}=\dfrac{x^2-a^2}{x-a}=\dfrac{(x+a)(x-a)}{x-a}=x+a$

b. Evaluating the formula:

$$\frac{f(x+h)-f(x)}{h}=\frac{(x+h)^2-x^2}{h}=\frac{x^2+2xh+h^2-x^2}{h}=\frac{h(2x+h)}{h}=2x+h$$

61. **a.** Evaluating the formula: $\dfrac{f(x)-f(a)}{x-a}=\dfrac{(x^2+1)-(a^2+1)}{x-a}=\dfrac{x^2-a^2}{x-a}=\dfrac{(x+a)(x-a)}{x-a}=x+a$

b. Evaluating the formula:

$$\frac{f(x+h)-f(x)}{h}=\frac{(x+h)^2+1-(x^2+1)}{h}=\frac{x^2+2xh+h^2+1-x^2-1}{h}=\frac{h(2x+h)}{h}=2x+h$$

63. **a.** Evaluating the formula:

$$\begin{aligned}\frac{f(x)-f(a)}{x-a}&=\frac{(x^2-3x+4)-(a^2-3a+4)}{x-a}\\&=\frac{x^2-a^2-3x+3a}{x-a}\\&=\frac{(x+a)(x-a)-3(x-a)}{x-a}\\&=\frac{(x-a)(x+a-3)}{x-a}\\&=x+a-3\end{aligned}$$

b. Evaluating the formula:

$$\begin{aligned}\frac{f(x+h)-f(x)}{h}&=\frac{(x+h)^2-3(x+h)+4-(x^2-3x+4)}{h}\\&=\frac{x^2+2xh+h^2-3x-3h+4-x^2+3x-4}{h}\\&=\frac{h(2x+h-3)}{h}\\&=2x+h-3\end{aligned}$$

65. **a.** From the graph: $f(2)=2$ **b.** From the graph: $f(-1)=-4$

c. From the graph: $f(0)$ is undefined **d.** From the graph: $g(3)=2$

67. Completing the table:

Weeks x	Weight (lb) $W(x)$
0	200
1	194
4	184
12	173
24	168

69. Multiplying: $\frac{6}{7}\bullet\frac{14}{18}=\frac{6}{7}\bullet\frac{2\bullet 7}{3\bullet 6}=\frac{2}{3}$

71. Multiplying: $5y^2\bullet 4x^2=20x^2y^2$

73. Multiplying: $9x^4\bullet 8y^5=72x^4y^5$

75. Factoring: $x^2-4=(x+2)(x-2)$

77. Factoring: $x^3-x^2y=x^2(x-y)$

79. Factoring: $2y^2-2=2(y^2-1)=2(y+1)(y-1)$

5.2 Multiplication and Division of Rational Expressions

1. Performing the operations: $\frac{2}{9}\bullet\frac{3}{4}=\frac{2}{3\bullet 3}\bullet\frac{3}{2\bullet 2}=\frac{1}{2\bullet 3}=\frac{1}{6}$

3. Performing the operations: $\frac{3}{4}\div\frac{1}{3}=\frac{3}{4}\bullet\frac{3}{1}=\frac{9}{4}$

5. Performing the operations: $\frac{3}{7}\bullet\frac{14}{24}\div\frac{1}{2}=\frac{1}{4}\div\frac{1}{2}=\frac{1}{4}\bullet\frac{2}{1}=\frac{2}{4}=\frac{1}{2}$

7. Performing the operations: $\frac{10x^2}{5y^2}\bullet\frac{15y^3}{2x^4}=\frac{150x^2y^3}{10x^4y^2}=\frac{15y}{x^2}$

9. Performing the operations: $\frac{11a^2b}{5ab^2}\div\frac{22a^3b^2}{10ab^4}=\frac{11a^2b}{5ab^2}\bullet\frac{10ab^4}{22a^3b^2}=\frac{110a^3b^5}{110a^4b^4}=\frac{b}{a}$

11. Performing the operations: $\frac{6x^2}{5y^3}\bullet\frac{11z^2}{2x^2}\div\frac{33z^5}{10y^8}=\frac{33z^2}{5y^3}\bullet\frac{10y^8}{33z^5}=\frac{2y^8z^2}{y^3z^5}=\frac{2y^5}{z^3}$

13. Performing the operations: $\frac{x^2-9}{x^2-4}\bullet\frac{x-2}{x-3}=\frac{(x+3)(x-3)}{(x+2)(x-2)}\bullet\frac{x-2}{x-3}=\frac{x+3}{x+2}$

15. Performing the operations: $\frac{y^2-1}{y+2}\bullet\frac{y^2+5y+6}{y^2+2y-3}=\frac{(y+1)(y-1)}{y+2}\bullet\frac{(y+2)(y+3)}{(y+3)(y-1)}=y+1$

17. Performing the operations: $\frac{3x-12}{x^2-4}\bullet\frac{x^2+6x+8}{x-4}=\frac{3(x-4)}{(x+2)(x-2)}\bullet\frac{(x+4)(x+2)}{x-4}=\frac{3(x+4)}{x-2}$

19. Performing the operations: $\frac{xy}{xy+1}\div\frac{x}{y}=\frac{xy}{xy+1}\bullet\frac{y}{x}=\frac{y^2}{xy+1}$

21. Performing the operations: $\frac{1}{x^2-9} \div \frac{1}{x^2+9} = \frac{1}{x^2-9} \bullet \frac{x^2+9}{1} = \frac{x^2+9}{x^2-9}$

23. Performing the operations: $\frac{y-3}{y^2-6y+9} \bullet \frac{y-3}{4} = \frac{y-3}{(y-3)^2} \bullet \frac{y-3}{4} = \frac{1}{4}$

25. Performing the operations:

$$\frac{5x+2y}{25x^2-5xy-6y^2} \bullet \frac{20x^2-7xy-3y^2}{4x+y} = \frac{5x+2y}{(5x+2y)(5x-3y)} \bullet \frac{(5x-3y)(4x+y)}{4x+y} = 1$$

27. Performing the operations:

$$\begin{aligned}\frac{a^2-5a+6}{a^2-2a-3} \div \frac{a-5}{a^2+3a+2} &= \frac{a^2-5a+6}{a^2-2a-3} \bullet \frac{a^2+3a+2}{a-5} \\ &= \frac{(a-3)(a-2)}{(a-3)(a+1)} \bullet \frac{(a+2)(a+1)}{a-5} \\ &= \frac{(a-2)(a+2)}{a-5}\end{aligned}$$

29. Performing the operations:

$$\begin{aligned}\frac{4t^2-1}{6t^2+t-2} \div \frac{8t^3+1}{27t^3+8} &= \frac{4t^2-1}{6t^2+t-2} \bullet \frac{27t^3+8}{8t^3+1} \\ &= \frac{(2t+1)(2t-1)}{(3t+2)(2t-1)} \bullet \frac{(3t+2)(9t^2-6t+4)}{(2t+1)(4t^2-2t+1)} \\ &= \frac{9t^2-6t+4}{4t^2-2t+1}\end{aligned}$$

31. Performing the operations:

$$\begin{aligned}\frac{2x^2-5x-12}{4x^2+8x+3} \div \frac{x^2-16}{2x^2+7x+3} &= \frac{2x^2-5x-12}{4x^2+8x+3} \bullet \frac{2x^2+7x+3}{x^2-16} \\ &= \frac{(2x+3)(x-4)}{(2x+1)(2x+3)} \bullet \frac{(2x+1)(x+3)}{(x+4)(x-4)} \\ &= \frac{x+3}{x+4}\end{aligned}$$

33. Performing the operations:

$$\begin{aligned}\frac{2a^2-21ab-36b^2}{a^2-11ab-12b^2} \div \frac{10a+15b}{a^2-b^2} &= \frac{2a^2-21ab-36b^2}{a^2-11ab-12b^2} \bullet \frac{a^2-b^2}{10a+15b} \\ &= \frac{(2a+3b)(a-12b)}{(a-12b)(a+b)} \bullet \frac{(a+b)(a-b)}{5(2a+3b)} \\ &= \frac{a-b}{5}\end{aligned}$$

35. Performing the operations:

$$\frac{6c^2-c-15}{9c^2-25} \bullet \frac{15c^2+22c-5}{6c^2+5c-6} = \frac{(3c-5)(2c+3)}{(3c+5)(3c-5)} \bullet \frac{(3c+5)(5c-1)}{(3c-2)(2c+3)} = \frac{5c-1}{3c-2}$$

37. Performing the operations:

$$\frac{6a^2b+2ab^2-20b^3}{4a^2b-16b^3}\bullet\frac{10a^2-22ab+4b^2}{27a^3-125b^3}$$
$$=\frac{2b\left(3a^2+ab-10b^2\right)}{4b\left(a^2-4b^2\right)}\bullet\frac{2\left(5a^2-11ab+2b^2\right)}{(3a-5b)\left(9a^2+15ab+25b^2\right)}$$
$$=\frac{2b(3a-5b)(a+2b)}{4b(a+2b)(a-2b)}\bullet\frac{2(5a-b)(a-2b)}{(3a-5b)\left(9a^2+15ab+25b^2\right)}$$
$$=\frac{5a-b}{9a^2+15ab+25b^2}$$

39. Performing the operations:

$$\frac{360x^3-490x}{36x^2+84x+49}\bullet\frac{30x^2+83x+56}{150x^3+65x^2-280x}=\frac{10x\left(36x^2-49\right)}{(6x+7)^2}\bullet\frac{(6x+7)(5x+8)}{5x\left(30x^2+13x-56\right)}$$
$$=\frac{10x(6x+7)(6x-7)}{(6x+7)^2}\bullet\frac{(6x+7)(5x+8)}{5x(6x-7)(5x+8)}$$
$$=2$$

41. Performing the operations:

$$\frac{x^5-x^2}{5x^2-5x}\bullet\frac{10x^4-10x^2}{2x^4+2x^3+2x^2}=\frac{x^2\left(x^3-1\right)}{5x(x-1)}\bullet\frac{10x^2\left(x^2-1\right)}{2x^2\left(x^2+x+1\right)}$$
$$=\frac{x^2(x-1)\left(x^2+x+1\right)}{5x(x-1)}\bullet\frac{10x^2(x+1)(x-1)}{2x^2\left(x^2+x+1\right)}$$
$$=x(x+1)(x-1)$$

43. Performing the operations:

$$\frac{a^2-16b^2}{a^2-8ab+16b^2}\bullet\frac{a^2-9ab+20b^2}{a^2-7ab+12b^2}\div\frac{a^2-25b^2}{a^2-6ab+9b^2}$$
$$=\frac{a^2-16b^2}{a^2-8ab+16b^2}\bullet\frac{a^2-9ab+20b^2}{a^2-7ab+12b^2}\bullet\frac{a^2-6ab+9b^2}{a^2-25b^2}$$
$$=\frac{(a+4b)(a-4b)}{(a-4b)^2}\bullet\frac{(a-5b)(a-4b)}{(a-3b)(a-4b)}\bullet\frac{(a-3b)^2}{(a+5b)(a-5b)}$$
$$=\frac{(a+4b)(a-3b)}{(a-4b)(a+5b)}$$

45. Performing the operations:

$$\frac{2y^2-7y-15}{42y^2-29y-5} \bullet \frac{12y^2-16y+5}{7y^2-36y+5} \div \frac{4y^2-9}{49y^2-1}$$

$$= \frac{2y^2-7y-15}{42y^2-29y-5} \bullet \frac{12y^2-16y+5}{7y^2-36y+5} \bullet \frac{49y^2-1}{4y^2-9}$$

$$= \frac{(2y+3)(y-5)}{(6y-5)(7y+1)} \bullet \frac{(6y-5)(2y-1)}{(7y-1)(y-5)} \bullet \frac{(7y+1)(7y-1)}{(2y+3)(2y-3)}$$

$$= \frac{2y-1}{2y-3}$$

47. Performing the operations:

$$\frac{xy-2x+3y-6}{xy+2x-4y-8} \bullet \frac{xy+x-4y-4}{xy-x+3y-3} = \frac{x(y-2)+3(y-2)}{x(y+2)-4(y+2)} \bullet \frac{x(y+1)-4(y+1)}{x(y-1)+3(y-1)}$$

$$= \frac{(y-2)(x+3)}{(y+2)(x-4)} \bullet \frac{(y+1)(x-4)}{(y-1)(x+3)}$$

$$= \frac{(y-2)(y+1)}{(y+2)(y-1)}$$

49. Performing the operations:

$$\frac{xy^2-y^2+4xy-4y}{xy-3y+4x-12} \div \frac{xy^3+2xy^2+y^3+2y^2}{xy^2-3y^2+2xy-6y} = \frac{xy^2-y^2+4xy-4y}{xy-3y+4x-12} \bullet \frac{xy^2-3y^2+2xy-6y}{xy^3+2xy^2+y^3+2y^2}$$

$$= \frac{y^2(x-1)+4y(x-1)}{y(x-3)+4(x-3)} \bullet \frac{y^2(x-3)+2y(x-3)}{xy^2(y+2)+y^2(y+2)}$$

$$= \frac{y(x-1)(y+4)}{(x-3)(y+4)} \bullet \frac{y(x-3)(y+2)}{y^2(y+2)(x+1)}$$

$$= \frac{x-1}{x+1}$$

51. Performing the operations:

$$\frac{2x^3+10x^2-8x-40}{x^3+4x^2-9x-36} \bullet \frac{x^2+x-12}{2x^2+14x+20} = \frac{2x^2(x+5)-8(x+5)}{x^2(x+4)-9(x+4)} \bullet \frac{(x+4)(x-3)}{2(x^2+7x+10)}$$

$$= \frac{2(x+5)(x^2-4)}{(x+4)(x^2-9)} \bullet \frac{(x+4)(x-3)}{2(x+5)(x+2)}$$

$$= \frac{2(x+5)(x+2)(x-2)}{(x+4)(x+3)(x-3)} \bullet \frac{(x+4)(x-3)}{2(x+5)(x+2)}$$

$$= \frac{x-2}{x+3}$$

53. Performing the operations:

$$\frac{w^3-w^2x}{wy-w}\div\left(\frac{w-x}{y-1}\right)^2=\frac{w^3-w^2x}{wy-w}\bullet\left(\frac{y-1}{w-x}\right)^2=\frac{w^2(w-x)}{w(y-1)}\bullet\frac{(y-1)^2}{(w-x)^2}=\frac{w(y-1)}{w-x}$$

55. Performing the operations:

$$\frac{mx+my+2x+2y}{6x^2-5xy-4y^2}\div\frac{2mx-4x+my-2y}{3mx-6x-4my+8y}=\frac{mx+my+2x+2y}{6x^2-5xy-4y^2}\bullet\frac{3mx-6x-4my+8y}{2mx-4x+my-2y}$$

$$=\frac{m(x+y)+2(x+y)}{(3x-4y)(2x+y)}\bullet\frac{3x(m-2)-4y(m-2)}{2x(m-2)+y(m-2)}$$

$$=\frac{(x+y)(m+2)}{(3x-4y)(2x+y)}\bullet\frac{(m-2)(3x-4y)}{(m-2)(2x+y)}$$

$$=\frac{(m+2)(x+y)}{(2x+y)^2}$$

57. Finding the product: $(3x-6)\bullet\frac{x}{x-2}=\frac{3(x-2)}{1}\bullet\frac{x}{x-2}=3x$

59. Finding the product: $\left(x^2-25\right)\bullet\frac{2}{x-5}=\frac{(x+5)(x-5)}{1}\bullet\frac{2}{x-5}=2(x+5)$

61. Finding the product: $\left(x^2-3x+2\right)\bullet\frac{3}{3x-3}=\frac{(x-2)(x-1)}{1}\bullet\frac{3}{3(x-1)}=x-2$

63. Finding the product:

$$(y-3)(y-4)(y+3)\bullet\frac{-1}{y^2-9}=\frac{(y-3)(y-4)(y+3)}{1}\bullet\frac{-1}{(y+3)(y-3)}=-(y-4)=4-y$$

65. Finding the product: $a(a+5)(a-5)\bullet\frac{a+1}{a^2+5a}=\frac{a(a+5)(a-5)}{1}\bullet\frac{a+1}{a(a+5)}=(a-5)(a+1)$

67. **a.** Simplifying: $\frac{16-1}{64-1}=\frac{15}{63}=\frac{5}{21}$

b. Reducing: $\frac{25x^2-9}{125x^3-27}=\frac{(5x-3)(5x+3)}{(5x-3)\left(25x^2+15x+9\right)}=\frac{5x+3}{25x^2+15x+9}$

c. Multiplying: $\frac{25x^2-9}{125x^3-27}\bullet\frac{5x-3}{5x+3}=\frac{(5x-3)(5x+3)}{(5x-3)\left(25x^2+15x+9\right)}\bullet\frac{5x-3}{5x+3}=\frac{5x-3}{25x^2+15x+9}$

d. Dividing:

$$\frac{25x^2-9}{125x^3-27}\div\frac{5x-3}{25x^2+15x+9}=\frac{(5x-3)(5x+3)}{(5x-3)\left(25x^2+15x+9\right)}\bullet\frac{25x^2+15x+9}{5x-3}=\frac{5x+3}{5x-3}$$

69. Combining: $\frac{4}{9}+\frac{2}{9}=\frac{6}{9}=\frac{2\bullet 3}{3\bullet 3}=\frac{2}{3}$

71. Combining: $\frac{3}{14}+\frac{7}{30}=\frac{3}{14}\bullet\frac{15}{15}+\frac{7}{30}\bullet\frac{7}{7}=\frac{45}{210}+\frac{49}{210}=\frac{94}{210}=\frac{47}{105}$

73. Multiplying: $-1(7-x)=-7+x=x-7$

75. Factoring: $x^2-1=(x+1)(x-1)$

77. Factoring: $2x+10=2(x+5)$

79. Factoring: $a^3-b^3=(a-b)(a^2+ab+b^2)$

5.3 Addition and Subtraction of Rational Expressions

1. Combining the fractions: $\frac{3}{4}+\frac{1}{2}=\frac{3}{4}+\frac{1}{2}\bullet\frac{2}{2}=\frac{3}{4}+\frac{2}{4}=\frac{5}{4}$

3. Combining the fractions: $\frac{2}{5}-\frac{1}{15}=\frac{2}{5}\bullet\frac{3}{3}-\frac{1}{15}=\frac{6}{15}-\frac{1}{15}=\frac{5}{15}=\frac{1}{3}$

5. Combining the fractions: $\frac{5}{6}+\frac{7}{8}=\frac{5}{6}\bullet\frac{4}{4}+\frac{7}{8}\bullet\frac{3}{3}=\frac{20}{24}+\frac{21}{24}=\frac{41}{24}$

7. Combining the fractions: $\frac{9}{48}-\frac{3}{54}=\frac{9}{48}\bullet\frac{9}{9}-\frac{3}{54}\bullet\frac{8}{8}=\frac{81}{432}-\frac{24}{432}=\frac{57}{432}=\frac{19}{144}$

9. Combining the fractions: $\frac{3}{4}-\frac{1}{8}+\frac{2}{3}=\frac{3}{4}\bullet\frac{6}{6}-\frac{1}{8}\bullet\frac{3}{3}+\frac{2}{3}\bullet\frac{8}{8}=\frac{18}{24}-\frac{3}{24}+\frac{16}{24}=\frac{31}{24}$

11. Combining the rational expressions: $\frac{x}{x+3}+\frac{3}{x+3}=\frac{x+3}{x+3}=1$

13. Combining the rational expressions: $\frac{4}{y-4}-\frac{y}{y-4}=\frac{4-y}{y-4}=\frac{-1(y-4)}{y-4}=-1$

15. Combining the rational expressions: $\frac{x}{x^2-y^2}-\frac{y}{x^2-y^2}=\frac{x-y}{x^2-y^2}=\frac{x-y}{(x+y)(x-y)}=\frac{1}{x+y}$

17. Combining the rational expressions: $\frac{2x-3}{x-2}-\frac{x-1}{x-2}=\frac{2x-3-x+1}{x-2}=\frac{x-2}{x-2}=1$

19. Combining the rational expressions: $\frac{1}{a}+\frac{2}{a^2}-\frac{3}{a^3}=\frac{1}{a}\bullet\frac{a^2}{a^2}+\frac{2}{a^2}\bullet\frac{a}{a}-\frac{3}{a^3}=\frac{a^2+2a-3}{a^3}$

21. Combining the rational expressions: $\frac{7x-2}{2x+1}-\frac{5x-3}{2x+1}=\frac{7x-2-5x+3}{2x+1}=\frac{2x+1}{2x+1}=1$

23. a. Multiplying: $\frac{3}{8}\bullet\frac{1}{6}=\frac{3}{8}\bullet\frac{1}{2\bullet3}=\frac{1}{16}$

b. Dividing: $\frac{3}{8}\div\frac{1}{6}=\frac{3}{8}\bullet\frac{6}{1}=\frac{3}{2\bullet4}\bullet\frac{2\bullet3}{1}=\frac{9}{4}$

c. Adding: $\frac{3}{8}+\frac{1}{6}=\frac{3}{8}\bullet\frac{3}{3}+\frac{1}{6}\bullet\frac{4}{4}=\frac{9}{24}+\frac{4}{24}=\frac{13}{24}$

d. Multiplying: $\frac{x+3}{x-3}\bullet\frac{5x+15}{x^2-9}=\frac{x+3}{x-3}\bullet\frac{5(x+3)}{(x+3)(x-3)}=\frac{5(x+3)}{(x-3)^2}$

e. Dividing: $\frac{x+3}{x-3}\div\frac{5x+15}{x^2-9}=\frac{x+3}{x-3}\bullet\frac{(x+3)(x-3)}{5(x+3)}=\frac{x+3}{5}$

f. Subtracting:

$$\frac{x+3}{x-3}-\frac{5x+15}{x^2-9}=\frac{x+3}{x-3}\bullet\frac{x+3}{x+3}-\frac{5x+15}{(x+3)(x-3)}$$
$$=\frac{x^2+6x+9}{(x+3)(x-3)}-\frac{5x+15}{(x+3)(x-3)}$$
$$=\frac{x^2+x-6}{(x+3)(x-3)}$$
$$=\frac{(x+3)(x-2)}{(x+3)(x-3)}$$
$$=\frac{x-2}{x-3}$$

25. Combining the rational expressions:

$$\frac{3x+1}{2x-6}-\frac{x+2}{x-3}=\frac{3x+1}{2(x-3)}-\frac{x+2}{x-3}\bullet\frac{2}{2}=\frac{3x+1}{2(x-3)}-\frac{2x+4}{2(x-3)}=\frac{3x+1-2x-4}{2(x-3)}=\frac{x-3}{2(x-3)}=\frac{1}{2}$$

27. Combining the rational expressions:

$$\frac{6x+5}{5x-25}-\frac{x+2}{x-5}=\frac{6x+5}{5(x-5)}-\frac{x+2}{x-5}\bullet\frac{5}{5}$$
$$=\frac{6x+5}{5(x-5)}-\frac{5x+10}{5(x-5)}$$
$$=\frac{6x+5-5x-10}{5(x-5)}$$
$$=\frac{x-5}{5(x-5)}$$
$$=\frac{1}{5}$$

29. Combining the rational expressions:

$$\frac{x+1}{2x-2}-\frac{2}{x^2-1}=\frac{x+1}{2(x-1)}\bullet\frac{x+1}{x+1}-\frac{2}{(x+1)(x-1)}\bullet\frac{2}{2}$$
$$=\frac{x^2+2x+1}{2(x+1)(x-1)}-\frac{4}{2(x+1)(x-1)}$$
$$=\frac{x^2+2x-3}{2(x+1)(x-1)}$$
$$=\frac{(x+3)(x-1)}{2(x+1)(x-1)}$$
$$=\frac{x+3}{2(x+1)}$$

31. Combining the rational expressions:

$$\begin{aligned}\frac{1}{a-b}-\frac{3ab}{a^3-b^3}&=\frac{1}{a-b}\bullet\frac{a^2+ab+b^2}{a^2+ab+b^2}-\frac{3ab}{a^3-b^3}\\&=\frac{a^2+ab+b^2}{a^3-b^3}-\frac{3ab}{a^3-b^3}\\&=\frac{a^2-2ab+b^2}{a^3-b^3}\\&=\frac{(a-b)^2}{(a-b)(a^2+ab+b^2)}\\&=\frac{a-b}{a^2+ab+b^2}\end{aligned}$$

33. Combining the rational expressions:

$$\begin{aligned}\frac{1}{2y-3}-\frac{18y}{8y^3-27}&=\frac{1}{2y-3}\bullet\frac{4y^2+6y+9}{4y^2+6y+9}-\frac{18y}{8y^3-27}\\&=\frac{4y^2+6y+9}{8y^3-27}-\frac{18y}{8y^3-27}\\&=\frac{4y^2-12y+9}{8y^3-27}\\&=\frac{(2y-3)^2}{(2y-3)(4y^2+6y+9)}\\&=\frac{2y-3}{4y^2+6y+9}\end{aligned}$$

35. Combining the rational expressions:

$$\begin{aligned}\frac{x}{x^2-5x+6}-\frac{3}{3-x}&=\frac{x}{(x-2)(x-3)}+\frac{3}{x-3}\bullet\frac{x-2}{x-2}\\&=\frac{x}{(x-2)(x-3)}+\frac{3x-6}{(x-2)(x-3)}\\&=\frac{4x-6}{(x-2)(x-3)}\\&=\frac{2(2x-3)}{(x-3)(x-2)}\end{aligned}$$

37. Combining the rational expressions:

$$\frac{2}{4t-5}+\frac{9}{8t^2-38t+35}=\frac{2}{4t-5}\bullet\frac{2t-7}{2t-7}+\frac{9}{(4t-5)(2t-7)}$$
$$=\frac{4t-14}{(4t-5)(2t-7)}+\frac{9}{(4t-5)(2t-7)}$$
$$=\frac{4t-5}{(4t-5)(2t-7)}$$
$$=\frac{1}{2t-7}$$

39. Combining the rational expressions:

$$\frac{1}{a^2-5a+6}+\frac{3}{a^2-a-2}=\frac{1}{(a-2)(a-3)}\bullet\frac{a+1}{a+1}+\frac{3}{(a-2)(a+1)}\bullet\frac{a-3}{a-3}$$
$$=\frac{a+1}{(a-2)(a-3)(a+1)}+\frac{3a-9}{(a-2)(a-3)(a+1)}$$
$$=\frac{4a-8}{(a-2)(a-3)(a+1)}$$
$$=\frac{4(a-2)}{(a-2)(a-3)(a+1)}$$
$$=\frac{4}{(a-3)(a+1)}$$

41. Combining the rational expressions:

$$\frac{1}{8x^3-1}-\frac{1}{4x^2-1}=\frac{1}{(2x-1)(4x^2+2x+1)}\bullet\frac{2x+1}{2x+1}-\frac{1}{(2x+1)(2x-1)}\bullet\frac{4x^2+2x+1}{4x^2+2x+1}$$
$$=\frac{2x+1}{(2x+1)(2x-1)(4x^2+2x+1)}-\frac{4x^2+2x+1}{(2x+1)(2x-1)(4x^2+2x+1)}$$
$$=\frac{2x+1-4x^2-2x-1}{(2x+1)(2x-1)(4x^2+2x+1)}$$
$$=\frac{-4x^2}{(2x+1)(2x-1)(4x^2+2x+1)}$$

43. Combining the rational expressions:

$$\begin{aligned}\frac{4}{4x^2-9}-\frac{6}{8x^2-6x-9}&=\frac{4}{(2x+3)(2x-3)}\bullet\frac{4x+3}{4x+3}-\frac{6}{(2x-3)(4x+3)}\bullet\frac{2x+3}{2x+3}\\&=\frac{16x+12}{(2x+3)(2x-3)(4x+3)}-\frac{12x+18}{(2x+3)(2x-3)(4x+3)}\\&=\frac{16x+12-12x-18}{(2x+3)(2x-3)(4x+3)}\\&=\frac{4x-6}{(2x+3)(2x-3)(4x+3)}\\&=\frac{2(2x-3)}{(2x+3)(2x-3)(4x+3)}\\&=\frac{2}{(2x+3)(4x+3)}\end{aligned}$$

45. Combining the rational expressions:

$$\begin{aligned}\frac{4a}{a^2+6a+5}-\frac{3a}{a^2+5a+4}&=\frac{4a}{(a+5)(a+1)}\bullet\frac{a+4}{a+4}-\frac{3a}{(a+4)(a+1)}\bullet\frac{a+5}{a+5}\\&=\frac{4a^2+16a}{(a+4)(a+5)(a+1)}-\frac{3a^2+15a}{(a+4)(a+5)(a+1)}\\&=\frac{4a^2+16a-3a^2-15a}{(a+4)(a+5)(a+1)}\\&=\frac{a^2+a}{(a+4)(a+5)(a+1)}\\&=\frac{a(a+1)}{(a+4)(a+5)(a+1)}\\&=\frac{a}{(a+4)(a+5)}\end{aligned}$$

47. Combining the rational expressions:

$$\frac{2x-1}{x^2+x-6}-\frac{x+2}{x^2+5x+6}=\frac{2x-1}{(x+3)(x-2)}\bullet\frac{x+2}{x+2}-\frac{x+2}{(x+3)(x+2)}\bullet\frac{x-2}{x-2}$$

$$=\frac{2x^2+3x-2}{(x+3)(x+2)(x-2)}-\frac{x^2-4}{(x+3)(x+2)(x-2)}$$

$$=\frac{2x^2+3x-2-x^2+4}{(x+3)(x+2)(x-2)}$$

$$=\frac{x^2+3x+2}{(x+3)(x+2)(x-2)}$$

$$=\frac{(x+2)(x+1)}{(x+3)(x+2)(x-2)}$$

$$=\frac{x+1}{(x-2)(x+3)}$$

49. Combining the rational expressions:

$$\frac{2x-8}{3x^2+8x+4}+\frac{x+3}{3x^2+5x+2}=\frac{2x-8}{(3x+2)(x+2)}+\frac{x+3}{(3x+2)(x+1)}$$

$$=\frac{2x-8}{(3x+2)(x+2)}\bullet\frac{x+1}{x+1}+\frac{x+3}{(3x+2)(x+1)}\bullet\frac{x+2}{x+2}$$

$$=\frac{2x^2-6x-8}{(3x+2)(x+2)(x+1)}+\frac{x^2+5x+6}{(3x+2)(x+2)(x+1)}$$

$$=\frac{3x^2-x-2}{(3x+2)(x+2)(x+1)}$$

$$=\frac{(3x+2)(x-1)}{(3x+2)(x+2)(x+1)}$$

$$=\frac{x-1}{(x+1)(x+2)}$$

51. Combining the rational expressions:

$$\frac{2}{x^2+5x+6}-\frac{4}{x^2+4x+3}+\frac{3}{x^2+3x+2}$$
$$=\frac{2}{(x+3)(x+2)}-\frac{4}{(x+3)(x+1)}+\frac{3}{(x+2)(x+1)}$$
$$=\frac{2}{(x+3)(x+2)}\bullet\frac{x+1}{x+1}-\frac{4}{(x+3)(x+1)}\bullet\frac{x+2}{x+2}+\frac{3}{(x+2)(x+1)}\bullet\frac{x+3}{x+3}$$
$$=\frac{2x+2}{(x+3)(x+2)(x+1)}-\frac{4x+8}{(x+3)(x+2)(x+1)}+\frac{3x+9}{(x+3)(x+2)(x+1)}$$
$$=\frac{2x+2-4x-8+3x+9}{(x+3)(x+2)(x+1)}$$
$$=\frac{x+3}{(x+3)(x+2)(x+1)}$$
$$=\frac{1}{(x+2)(x+1)}$$

53. Combining the rational expressions:

$$\frac{2x+8}{x^2+5x+6}-\frac{x+5}{x^2+4x+3}-\frac{x-1}{x^2+3x+2}$$
$$=\frac{2x+8}{(x+3)(x+2)}-\frac{x+5}{(x+3)(x+1)}-\frac{x-1}{(x+2)(x+1)}$$
$$=\frac{2x+8}{(x+3)(x+2)}\bullet\frac{x+1}{x+1}-\frac{x+5}{(x+3)(x+1)}\bullet\frac{x+2}{x+2}-\frac{x-1}{(x+2)(x+1)}\bullet\frac{x+3}{x+3}$$
$$=\frac{2x^2+10x+8}{(x+3)(x+2)(x+1)}-\frac{x^2+7x+10}{(x+3)(x+2)(x+1)}-\frac{x^2+2x-3}{(x+3)(x+2)(x+1)}$$
$$=\frac{2x^2+10x+8-x^2-7x-10-x^2-2x+3}{(x+3)(x+2)(x+1)}$$
$$=\frac{x+1}{(x+3)(x+2)(x+1)}$$
$$=\frac{1}{(x+2)(x+3)}$$

55. Combining the rational expressions: $2+\frac{3}{2x+1}=\frac{2}{1}\bullet\frac{2x+1}{2x+1}+\frac{3}{2x+1}=\frac{4x+2}{2x+1}+\frac{3}{2x+1}=\frac{4x+5}{2x+1}$

57. Combining the rational expressions: $5+\frac{2}{4-t}=\frac{5}{1}\bullet\frac{4-t}{4-t}+\frac{2}{4-t}=\frac{20-5t}{4-t}+\frac{2}{4-t}=\frac{22-5t}{4-t}$

59. Combining the rational expressions:

$$x-\frac{4}{2x+3}=\frac{x}{1}\bullet\frac{2x+3}{2x+3}-\frac{4}{2x+3}=\frac{2x^2+3x}{2x+3}-\frac{4}{2x+3}=\frac{2x^2+3x-4}{2x+3}$$

61. Combining the rational expressions:

$$\frac{x}{x+2}+\frac{1}{2x+4}-\frac{3}{x^2+2x}=\frac{x}{x+2}\bullet\frac{2x}{2x}+\frac{1}{2(x+2)}\bullet\frac{x}{x}-\frac{3}{x(x+2)}\bullet\frac{2}{2}$$

$$=\frac{2x^2}{2x(x+2)}+\frac{x}{2x(x+2)}-\frac{6}{2x(x+2)}$$

$$=\frac{2x^2+x-6}{2x(x+2)}$$

$$=\frac{(2x-3)(x+2)}{2x(x+2)}$$

$$=\frac{2x-3}{2x}$$

63. Combining the rational expressions:

$$\frac{1}{x}+\frac{x}{2x+4}-\frac{2}{x^2+2x}=\frac{1}{x}\bullet\frac{2(x+2)}{2(x+2)}+\frac{x}{2(x+2)}\bullet\frac{x}{x}-\frac{2}{x(x+2)}\bullet\frac{2}{2}$$

$$=\frac{2x+4}{2x(x+2)}+\frac{x^2}{2x(x+2)}-\frac{4}{2x(x+2)}$$

$$=\frac{x^2+2x}{2x(x+2)}$$

$$=\frac{x(x+2)}{2x(x+2)}$$

$$=\frac{1}{2}$$

65. Finding the sum:

$$f(x)+g(x)=\frac{2}{x+4}+\frac{x-1}{x^2+3x-4}$$

$$=\frac{2}{x+4}\bullet\frac{x-1}{x-1}+\frac{x-1}{(x+4)(x-1)}$$

$$=\frac{2x-2}{(x+4)(x-1)}+\frac{x-1}{(x+4)(x-1)}$$

$$=\frac{3x-3}{(x+4)(x-1)}$$

$$=\frac{3(x-1)}{(x+4)(x-1)}$$

$$=\frac{3}{x+4}$$

67. Finding the sum:

$$\begin{aligned} f(x)+g(x) &= \frac{2x}{x^2-x-2}+\frac{5}{x^2+x-6} \\ &= \frac{2x}{(x-2)(x+1)}\bullet\frac{x+3}{x+3}+\frac{5}{(x+3)(x-2)}\bullet\frac{x+1}{x+1} \\ &= \frac{2x^2+6x}{(x-2)(x+1)(x+3)}+\frac{5x+5}{(x-2)(x+1)(x+3)} \\ &= \frac{2x^2+11x+5}{(x-2)(x+1)(x+3)} \\ &= \frac{(2x+1)(x+5)}{(x-2)(x+1)(x+3)} \end{aligned}$$

69. Substituting the values: $P=\frac{1}{10}+\frac{1}{0.2}=0.1+5=5.1$

71. Writing the expression and simplifying: $x+\frac{4}{x}=\frac{x^2+4}{x}$

73. Writing the expression and simplifying:

$$\frac{1}{x}+\frac{1}{x+1}=\frac{1}{x}\bullet\frac{x+1}{x+1}+\frac{1}{x+1}\bullet\frac{x}{x}=\frac{x+1}{x(x+1)}+\frac{x}{x(x+1)}=\frac{2x+1}{x(x+1)}$$

75. Dividing: $\frac{3}{4}\div\frac{5}{8}=\frac{3}{4}\bullet\frac{8}{5}=\frac{24}{20}=\frac{4\bullet 6}{4\bullet 5}=\frac{6}{5}$

77. Multiplying: $x\left(1+\frac{2}{x}\right)=x\bullet 1+x\bullet\frac{2}{x}=x+2$

79. Multiplying: $3x\left(\frac{1}{x}-\frac{1}{3}\right)=3x\bullet\frac{1}{x}-3x\bullet\frac{1}{3}=3-x$

81. Factoring: $x^2-4=(x+2)(x-2)$

5.4 Complex Fractions

1. Simplifying the complex fraction: $\dfrac{\frac{3}{4}}{\frac{2}{3}}=\dfrac{\frac{3}{4}\bullet 12}{\frac{2}{3}\bullet 12}=\frac{9}{8}$

3. Simplifying the complex fraction: $\dfrac{\frac{1}{3}-\frac{1}{4}}{\frac{1}{2}+\frac{1}{8}}=\dfrac{\left(\frac{1}{3}-\frac{1}{4}\right)\bullet 24}{\left(\frac{1}{2}+\frac{1}{8}\right)\bullet 24}=\frac{8-6}{12+3}=\frac{2}{15}$

5. Simplifying the complex fraction: $\dfrac{3+\frac{2}{5}}{1-\frac{3}{7}}=\dfrac{\left(3+\frac{2}{5}\right)\bullet 35}{\left(1-\frac{3}{7}\right)\bullet 35}=\frac{105+14}{35-15}=\frac{119}{20}$

7. Simplifying the complex fraction: $\dfrac{\frac{1}{x}}{1+\frac{1}{x}}=\dfrac{\left(\frac{1}{x}\right)\bullet x}{\left(1+\frac{1}{x}\right)\bullet x}=\dfrac{1}{x+1}$

9. Simplifying the complex fraction: $\dfrac{1+\frac{1}{a}}{1-\frac{1}{a}}=\dfrac{\left(1+\frac{1}{a}\right)\bullet a}{\left(1-\frac{1}{a}\right)\bullet a}=\dfrac{a+1}{a-1}$

11. Simplifying the complex fraction: $\dfrac{\frac{1}{x}-\frac{1}{y}}{\frac{1}{x}+\frac{1}{y}}=\dfrac{\left(\frac{1}{x}-\frac{1}{y}\right)\bullet xy}{\left(\frac{1}{x}+\frac{1}{y}\right)\bullet xy}=\dfrac{y-x}{y+x}$

13. Simplifying the complex fraction:

$$\dfrac{\frac{x-5}{x^2-4}}{\frac{x^2-25}{x+2}}=\dfrac{\frac{x-5}{(x+2)(x-2)}\bullet(x+2)(x-2)}{\frac{(x+5)(x-5)}{x+2}\bullet(x+2)(x-2)}=\dfrac{x-5}{(x+5)(x-5)(x-2)}=\dfrac{1}{(x+5)(x-2)}$$

15. Simplifying the complex fraction:

$$\dfrac{\frac{4a}{2a^3+2}}{\frac{8a}{4a+4}}=\dfrac{\frac{4a}{2(a+1)(a^2-a+1)}\bullet 2(a+1)(a^2-a+1)}{\frac{8a}{4(a+1)}\bullet 2(a+1)(a^2-a+1)}=\dfrac{4a}{4a(a^2-a+1)}=\dfrac{1}{a^2-a+1}$$

17. Simplifying the complex fraction:

$$\dfrac{1-\frac{9}{x^2}}{1-\frac{1}{x}-\frac{6}{x^2}}=\dfrac{\left(1-\frac{9}{x^2}\right)\bullet x^2}{\left(1-\frac{1}{x}-\frac{6}{x^2}\right)\bullet x^2}=\dfrac{x^2-9}{x^2-x-6}=\dfrac{(x+3)(x-3)}{(x+2)(x-3)}=\dfrac{x+3}{x+2}$$

19. Simplifying the complex fraction:

$$\dfrac{2+\frac{5}{a}-\frac{3}{a^2}}{2-\frac{5}{a}+\frac{2}{a^2}}=\dfrac{\left(2+\frac{5}{a}-\frac{3}{a^2}\right)\bullet a^2}{\left(2-\frac{5}{a}+\frac{2}{a^2}\right)\bullet a^2}=\dfrac{2a^2+5a-3}{2a^2-5a+2}=\dfrac{(2a-1)(a+3)}{(2a-1)(a-2)}=\dfrac{a+3}{a-2}$$

21. Simplifying the complex fraction:

$$\frac{2+\frac{3}{x}-\frac{18}{x^2}-\frac{27}{x^3}}{2+\frac{9}{x}+\frac{9}{x^2}}=\frac{\left(2+\frac{3}{x}-\frac{18}{x^2}-\frac{27}{x^3}\right)\bullet x^3}{\left(2+\frac{9}{x}+\frac{9}{x^2}\right)\bullet x^3}$$

$$=\frac{2x^3+3x^2-18x-27}{2x^3+9x^2+9x}$$

$$=\frac{x^2(2x+3)-9(2x+3)}{x(2x^2+9x+9)}$$

$$=\frac{(2x+3)(x^2-9)}{x(2x+3)(x+3)}$$

$$=\frac{(2x+3)(x+3)(x-3)}{x(2x+3)(x+3)}$$

$$=\frac{x-3}{x}$$

23. Simplifying the complex fraction: $\frac{1+\frac{1}{x+3}}{1-\frac{1}{x+3}}=\frac{\left(1+\frac{1}{x+3}\right)\bullet(x+3)}{\left(1-\frac{1}{x+3}\right)\bullet(x+3)}=\frac{x+3+1}{x+3-1}=\frac{x+4}{x+2}$

25. Simplifying the complex fraction:

$$\frac{1+\frac{1}{x+3}}{1+\frac{7}{x-3}}=\frac{\left(1+\frac{1}{x+3}\right)\bullet(x+3)(x-3)}{\left(1+\frac{7}{x-3}\right)\bullet(x+3)(x-3)}$$

$$=\frac{(x+3)(x-3)+(x-3)}{(x+3)(x-3)+7(x+3)}$$

$$=\frac{(x-3)(x+3+1)}{(x+3)(x-3+7)}$$

$$=\frac{(x-3)(x+4)}{(x+3)(x+4)}$$

$$=\frac{x-3}{x+3}$$

27. Simplifying the complex fraction:

$$\frac{1-\frac{1}{a+1}}{1+\frac{1}{a-1}}=\frac{\left(1-\frac{1}{a+1}\right)\bullet(a+1)(a-1)}{\left(1+\frac{1}{a-1}\right)\bullet(a+1)(a-1)}$$
$$=\frac{(a+1)(a-1)-(a-1)}{(a+1)(a-1)+(a+1)}$$
$$=\frac{(a-1)(a+1-1)}{(a+1)(a-1+1)}$$
$$=\frac{a(a-1)}{a(a+1)}$$
$$=\frac{a-1}{a+1}$$

29. Simplifying the complex fraction:

$$\frac{\frac{1}{x+3}+\frac{1}{x-3}}{\frac{1}{x+3}-\frac{1}{x-3}}=\frac{\left(\frac{1}{x+3}+\frac{1}{x-3}\right)\bullet(x+3)(x-3)}{\left(\frac{1}{x+3}-\frac{1}{x-3}\right)\bullet(x+3)(x-3)}=\frac{(x-3)+(x+3)}{(x-3)-(x+3)}=\frac{2x}{-6}=-\frac{x}{3}$$

31. Simplifying the complex fraction:

$$\frac{\frac{y+1}{y-1}+\frac{y-1}{y+1}}{\frac{y+1}{y-1}-\frac{y-1}{y+1}}=\frac{\left(\frac{y+1}{y-1}+\frac{y-1}{y+1}\right)\bullet(y+1)(y-1)}{\left(\frac{y+1}{y-1}-\frac{y-1}{y+1}\right)\bullet(y+1)(y-1)}$$
$$=\frac{(y+1)^2+(y-1)^2}{(y+1)^2-(y-1)^2}$$
$$=\frac{y^2+2y+1+y^2-2y+1}{y^2+2y+1-y^2+2y-1}$$
$$=\frac{2y^2+2}{4y}$$
$$=\frac{2(y^2+1)}{4y}$$
$$=\frac{y^2+1}{2y}$$

33. Simplifying the complex fraction: $1-\frac{x}{1-\frac{1}{x}}=1-\frac{x\bullet x}{\left(1-\frac{1}{x}\right)\bullet x}=1-\frac{x^2}{x-1}=\frac{x-1-x^2}{x-1}=\frac{-x^2+x-1}{x-1}$

35. Simplifying the complex fraction: $1+\dfrac{1}{1+\dfrac{1}{1+1}}=1+\dfrac{1}{1+\dfrac{1}{2}}=1+\dfrac{1}{\frac{3}{2}}=1+\dfrac{2}{3}=\dfrac{5}{3}$

37. Simplifying the complex fraction:

$$\frac{1-\dfrac{1}{x+\frac{1}{2}}}{1+\dfrac{1}{x+\frac{1}{2}}}=\frac{1-\dfrac{1\bullet 2}{\left(x+\frac{1}{2}\right)\bullet 2}}{1+\dfrac{1\bullet 2}{\left(x+\frac{1}{2}\right)\bullet 2}}=\frac{1-\dfrac{2}{2x+1}}{1+\dfrac{2}{2x+1}}=\frac{\left(1-\dfrac{2}{2x+1}\right)(2x+1)}{\left(1+\dfrac{2}{2x+1}\right)(2x+1)}=\frac{2x+1-2}{2x+1+2}=\frac{2x-1}{2x+3}$$

39. Simplifying the complex fraction:

$$\frac{\dfrac{1}{x+h}-\dfrac{1}{x}}{h}=\frac{\left(\dfrac{1}{x+h}-\dfrac{1}{x}\right)\bullet x(x+h)}{h\bullet x(x+h)}=\frac{x-(x+h)}{hx(x+h)}=\frac{x-x-h}{hx(x+h)}=\frac{-h}{hx(x+h)}=-\frac{1}{x(x+h)}$$

41. Simplifying the complex fraction: $\dfrac{\dfrac{3}{ab}+\dfrac{4}{bc}-\dfrac{2}{ac}}{\dfrac{5}{abc}}=\dfrac{\left(\dfrac{3}{ab}+\dfrac{4}{bc}-\dfrac{2}{ac}\right)\bullet abc}{\left(\dfrac{5}{abc}\right)\bullet abc}=\dfrac{3c+4a-2b}{5}$

43. Simplifying the complex fraction:

$$\frac{\dfrac{t^2-2t-8}{t^2+7t+6}}{\dfrac{t^2-t-6}{t^2+2t+1}}=\frac{\dfrac{(t-4)(t+2)}{(t+6)(t+1)}\bullet(t+6)(t+1)^2}{\dfrac{(t-3)(t+2)}{(t+1)^2}\bullet(t+6)(t+1)^2}=\frac{(t-4)(t+2)(t+1)}{(t-3)(t+2)(t+6)}=\frac{(t-4)(t+1)}{(t+6)(t-3)}$$

45. Simplifying the complex fraction:

$$\frac{5+\dfrac{4}{b-1}}{\dfrac{7}{b+5}-\dfrac{3}{b-1}}=\frac{\left(5+\dfrac{4}{b-1}\right)\bullet(b+5)(b-1)}{\left(\dfrac{7}{b+5}-\dfrac{3}{b-1}\right)\bullet(b+5)(b-1)}$$

$$=\frac{5(b+5)(b-1)+4(b+5)}{7(b-1)-3(b+5)}$$

$$=\frac{(b+5)(5b-5+4)}{7b-7-3b-15}$$

$$=\frac{(b+5)(5b-1)}{4b-22}$$

$$=\frac{(5b-1)(b+5)}{2(2b-11)}$$

47. Simplifying the complex fraction:

$$\frac{\frac{3}{x^2-x-6}}{\frac{2}{x+2}-\frac{4}{x-3}}=\frac{\frac{3}{(x-3)(x+2)}\bullet(x-3)(x+2)}{\left(\frac{2}{x+2}-\frac{4}{x-3}\right)\bullet(x-3)(x+2)}$$

$$=\frac{3}{2(x-3)-4(x+2)}$$

$$=\frac{3}{2x-6-4x-8}$$

$$=\frac{3}{-2x-14}$$

$$=-\frac{3}{2x+14}$$

49. Simplifying the complex fraction:

$$\frac{\frac{1}{m-4}+\frac{1}{m-5}}{\frac{1}{m^2-9m+20}}=\frac{\left(\frac{1}{m-4}+\frac{1}{m-5}\right)\bullet(m-4)(m-5)}{\frac{1}{(m-4)(m-5)}\bullet(m-4)(m-5)}=\frac{(m-5)+(m-4)}{1}=2m-9$$

51. **a.** Simplifying the difference quotient:

$$\frac{f(x)-f(a)}{x-a}=\frac{\frac{4}{x}-\frac{4}{a}}{x-a}=\frac{\left(\frac{4}{x}-\frac{4}{a}\right)ax}{(x-a)ax}=\frac{4a-4x}{ax(x-a)}=\frac{-4(x-a)}{ax(x-a)}=-\frac{4}{ax}$$

b. Simplifying the difference quotient:

$$\frac{f(x)-f(a)}{x-a}=\frac{\frac{1}{x+1}-\frac{1}{a+1}}{x-a}$$

$$=\frac{\left(\frac{1}{x+1}-\frac{1}{a+1}\right)(x+1)(a+1)}{(x-a)(x+1)(a+1)x}$$

$$=\frac{a+1-x-1}{x(x-a)(x+1)(a+1)}$$

$$=\frac{a-x}{x(x-a)(x+1)(a+1)}$$

$$=-\frac{1}{x(x+1)(a+1)}$$

c. Simplifying the difference quotient:

$$\frac{f(x)-f(a)}{x-a}=\frac{\frac{1}{x^2}-\frac{1}{a^2}}{x-a}=\frac{\left(\frac{1}{x^2}-\frac{1}{a^2}\right)a^2x^2}{a^2x^2(x-a)}=\frac{a^2-x^2}{a^2x^2(x-a)}=\frac{(a+x)(a-x)}{a^2x^2(x-a)}=-\frac{a+x}{a^2x^2}$$

53. **a.** As v approaches 0, the denominator approaches 1.

b. Solving v:

$$h=\frac{f}{1+\frac{v}{s}}$$
$$h=\frac{f\bullet s}{\left(1+\frac{v}{s}\right)s}$$
$$h=\frac{fs}{s+v}$$
$$h(s+v)=fs$$
$$s+v=\frac{fs}{h}$$
$$v=\frac{fs}{h}-s$$

55. Multiplying: $x(y-2)=xy-2x$

57. Multiplying: $6\left(\frac{x}{2}-3\right)=6\bullet\frac{x}{2}-6\bullet3=3x-18$

59. Multiplying: $xab\bullet\frac{1}{x}=ab$

61. Factoring: $y^2-25=(y+5)(y-5)$

63. Factoring: $xa+xb=x(a+b)$

65. Solving the equation:

$$5x-4=6$$
$$5x=10$$
$$x=2$$

5.5 Equations With Rational Expressions

1. Solving the equation:

$$\frac{x}{5}+4=\frac{5}{3}$$
$$15\left(\frac{x}{5}+4\right)=15\left(\frac{5}{3}\right)$$
$$3x+60=25$$
$$3x=-35$$
$$x=-\frac{35}{3}$$

3. Solving the equation:

$$\frac{a}{3}+2=\frac{4}{5}$$
$$15\left(\frac{a}{3}+2\right)=15\left(\frac{4}{5}\right)$$
$$5a+30=12$$
$$5a=-18$$
$$a=-\frac{18}{5}$$

5. Solving the equation:

$$\frac{y}{2}+\frac{y}{4}+\frac{y}{6}=3$$
$$12\left(\frac{y}{2}+\frac{y}{4}+\frac{y}{6}\right)=12(3)$$
$$6y+3y+2y=36$$
$$11y=36$$
$$y=\frac{36}{11}$$

7. Solving the equation:

$$\frac{5}{2x}=\frac{1}{x}+\frac{3}{4}$$
$$4x\left(\frac{5}{2x}\right)=4x\left(\frac{1}{x}+\frac{3}{4}\right)$$
$$10=4+3x$$
$$3x=6$$
$$x=2$$

9. Solving the equation:

$$\frac{1}{x}=\frac{1}{3}-\frac{2}{3x}$$
$$3x\left(\frac{1}{x}\right)=3x\left(\frac{1}{3}-\frac{2}{3x}\right)$$
$$3=x-2$$
$$x=5$$

11. Solving the equation:

$$\frac{2x}{x-3}+2=\frac{2}{x-3}$$
$$(x-3)\left(\frac{2x}{x-3}+2\right)=(x-3)\left(\frac{2}{x-3}\right)$$
$$2x+2(x-3)=2$$
$$2x+2x-6=2$$
$$4x=8$$
$$x=2$$

13. Solving the equation:

$$1-\frac{1}{x}=\frac{12}{x^2}$$
$$x^2\left(1-\frac{1}{x}\right)=x^2\left(\frac{12}{x^2}\right)$$
$$x^2-x=12$$
$$x^2-x-12=0$$
$$(x+3)(x-4)=0$$
$$x=-3,4$$

15. Solving the equation:

$$y-\frac{4}{3y}=-\frac{1}{3}$$
$$3y\left(y-\frac{4}{3y}\right)=3y\left(-\frac{1}{3}\right)$$
$$3y^2-4=-y$$
$$3y^2+y-4=0$$
$$(3y+4)(y-1)=0$$
$$x=-\frac{4}{3},1$$

17. Solving the equation:

$$\frac{x+2}{x+1}=\frac{1}{x+1}+2$$
$$(x+1)\left(\frac{x+2}{x+1}\right)=(x+1)\left(\frac{1}{x+1}+2\right)$$
$$x+2=1+2(x+1)$$
$$x+2=1+2x+2$$
$$x+2=2x+3$$
$$x=-1 \text{ (does not check)}$$

There is no solution (–1 does not check).

19. Solving the equation:

$$\frac{3}{a-2}=\frac{2}{a-3}$$

$$(a-2)(a-3)\left(\frac{3}{a-2}\right)=(a-2)(a-3)\left(\frac{2}{a-3}\right)$$

$$3(a-3)=2(a-2)$$

$$3a-9=2a-4$$

$$a=5$$

21. Solving the equation:

$$6-\frac{5}{x^2}=\frac{7}{x}$$

$$x^2\left(6-\frac{5}{x^2}\right)=x^2\left(\frac{7}{x}\right)$$

$$6x^2-5=7x$$

$$6x^2-7x-5=0$$

$$(2x+1)(3x-5)=0$$

$$x=-\frac{1}{2},\frac{5}{3}$$

23. Solving the equation:

$$\frac{1}{x-1}-\frac{1}{x+1}=\frac{3x}{x^2-1}$$

$$(x+1)(x-1)\left(\frac{1}{x-1}-\frac{1}{x+1}\right)=(x+1)(x-1)\left(\frac{3x}{(x+1)(x-1)}\right)$$

$$(x+1)-(x-1)=3x$$

$$x+1-x+1=3x$$

$$3x=2$$

$$x=\frac{2}{3}$$

25. Solving the equation:

$$\frac{2}{x-3}+\frac{x}{x^2-9}=\frac{4}{x+3}$$

$$(x+3)(x-3)\left(\frac{2}{x-3}+\frac{x}{(x+3)(x-3)}\right)=(x+3)(x-3)\left(\frac{4}{x+3}\right)$$

$$2(x+3)+x=4(x-3)$$

$$2x+6+x=4x-12$$

$$3x+6=4x-12$$

$$-x=-18$$

$$x=18$$

27. Solving the equation:

$$\frac{3}{2}-\frac{1}{x-4}=\frac{-2}{2x-8}$$

$$2(x-4)\left(\frac{3}{2}-\frac{1}{x-4}\right)=2(x-4)\left(\frac{-2}{2(x-4)}\right)$$

$$3(x-4)-2=-2$$

$$3x-12-2=-2$$

$$3x-14=-2$$

$$3x=12$$

$$x=4 \quad (\text{does not check})$$

There is no solution (4 does not check).

29. Solving the equation:

$$\frac{t-4}{t^2-3t}=\frac{-2}{t^2-9}$$

$$t(t+3)(t-3)\bullet\frac{t-4}{t(t-3)}=t(t+3)(t-3)\bullet\frac{-2}{(t+3)(t-3)}$$

$$(t+3)(t-4)=-2t$$

$$t^2-t-12=-2t$$

$$t^2+t-12=0$$

$$(t+4)(t-3)=0$$

$$t=-4 \quad (t=3 \text{ does not check})$$

The solution is –4 (3 does not check).

31. Solving the equation:

$$\frac{3}{y-4}-\frac{2}{y+1}=\frac{5}{y^2-3y-4}$$

$$(y-4)(y+1)\left(\frac{3}{y-4}-\frac{2}{y+1}\right)=(y-4)(y+1)\left(\frac{5}{(y-4)(y+1)}\right)$$

$$3(y+1)-2(y-4)=5$$

$$3y+3-2y+8=5$$

$$y+11=5$$

$$y=-6$$

33. Solving the equation:

$$\frac{2}{1+a}=\frac{3}{1-a}+\frac{5}{a}$$

$$a(1+a)(1-a)\left(\frac{2}{1+a}\right)=a(1+a)(1-a)\left(\frac{3}{1-a}+\frac{5}{a}\right)$$

$$2a(1-a)=3a(1+a)+5(1+a)(1-a)$$

$$2a-2a^2=3a+3a^2+5-5a^2$$

$$-2a^2+2a=-2a^2+3a+5$$

$$2a=3a+5$$

$$-a=5$$

$$a=-5$$

35. Solving the equation:

$$\frac{3}{2x-6}-\frac{x+1}{4x-12}=4$$

$$4(x-3)\left(\frac{3}{2(x-3)}-\frac{x+1}{4(x-3)}\right)=4(x-3)(4)$$

$$6-(x+1)=16x-48$$

$$5-x=16x-48$$

$$-17x=-53$$

$$x=\frac{53}{17}$$

37. Solving the equation:

$$\frac{y+2}{y^2-y}-\frac{6}{y^2-1}=0$$

$$y(y+1)(y-1)\left(\frac{y+2}{y(y-1)}-\frac{6}{(y+1)(y-1)}\right)=y(y+1)(y-1)(0)$$

$$(y+1)(y+2)-6y=0$$

$$y^2+3y+2-6y=0$$

$$y^2-3y+2=0$$

$$(y-1)(y-2)=0$$

$$y=2 \quad (y=1 \text{ does not check})$$

The solution is 2 (1 does not check).

39. Solving the equation:

$$\frac{4}{2x-6}-\frac{12}{4x+12}=\frac{12}{x^2-9}$$

$$4(x+3)(x-3)\left(\frac{4}{2(x-3)}-\frac{12}{4(x+3)}\right)=4(x+3)(x-3)\left(\frac{12}{(x+3)(x-3)}\right)$$

$$8(x+3)-12(x-3)=48$$

$$8x+24-12x+36=48$$

$$-4x+60=48$$

$$-4x=-12$$

$$x=3 \quad (x=3 \text{ does not check})$$

There is no solution (3 does not check).

41. Solving the equation:

$$\frac{2}{y^2-7y+12}-\frac{1}{y^2-9}=\frac{4}{y^2-y-12}$$

$$(y+3)(y-3)(y-4)\left(\frac{2}{(y-3)(y-4)}-\frac{1}{(y+3)(y-3)}\right)=(y+3)(y-3)(y-4)\left(\frac{4}{(y-4)(y+3)}\right)$$

$$2(y+3)-(y-4)=4(y-3)$$

$$2y+6-y+4=4y-12$$

$$y+10=4y-12$$

$$-3y=-22$$

$$y=\frac{22}{3}$$

43. **a.** Solving the equation:

$$f(x)+g(x)=\frac{5}{8}$$

$$\frac{1}{x-3}+\frac{1}{x+3}=\frac{5}{8}$$

$$8(x-3)(x+3)\left(\frac{1}{x-3}+\frac{1}{x+3}\right)=8(x-3)(x+3)\left(\frac{5}{8}\right)$$

$$8(x+3)+8(x-3)=5(x-3)(x+3)$$

$$8x+24+8x-24=5x^2-45$$

$$16x=5x^2-45$$

$$0=5x^2-16x-45$$

$$0=(5x+9)(x-5)$$

$$x=-\frac{9}{5},5$$

b. Solving the equation:

$$\frac{f(x)}{g(x)}=5$$
$$\frac{\frac{1}{x-3}}{\frac{1}{x+3}}=5$$
$$\frac{x+3}{x-3}=5$$
$$x+3=5(x-3)$$
$$x+3=5x-15$$
$$18=4x$$
$$x=\frac{9}{2}$$

c. Solving the equation:

$$f(x)=g(x)$$
$$\frac{1}{x-3}=\frac{1}{x+3}$$
$$x+3=x-3$$
$$3=-3 \text{ (false)}$$

There is no solution $(\varnothing)$.

45. **a.** Solving the equation:

$$6x-2=0$$
$$6x=2$$
$$x=\frac{1}{3}$$

b. Solving the equation:

$$\frac{6}{x}-2=0$$
$$x\left(\frac{6}{x}-2\right)=x(0)$$
$$6-2x=0$$
$$6=2x$$
$$x=3$$

c. Solving the equation:

$$\frac{x}{6}-2=-\frac{1}{2}$$
$$6\left(\frac{x}{6}-2\right)=6\left(-\frac{1}{2}\right)$$
$$x-12=-3$$
$$x=9$$

d. Solving the equation:

$$\frac{6}{x}-2=-\frac{1}{2}$$
$$2x\left(\frac{6}{x}-2\right)=2x\left(-\frac{1}{2}\right)$$
$$12-4x=-x$$
$$12=3x$$
$$x=4$$

e. Solving the equation:

$$\frac{6}{x^2}+6=\frac{20}{x}$$
$$x^2\left(\frac{6}{x^2}+6\right)=x^2\left(\frac{20}{x}\right)$$
$$6+6x^2=20x$$
$$6x^2-20x+6=0$$
$$3x^2-10x+3=0$$
$$(3x-1)(x-3)=0$$
$$x=\frac{1}{3},3$$

47. a. Dividing: $\frac{6}{x^2-2x-8}\div\frac{x+3}{x+2}=\frac{6}{(x-4)(x+2)}\bullet\frac{x+2}{x+3}=\frac{6}{(x-4)(x+3)}$

b. Adding:

$$\frac{6}{x^2-2x-8}+\frac{x+3}{x+2}=\frac{6}{(x-4)(x+2)}+\frac{x+3}{x+2}\bullet\frac{x-4}{x-4}$$
$$=\frac{6}{(x-4)(x+2)}+\frac{x^2-x-12}{(x-4)(x+2)}$$
$$=\frac{x^2-x-6}{(x-4)(x+2)}$$
$$=\frac{(x-3)(x+2)}{(x-4)(x+2)}$$
$$=\frac{x-3}{x-4}$$

c. Solving the equation:

$$\frac{6}{x^2-2x-8}+\frac{x+3}{x+2}=2$$
$$(x-4)(x+2)\left(\frac{6}{(x-4)(x+2)}+\frac{x+3}{x+2}\right)=(x-4)(x+2)(2)$$
$$6+(x-4)(x+3)=2(x-4)(x+2)$$
$$6+x^2-x-12=2x^2-4x-16$$
$$0=x^2-3x-10$$
$$0=(x-5)(x+2)$$
$$x=-2,5$$

Note that $x=-2$ does not check in the original equation, so the solution is $x=5$.

49. Solving for x:

$$\begin{aligned}\frac{1}{x}&=\frac{1}{b}-\frac{1}{a}\\ abx\left(\frac{1}{x}\right)&=abx\left(\frac{1}{b}-\frac{1}{a}\right)\\ ab&=ax-bx\\ ab&=x(a-b)\\ x&=\frac{ab}{a-b}\end{aligned}$$

51. Solving for y:

$$\begin{aligned}x&=\frac{y-3}{y-1}\\ x(y-1)&=y-3\\ xy-x&=y-3\\ xy-y&=x-3\\ y(x-1)&=x-3\\ y&=\frac{x-3}{x-1}\end{aligned}$$

53. Solving for y:

$$\begin{aligned}x&=\frac{2y+1}{3y+1}\\ x(3y+1)&=2y+1\\ 3xy+x&=2y+1\\ 3xy-2y&=-x+1\\ y(3x-2)&=-x+1\\ y&=\frac{1-x}{3x-2}\end{aligned}$$

55. Graphing the function:

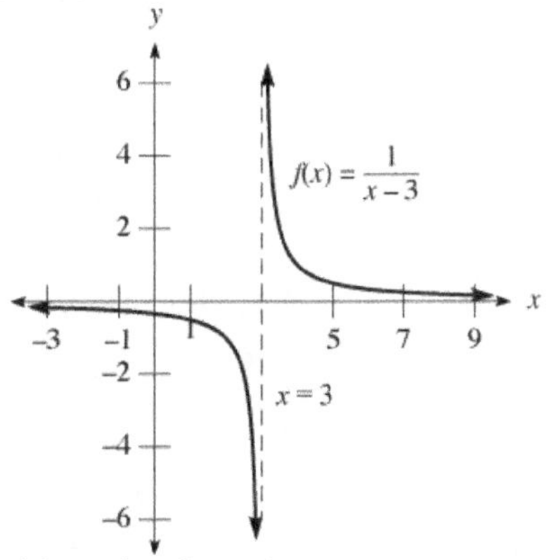

57. Graphing the function:

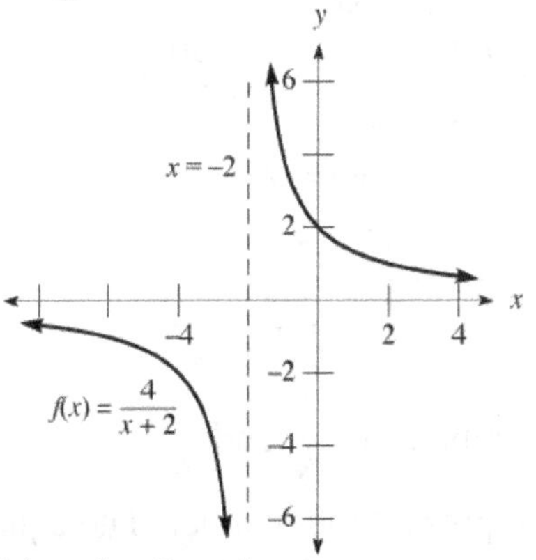

59. Graphing the function:

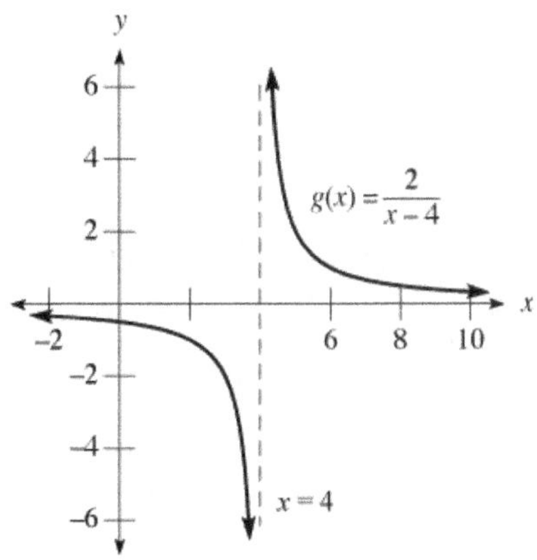

61. Graphing the function:

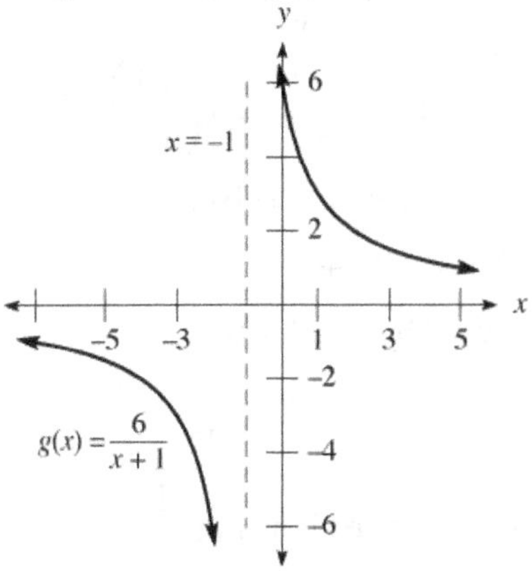

63. Substituting $y_1 = 12$ and $y_2 = 8$:

$$\frac{1}{h} = \frac{1}{12} + \frac{1}{8} = \frac{2}{24} + \frac{3}{24} = \frac{5}{24}$$

$$h = \frac{24}{5} \text{ feet}$$

65. Multiplying: $39.3 \bullet 60 = 2,358$

67. Dividing: $65,000 \div 5,280 \approx 12.3$

69. Multiplying: $2x\left(\frac{1}{x} + \frac{1}{2x}\right) = 2x \bullet \frac{1}{x} + 2x \bullet \frac{1}{2x} = 2 + 1 = 3$

71. Solving the equation:

$$\begin{aligned} 12(x+3) + 12(x-3) &= 3(x^2 - 9) \\ 12x + 36 + 12x - 36 &= 3x^2 - 27 \\ 24x &= 3x^2 - 27 \\ 3x^2 - 24x - 27 &= 0 \\ 3(x^2 - 8x - 9) &= 0 \\ 3(x-9)(x+1) &= 0 \\ x &= -1, 9 \end{aligned}$$

73. Solving the equation:

$$\begin{aligned} \frac{1}{10} - \frac{1}{12} &= \frac{1}{x} \\ 60x\left(\frac{1}{10} - \frac{1}{12}\right) &= 60x\left(\frac{1}{x}\right) \\ 6x - 5x &= 60 \\ x &= 60 \end{aligned}$$

5.6 Applications

1. Let x and $3x$ represent the two numbers. The equation is:

$$\begin{aligned} \frac{1}{x} + \frac{1}{3x} &= \frac{20}{3} \\ 3x\left(\frac{1}{x} + \frac{1}{3x}\right) &= 3x\left(\frac{20}{3}\right) \\ 3 + 1 &= 20x \\ 20x &= 4 \\ x &= \frac{1}{5} \end{aligned}$$

The numbers are $\frac{1}{5}$ and $\frac{3}{5}$.

3. Let x represent the number. The equation is:

$$\begin{aligned} x + \frac{1}{x} &= \frac{10}{3} \\ 3x\left(x + \frac{1}{x}\right) &= 3x\left(\frac{10}{3}\right) \\ 3x^2 + 3 &= 10x \\ 3x^2 - 10x + 3 &= 0 \\ (3x-1)(x-3) &= 0 \\ x &= \frac{1}{3}, 3 \end{aligned}$$

The number is either 3 or $\frac{1}{3}$.

5. Let x and $x + 1$ represent the two integers. The equation is:

$$\frac{1}{x}+\frac{1}{x+1}=\frac{7}{12}$$

$$12x(x+1)\left(\frac{1}{x}+\frac{1}{x+1}\right)=12x(x+1)\left(\frac{7}{12}\right)$$

$$12(x+1)+12x=7x(x+1)$$

$$12x+12+12x=7x^2+7x$$

$$0=7x^2-17x-12$$

$$0=(7x+4)(x-3)$$

$$x=3 \quad \left(x=-\frac{4}{7}\text{ is not an integer}\right)$$

The two integers are 3 and 4.

7. Let x represent the number. The equation is:

$$\frac{7+x}{9+x}=\frac{5}{6}$$

$$6(9+x)\left(\frac{7+x}{9+x}\right)=6(9+x)\left(\frac{5}{6}\right)$$

$$6(7+x)=5(9+x)$$

$$42+6x=45+5x$$

$$x=3$$

The number is 3.

9. **a.** Completing the table:

	d	r	t
Upstream	1.5	$5-x$	$\frac{1.5}{5-x}$
Downstream	3	$5+x$	$\frac{3}{5+x}$

b. The time column follows from dividing distance by rate.

c. The two times are equal. This yields the equation: $\frac{3}{5+x}=\frac{1.5}{5-x}$

d. Solving the equation:

$$\frac{3}{5+x}=\frac{1.5}{5-x}$$

$$3(5-x)=1.5(5+x)$$

$$15-3x=7.5+1.5x$$

$$7.5=4.5x$$

$$x=\frac{75}{45}=\frac{5}{3}$$

The speed of the current is $\frac{5}{3}$ mph.

11. Let x represent the speed of the boat. Since the total time is 3 hours:

$$\frac{8}{x-2}+\frac{8}{x+2}=3$$

$$(x+2)(x-2)\left(\frac{8}{x-2}+\frac{8}{x+2}\right)=3(x+2)(x-2)$$

$$8(x+2)+8(x-2)=3x^2-12$$

$$16x=3x^2-12$$

$$0=3x^2-16x-12$$

$$0=(3x+2)(x-6)$$

$$x=6 \quad \left(x=-\frac{2}{3} \text{ is impossible}\right)$$

The speed of the boat is 6 mph.

13. **a.** Completing the table:

	d	r	t
Train A	150	$x+15$	$\frac{150}{x+15}$
Train B	120	x	$\frac{120}{x}$

b. The time column follows from dividing distance by rate.

c. The two times are equal. This yields the equation: $\frac{150}{x+15}=\frac{120}{x}$

d. Solving the equation:

$$\frac{150}{x+15}=\frac{120}{x}$$

$$150x=120(x+15)$$

$$150x=120x+1800$$

$$30x=1800$$

$$x=60$$

The speed of train A is 75 mph and the speed of train B is 60 mph.

15. The smaller plane makes the trip in 3 hours, so the 747 must take $1\frac{1}{2}$ hours to complete the trip.

Thus the average speed is given by: $\dfrac{810 \text{ miles}}{1\frac{1}{2} \text{ hours}}=540$ miles per hour

17. Let r represent the bus's usual speed. The difference of the two times is $\frac{1}{2}$ hour, therefore:

$$\frac{270}{r} - \frac{270}{r+6} = \frac{1}{2}$$

$$2r(r+6)\left(\frac{270}{r} - \frac{270}{r+6}\right) = 2r(r+6)\left(\frac{1}{2}\right)$$

$$540(r+6) - 540(r) = r(r+6)$$

$$540r + 3240 - 540r = r^2 + 6r$$

$$0 = r^2 + 6r - 3240$$

$$0 = (r-54)(r+60)$$

$$r = 54 \quad (r = -60 \text{ is impossible})$$

The usual speed is 54 mph.

19. Let x represent the time to fill the tank if both pipes are open. The rate equation is:

$$\frac{1}{8} - \frac{1}{16} = \frac{1}{x}$$

$$16x\left(\frac{1}{8} - \frac{1}{16}\right) = 16x\left(\frac{1}{x}\right)$$

$$2x - x = 16$$

$$x = 16$$

It will take 16 hours to fill the tank if both pipes are open.

21. Let x represent the time to fill the pool with both pipes open. The rate equation is:

$$\frac{1}{10} - \frac{1}{15} = \frac{1}{2} \bullet \frac{1}{x}$$

$$30x\left(\frac{1}{10} - \frac{1}{15}\right) = 30x\left(\frac{1}{2x}\right)$$

$$3x - 2x = 15$$

$$x = 15$$

It will take 15 hours to fill the pool with both pipes open.

23. Let x represent the time to fill the sink with the hot water faucet. The rate equation is:

$$\frac{1}{3.5} + \frac{1}{x} = \frac{1}{2.1}$$

$$7.35x\left(\frac{1}{3.5} + \frac{1}{x}\right) = 7.35x\left(\frac{1}{2.1}\right)$$

$$2.1x + 7.35 = 3.5x$$

$$7.35 = 1.4x$$

$$x = 5.25$$

It will take $5\frac{1}{4}$ minutes to fill the sink with the hot water faucet.

25. Solving the equation:

$$\frac{1}{3}\left[\left(x+\frac{2}{3}x\right)+\frac{1}{3}\left(x+\frac{2}{3}x\right)\right]=10$$

$$\left(x+\frac{2}{3}x\right)+\frac{1}{3}\left(x+\frac{2}{3}x\right)=30$$

$$x+\frac{2}{3}x+\frac{1}{3}x+\frac{2}{9}x=30$$

$$\frac{20}{9}x=30$$

$$20x=270$$

$$x=\frac{27}{2}$$

27. **a.** Converting to grams: $2.5 \text{ moles} \bullet \frac{12.01 \text{ grams}}{1 \text{ mole}} \approx 30 \text{ grams}$

b. Converting to moles: $39 \text{ grams} \bullet \frac{1 \text{ mole}}{12.01 \text{ grams}} \approx 3.25 \text{ moles}$

29. Graphing the function:

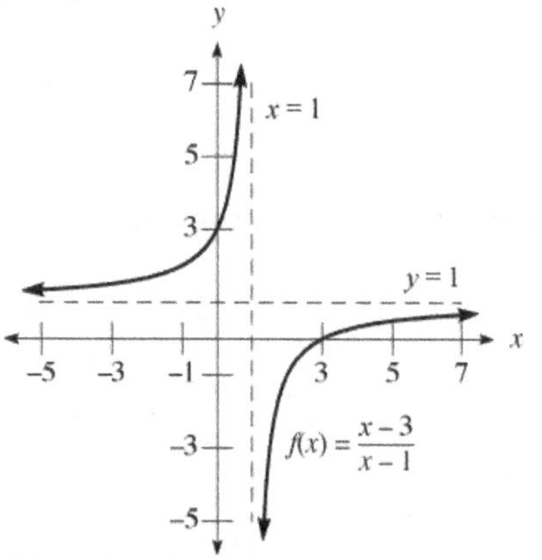

31. Graphing the function:

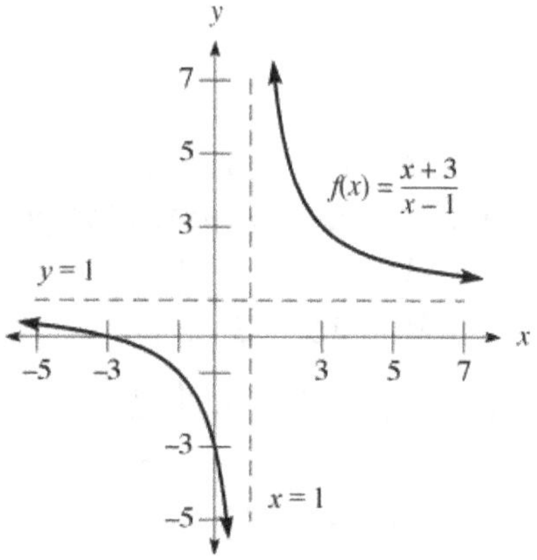

33. Graphing the function:

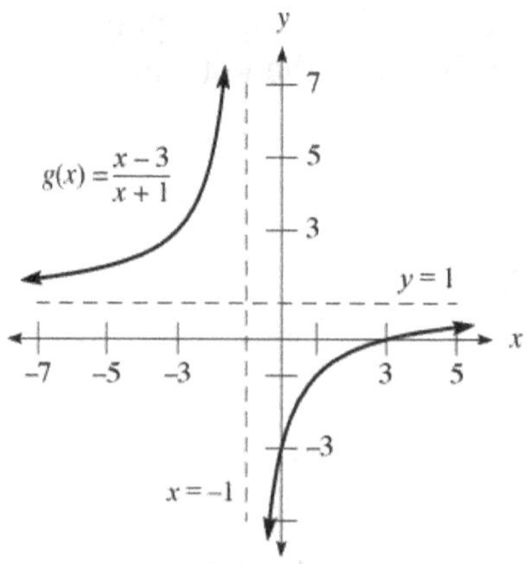

35. Dividing: $\frac{10x^2}{5x^2} = 2x^{2-2} = 2$

37. Dividing: $\frac{4x^4y^3}{-2x^2y} = -2x^{4-2}y^{3-1} = -2x^2y^2$

39. Dividing: $4,628 \div 25 = 185.12$

41. Multiplying: $2x^2(2x-4) = 4x^3 - 8x^2$

43. Multiplying:
$$\begin{aligned}(2x-4)(2x^2+4x+5) &= 2x(2x^2+4x+5) - 4(2x^2+4x+5)\\ &= 4x^3+8x^2+10x-8x^2-16x-20\\ &= 4x^3-6x-20\end{aligned}$$

45. Subtracting: $(2x^2-7x+9)-(2x^2-4x) = 2x^2-7x+9-2x^2+4x = -3x+9$

47. Factoring: $x^2 - a^2 = (x+a)(x-a)$

49. Factoring: $x^2 - 6xy - 7y^2 = (x-7y)(x+y)$

5.7 Division of Polynomials

1. Dividing: $\frac{4x^3-8x^2+6x}{2x} = \frac{4x^3}{2x} - \frac{8x^2}{2x} + \frac{6x}{2x} = 2x^2 - 4x + 3$

3. Dividing: $\frac{10x^4+15x^3-20x^2}{-5x^2} = \frac{10x^4}{-5x^2} + \frac{15x^3}{-5x^2} - \frac{20x^2}{-5x^2} = -2x^2 - 3x + 4$

5. Dividing: $\frac{8y^5+10y^3-6y}{4y^3} = \frac{8y^5}{4y^3} + \frac{10y^3}{4y^3} - \frac{6y}{4y^3} = 2y^2 + \frac{5}{2} - \frac{3}{2y^2}$

7. Dividing: $\frac{5x^3-8x^2-6x}{-2x^2} = \frac{5x^3}{-2x^2} - \frac{8x^2}{-2x^2} - \frac{6x}{-2x^2} = -\frac{5}{2}x + 4 + \frac{3}{x}$

9. Dividing: $\frac{28a^3b^5+42a^4b^3}{7a^2b^2} = \frac{28a^3b^5}{7a^2b^2} + \frac{42a^4b^3}{7a^2b^2} = 4ab^3 + 6a^2b$

11. Dividing: $\frac{10x^3y^2-20x^2y^3-30x^3y^3}{-10x^2y} = \frac{10x^3y^2}{-10x^2y} - \frac{20x^2y^3}{-10x^2y} - \frac{30x^3y^3}{-10x^2y} = -xy + 2y^2 + 3xy^2$

13. Dividing by factoring: $\frac{x^2-x-6}{x-3} = \frac{(x-3)(x+2)}{x-3} = x+2$

15. Dividing by factoring: $\frac{2a^2-3a-9}{2a+3} = \frac{(2a+3)(a-3)}{2a+3} = a-3$

17. Dividing by factoring: $\frac{5x^2-14xy-24y^2}{x-4y} = \frac{(5x+6y)(x-4y)}{x-4y} = 5x+6y$

19. Dividing by factoring: $\frac{x^3-y^3}{x-y} = \frac{(x-y)(x^2+xy+y^2)}{x-y} = x^2+xy+y^2$

21. Dividing by factoring: $\frac{y^4-16}{y-2} = \frac{(y^2+4)(y^2-4)}{y-2} = \frac{(y^2+4)(y+2)(y-2)}{y-2} = (y^2+4)(y+2)$

23. Dividing by factoring:

$$\begin{aligned}\frac{x^3+2x^2-25x-50}{x-5}&=\frac{x^2(x+2)-25(x+2)}{x-5}\\&=\frac{(x+2)(x^2-25)}{x-5}\\&=\frac{(x+2)(x+5)(x-5)}{x-5}\\&=(x+2)(x+5)\end{aligned}$$

25. Dividing by factoring:

$$\begin{aligned}\frac{4x^3+12x^2-9x-27}{x+3}&=\frac{4x^2(x+3)-9(x+3)}{x+3}\\&=\frac{(x+3)(4x^2-9)}{x+3}\\&=\frac{(x+3)(2x+3)(2x-3)}{x+3}\\&=(2x+3)(2x-3)\end{aligned}$$

27. Dividing using long division:

$$\begin{array}{r}x-7\\x+2\overline{)x^2-5x-7}\\\underline{x^2+2x}\\-7x-7\\\underline{-7x-14}\\7\end{array}$$

The quotient is $x-7+\frac{7}{x+2}$.

29. Dividing using long division:

$$\begin{array}{r}2x+5\\3x-4\overline{)6x^2+7x-18}\\\underline{6x^2-8x}\\15x-18\\\underline{15x-20}\\2\end{array}$$

The quotient is $2x+5+\frac{2}{3x-4}$.

31. Dividing using long division:

$$\begin{array}{r}2x^2-5x+1\\x+1\overline{)2x^3-3x^2-4x+5}\\\underline{2x^3+2x^2}\\-5x^2-4x\\\underline{-5x^2-5x}\\x+5\\\underline{x+1}\\4\end{array}$$

The quotient is $2x^2-5x+1+\frac{4}{x+1}$.

33. Dividing using long division:

$$\begin{array}{r}y^2-3y-13\\2y-3\overline{)2y^3-9y^2-17y+39}\\\underline{2y^3-3y^2}\\-6y^2-17y\\\underline{-6y^2+9y}\\-26y+39\\\underline{-26y+39}\\0\end{array}$$

The quotient is $y^2-3y-13$.

35. Dividing using long division:

$$\begin{array}{r} x-3 \\ 2x^2-3x+2\overline{)2x^3-9x^2+11x-6} \\ \underline{2x^3-3x^2+2x} \\ -6x^2+9x-6 \\ \underline{-6x^2+9x-6} \\ 0 \end{array}$$

The quotient is $x-3$.

37. Dividing using long division:

$$\begin{array}{r} 3y^2+6y+8 \\ 2y-4\overline{)6y^3+0y^2-8y+5} \\ \underline{6y^3-12y^2} \\ 12y^2-8y \\ \underline{12y^2-24y} \\ 16y+5 \\ \underline{16y-32} \\ 37 \end{array}$$

The quotient is $3y^2+6y+8+\dfrac{37}{2y-4}$.

39. Dividing using long division:

$$\begin{array}{r} a^3+2a^2+4a+6 \\ a-2\overline{)a^4+0a^3+0a^2-2a+5} \\ \underline{a^4-2a^3} \\ 2a^3+0a^2 \\ \underline{2a^3-4a^2} \\ 4a^2-2a \\ \underline{4a^2-8a} \\ 6a+5 \\ \underline{6a-12} \\ 17 \end{array}$$

The quotient is $a^3+2a^2+4a+6+\dfrac{17}{a-2}$.

41. Dividing using long division:

$$\begin{array}{r} y^3+2y^2+4y+8 \\ y-2\overline{)y^4+0y^3+0y^2+0y-16} \\ \underline{y^4-2y^3} \\ 2y^3+0y^2 \\ \underline{2y^3-4y^2} \\ 4y^2+0y \\ \underline{4y^2-8y} \\ 8y-16 \\ \underline{8y-16} \\ 0 \end{array}$$

The quotient is y^3+2y^2+4y+8.

43. Dividing using long division:

$$\begin{array}{r} x^2-2x+1 \\ x^2+3x+2\overline{)x^4+x^3-3x^2-x+2} \\ \underline{x^4+3x^3+2x^2} \\ -2x^3-5x^2-x \\ \underline{-2x^3-6x^2-4x} \\ x^2+3x+2 \\ \underline{x^2+3x+2} \\ 0 \end{array}$$

The quotient is x^2-2x+1.

45. First use long division to find the remaining factors:

$$\begin{array}{r} x^2+3x+2 \\ x+3\overline{\smash{)}x^3+6x^2+11x+6} \\ \underline{x^3+3x^2} \\ 3x^2+11x \\ \underline{3x^2+\ 9x} \\ 2x+6 \\ \underline{2x+6} \\ 0 \end{array}$$

Thus $x^3+6x^2+11x+6=(x+3)(x^2+3x+2)=(x+3)(x+2)(x+1)$.

47. First use long division to find the remaining factors:

$$\begin{array}{r} x^2+2x-8 \\ x+3\overline{\smash{)}x^3+5x^2-2x-24} \\ \underline{x^3+3x^2} \\ 2x^2-2x \\ \underline{2x^2+6x} \\ -8x-24 \\ \underline{-8x-24} \\ 0 \end{array}$$

Thus $x^3+5x^2-2x-24=(x+3)(x^2+2x-8)=(x+3)(x+4)(x-2)$.

49. Yes, both answers are identical.

51. $P(-2)=(-2)^2-5(-2)-7=4+10-7=7$

The answer is the same.

53. **a.** Using long division:

$$\begin{array}{r} x^2-x+3 \\ x-2\overline{\smash{)}x^3-3x^2+5x-6} \\ \underline{x^3-2x^2} \\ -x^2+5x \\ \underline{-x^2+2x} \\ 3x-6 \\ \underline{3x-6} \\ 0 \end{array}$$

Since the remainder is 0, $x-2$ is a factor of x^3-3x^2+5x-6.

Also note that: $P(2)=(2)^3-3(2)^2+5(2)-6=8-12+10-6=0$

b. Using long division:

$$\begin{array}{r} x^3 - x + 1 \\ x-5\overline{)x^4 - 5x^3 - x^2 + 6x - 5} \\ \underline{x^4 - 5x^3} \quad\quad\quad\quad\quad \\ -x^2 + 6x \quad\;\; \\ \underline{-x^2 + 5x} \quad\;\; \\ x - 5 \\ \underline{x - 5} \\ 0 \end{array}$$

Since the remainder is 0, $x-5$ is a factor of $x^4 - 5x^3 - x^2 + 6x - 5$.
Also note that: $P(5) = (5)^4 - 5(5)^3 - (5)^2 + 6(5) - 5 = 625 - 625 - 25 + 30 - 5 = 0$

55. **a.** Completing the table:

x	1	5	10	15	20
$C(x)$	2.15	2.75	3.50	4.25	5.00

b. The average cost function is $\overline{C}(x) = \dfrac{2}{x} + 0.15$.

c. Completing the table:

x	1	5	10	15	20
$\overline{C}(x)$	2.15	0.55	0.35	0.28	0.25

d. The average cost function decreases.

e. For $y = C(x)$, the domain is $\{x \mid 1 \le x \le 20\}$ and the range is $\{y \mid 2.15 \le y \le 5.00\}$.
For $y = \overline{C}(x)$, the domain is $\{x \mid 1 \le x \le 20\}$ and the range is $\{y \mid 0.25 \le y \le 2.15\}$.

57. **a.** Substituting the values:

$$T(100) = 4.95 + 0.07(100) = \$11.95$$
$$T(400) = 4.95 + 0.07(400) = \$32.95$$
$$T(500) = 4.95 + 0.07(500) = \$39.95$$

b. The average cost function is $\overline{T}(m) = \dfrac{4.95}{m} + 0.07$.

c. Substituting the values:

$$\overline{T}(100) = \frac{4.95}{100} + 0.07 = \$0.1195 \text{ per minute}$$
$$\overline{T}(400) = \frac{4.95}{400} + 0.07 = \$0.0824 \text{ per minute}$$
$$\overline{T}(500) = \frac{4.95}{500} + 0.07 = \$0.0799 \text{ per minute}$$

59. Performing the operations: $\dfrac{2a+10}{a^3} \bullet \dfrac{a^2}{3a+15} = \dfrac{2(a+5)}{a^3} \bullet \dfrac{a^2}{3(a+5)} = \dfrac{2}{3a}$

61. Performing the operations: $(x^2-9)\left(\dfrac{x+2}{x+3}\right)=(x+3)(x-3)\left(\dfrac{x+2}{x+3}\right)=(x-3)(x+2)$

63. Performing the operations: $\dfrac{2x-7}{x-2}-\dfrac{x-5}{x-2}=\dfrac{2x-7-x+5}{x-2}=\dfrac{x-2}{x-2}=1$

65. Simplifying the expression: $\dfrac{\frac{1}{x}-\frac{1}{3}}{\frac{1}{x}+\frac{1}{3}}=\dfrac{\left(\frac{1}{x}-\frac{1}{3}\right)\bullet 3x}{\left(\frac{1}{x}+\frac{1}{3}\right)\bullet 3x}=\dfrac{3-x}{3+x}$

67. Solving the equation:

$$\begin{aligned}
\frac{x}{x-3}+\frac{3}{2}&=\frac{3}{x-3}\\
2(x-3)\left(\frac{x}{x-3}+\frac{3}{2}\right)&=2(x-3)\left(\frac{3}{x-3}\right)\\
2x+3(x-3)&=6\\
2x+3x-9&=6\\
5x&=15\\
x&=3 \quad (\text{does not check})
\end{aligned}$$

There is no solution (3 does not check).

Chapter 5 Test

See www.mathtv.com for video solutions to all problems in this chapter test.

Chapter 6
Rational Exponents and Roots

6.1 Rational Exponents

1. Finding the root: $\sqrt{144} = 12$
3. $\sqrt{-144}$ is not a real number
5. Finding the root: $-\sqrt{49} = -7$
7. Finding the root: $\sqrt[3]{-27} = -3$
9. Finding the root: $\sqrt[4]{16} = 2$
11. $\sqrt[4]{-16}$ is not a real number
13. Finding the root: $\sqrt{0.04} = 0.2$
15. Finding the root: $\sqrt[3]{0.008} = 0.2$
17. Simplifying: $\sqrt{36a^8} = 6a^4$
19. Simplifying: $\sqrt[3]{27a^{12}} = 3a^4$
21. Simplifying: $\sqrt[3]{x^3y^6} = xy^2$
23. Simplifying: $\sqrt[5]{32x^{10}y^5} = 2x^2y$
25. Simplifying: $\sqrt[4]{16a^{12}b^{20}} = 2a^3b^5$
27. Writing as a root and simplifying: $36^{1/2} = \sqrt{36} = 6$
29. Writing as a root and simplifying: $-9^{1/2} = -\sqrt{9} = -3$
31. Writing as a root and simplifying: $8^{1/3} = \sqrt[3]{8} = 2$
33. Writing as a root and simplifying: $(-8)^{1/3} = \sqrt[3]{-8} = -2$
35. Writing as a root and simplifying: $32^{1/5} = \sqrt[5]{32} = 2$
37. Writing as a root and simplifying: $\left(\frac{81}{25}\right)^{1/2} = \sqrt{\frac{81}{25}} = \frac{9}{5}$
39. Writing as a root and simplifying: $\left(\frac{64}{125}\right)^{1/3} = \sqrt[3]{\frac{64}{125}} = \frac{4}{5}$
41. Simplifying: $27^{2/3} = \left(27^{1/3}\right)^2 = 3^2 = 9$
43. Simplifying: $25^{3/2} = \left(25^{1/2}\right)^3 = 5^3 = 125$
45. Simplifying: $16^{3/4} = \left(16^{1/4}\right)^3 = 2^3 = 8$
47. Simplifying: $27^{-1/3} = \left(27^{1/3}\right)^{-1} = 3^{-1} = \frac{1}{3}$
49. Simplifying: $81^{-3/4} = \left(81^{1/4}\right)^{-3} = 3^{-3} = \frac{1}{3^3} = \frac{1}{27}$
51. Simplifying: $\left(\frac{25}{26}\right)^{-1/2} = \left(\frac{36}{25}\right)^{1/2} = \frac{6}{5}$
53. Simplifying: $\left(\frac{81}{16}\right)^{-3/4} = \left(\frac{16}{81}\right)^{3/4} = \left[\left(\frac{16}{81}\right)^{1/4}\right]^3 = \left(\frac{2}{3}\right)^3 = \frac{8}{27}$
55. Simplifying: $16^{1/2} + 27^{1/3} = 4 + 3 = 7$
57. Simplifying: $8^{-2/3} + 4^{-1/2} = \left(8^{1/3}\right)^{-2} + \left(4^{1/2}\right)^{-1} = 2^{-2} + 2^{-1} = \frac{1}{4} + \frac{1}{2} = \frac{3}{4}$
59. Using properties of exponents: $x^{3/5} \bullet x^{1/5} = x^{3/5+1/5} = x^{4/5}$

61. Using properties of exponents: $\left(a^{3/4}\right)^{4/3} = a^{3/4 \bullet 4/3} = a$

63. Using properties of exponents: $\dfrac{x^{1/5}}{x^{3/5}} = x^{1/5-3/5} = x^{-2/5} = \dfrac{1}{x^{2/5}}$

65. Using properties of exponents: $\dfrac{x^{5/6}}{x^{2/3}} = x^{5/6-2/3} = x^{5/6-4/6} = x^{1/6}$

67. Using properties of exponents: $\left(x^{3/5}y^{5/6}z^{1/3}\right)^{3/5} = x^{3/5 \bullet 3/5}y^{5/6 \bullet 3/5}z^{1/3 \bullet 3/5} = x^{9/25}y^{1/2}z^{1/5}$

69. Using properties of exponents: $\dfrac{a^{3/4}b^2}{a^{7/8}b^{1/4}} = a^{3/4-7/8}b^{2-1/4} = a^{6/8-7/8}b^{8/4-1/4} = a^{-1/8}b^{7/4} = \dfrac{b^{7/4}}{a^{1/8}}$

71. Using properties of exponents: $\dfrac{\left(y^{2/3}\right)^{3/4}}{\left(y^{1/3}\right)^{3/5}} = \dfrac{y^{1/2}}{y^{1/5}} = y^{1/2-1/5} = y^{5/10-2/10} = y^{3/10}$

73. Using properties of exponents: $\left(\dfrac{a^{-1/4}}{b^{1/2}}\right)^8 = \dfrac{a^{-1/4 \bullet 8}}{b^{1/2 \bullet 8}} = \dfrac{a^{-2}}{b^4} = \dfrac{1}{a^2b^4}$

75. **a.** Simplifying: $\sqrt{25} = \sqrt{5^2} = 5$

b. Simplifying: $\sqrt{0.25} = \sqrt{0.5^2} = 0.5$

c. Simplifying: $\sqrt{2500} = \sqrt{50^2} = 50$

d. Simplifying: $\sqrt{0.0025} = \sqrt{0.05^2} = 0.05$

77. **a.** Simplifying: $\sqrt{16a^4b^8} = \sqrt{\left(4a^2b^4\right)^2} = 4a^2b^4$

b. Simplifying: $\sqrt[3]{16a^4b^8} = \sqrt[3]{\left(2ab^2\right)^3\left(2ab^2\right)} = 2ab^2\sqrt[3]{2ab^2}$

c. Simplifying: $\sqrt[4]{16a^4b^8} = \sqrt[4]{\left(2ab^2\right)^4} = 2ab^2$

79. Simplifying each expression:

$$\left(9^{1/2} + 4^{1/2}\right)^2 = (3+2)^2 = 5^2 = 25$$
$$9 + 4 = 13$$

Note that the values are not equal.

81. Rewriting with exponents: $\sqrt{\sqrt{a}} = \sqrt{a^{1/2}} = \left(a^{1/2}\right)^{1/2} = a^{1/4} = \sqrt[4]{a}$

83. Substituting $r = 250$: $v = \left(\dfrac{5 \bullet 250}{2}\right)^{1/2} = 625^{1/2} = 25$. The maximum speed is 25 mph.

85. Using a calculator: $\dfrac{1+\sqrt{5}}{2} \approx 1.618$

87. The next term is $\dfrac{13}{8}$. The numerator and denominator are consecutive members of the Fibonacci sequence.

89. **a.** The length of the side is 60 + 150 + 150 + 60 = 420 pm

b. Let d represent the diagonal. Using the Pythagorean theorem:

$$d^2 = 420^2 + 420^2 = 352{,}800$$
$$d = \sqrt{352{,}800} \approx 594.0 \text{ pm}$$

c. Converting to meters: $594 \text{ pm} \bullet \dfrac{1 \text{ m}}{10^{12} \text{ pm}} = 5.94 \times 10^{-10} \text{ m}$

91. Simplifying: $\sqrt{25}=5$

93. Simplifying: $\sqrt{6^2}=6$

95. Simplifying: $\sqrt{16x^4y^2}=4x^2y$

97. Simplifying: $\sqrt{(5y)^2}=5y$

99. Simplifying: $\sqrt[3]{27}=3$

101. Simplifying: $\sqrt[3]{2^3}=2$

103. Simplifying: $\sqrt[3]{8a^3b^3}=2ab$

105. Filling in the blank: $50=25\bullet 2$

107. Filling in the blank: $48x^4y^3=48x^4y^2\bullet y$

109. Filling in the blank: $12x^7y^6=4x^6y^6\bullet 3x$

6.2 Simplified Form for Radicals

1. Simplifying the radical: $\sqrt{8}=\sqrt{4\bullet 2}=2\sqrt{2}$

3. Simplifying the radical: $\sqrt{98}=\sqrt{49\bullet 2}=7\sqrt{2}$

5. Simplifying the radical: $\sqrt{288}=\sqrt{144\bullet 2}=12\sqrt{2}$

7. Simplifying the radical: $\sqrt{80}=\sqrt{16\bullet 5}=4\sqrt{5}$

9. Simplifying the radical: $\sqrt{48}=\sqrt{16\bullet 3}=4\sqrt{3}$

11. Simplifying the radical: $\sqrt{675}=\sqrt{225\bullet 3}=15\sqrt{3}$

13. Simplifying the radical: $\sqrt[3]{54}=\sqrt[3]{27\bullet 2}=3\sqrt[3]{2}$

15. Simplifying the radical: $\sqrt[3]{128}=\sqrt[3]{64\bullet 2}=4\sqrt[3]{2}$

17. Simplifying the radical: $\sqrt[3]{432}=\sqrt[3]{216\bullet 2}=6\sqrt[3]{2}$

19. Simplifying the radical: $\sqrt[5]{64}=\sqrt[5]{32\bullet 2}=2\sqrt[5]{2}$

21. Simplifying the radical: $\sqrt{18x^3}=\sqrt{9x^2\bullet 2x}=3x\sqrt{2x}$

23. Simplifying the radical: $\sqrt[4]{32y^7}=\sqrt[4]{16y^4\bullet 2y^3}=2y\sqrt[4]{2y^3}$

25. Simplifying the radical: $\sqrt[3]{40x^4y^7}=\sqrt[3]{8x^3y^6\bullet 5xy}=2xy^2\sqrt[3]{5xy}$

27. Simplifying the radical: $\sqrt{48a^2b^3c^4}=\sqrt{16a^2b^2c^4\bullet 3b}=4abc^2\sqrt{3b}$

29. Simplifying the radical: $\sqrt[3]{48a^2b^3c^4}=\sqrt[3]{8b^3c^3\bullet 6a^2c}=2bc\sqrt[3]{6a^2c}$

31. Simplifying the radical: $\sqrt[5]{64x^8y^{12}}=\sqrt[5]{32x^5y^{10}\bullet 2x^3y^2}=2xy^2\sqrt[5]{2x^3y^2}$

33. Simplifying the radical: $\sqrt[5]{243x^7y^{10}z^5}=\sqrt[5]{243x^5y^{10}z^5\bullet x^2}=3xy^2z\sqrt[5]{x^2}$

35. Substituting into the expression: $\sqrt{b^2-4ac}=\sqrt{(-6)^2-4(2)(3)}=\sqrt{36-24}=\sqrt{12}=2\sqrt{3}$

37. Substituting into the expression: $\sqrt{b^2-4ac}=\sqrt{(2)^2-4(1)(6)}=\sqrt{4-24}=\sqrt{-20}$, which is not a real number

39. Substituting into the expression: $\sqrt{b^2-4ac}=\sqrt{\left(-\frac{1}{2}\right)^2-4\left(\frac{1}{2}\right)\left(-\frac{5}{4}\right)}=\sqrt{\frac{1}{4}+\frac{5}{2}}=\sqrt{\frac{11}{4}}=\frac{\sqrt{11}}{2}$

41. **a.** Simplifying: $\frac{\sqrt{20}}{4}=\frac{\sqrt{4 \bullet 5}}{4}=\frac{2\sqrt{5}}{4}=\frac{\sqrt{5}}{2}$

b. Simplifying: $\frac{3\sqrt{20}}{15}=\frac{3\sqrt{4 \bullet 5}}{15}=\frac{6\sqrt{5}}{15}=\frac{2\sqrt{5}}{5}$

c. Simplifying: $\frac{4+\sqrt{12}}{2}=\frac{4+\sqrt{4 \bullet 3}}{2}=\frac{4+2\sqrt{3}}{2}=\frac{2\left(2+\sqrt{3}\right)}{2}=2+\sqrt{3}$

d. Simplifying: $\frac{2+\sqrt{9}}{5}=\frac{2+3}{5}=\frac{5}{5}=1$

43. **a.** Simplifying: $\frac{10+\sqrt{75}}{5}=\frac{10+\sqrt{25 \bullet 3}}{5}=\frac{10+5\sqrt{3}}{5}=\frac{5\left(2+\sqrt{3}\right)}{5}=2+\sqrt{3}$

b. Simplifying: $\frac{-6+\sqrt{45}}{3}=\frac{-6+\sqrt{9 \bullet 5}}{3}=\frac{-6+3\sqrt{5}}{3}=\frac{3\left(-2+\sqrt{5}\right)}{3}=-2+\sqrt{5}$

c. Simplifying: $\frac{-2-\sqrt{27}}{6}=\frac{-2-\sqrt{9 \bullet 3}}{6}=\frac{-2-3\sqrt{3}}{6}$

45. Rationalizing the denominator: $\frac{2}{\sqrt{3}}=\frac{2}{\sqrt{3}} \bullet \frac{\sqrt{3}}{\sqrt{3}}=\frac{2\sqrt{3}}{3}$

47. Rationalizing the denominator: $\frac{5}{\sqrt{6}}=\frac{5}{\sqrt{6}} \bullet \frac{\sqrt{6}}{\sqrt{6}}=\frac{5\sqrt{6}}{6}$

49. Rationalizing the denominator: $\sqrt{\frac{1}{2}}=\frac{1}{\sqrt{2}} \bullet \frac{\sqrt{2}}{\sqrt{2}}=\frac{\sqrt{2}}{2}$

51. Rationalizing the denominator: $\sqrt{\frac{1}{5}}=\frac{1}{\sqrt{5}} \bullet \frac{\sqrt{5}}{\sqrt{5}}=\frac{\sqrt{5}}{5}$

53. Rationalizing the denominator: $\frac{4}{\sqrt[3]{2}}=\frac{4}{\sqrt[3]{2}} \bullet \frac{\sqrt[3]{4}}{\sqrt[3]{4}}=\frac{4\sqrt[3]{4}}{2}=2\sqrt[3]{4}$

55. Rationalizing the denominator: $\frac{2}{\sqrt[3]{9}}=\frac{2}{\sqrt[3]{9}} \bullet \frac{\sqrt[3]{3}}{\sqrt[3]{3}}=\frac{2\sqrt[3]{3}}{3}$

57. Rationalizing the denominator: $\sqrt[4]{\frac{3}{2x^2}}=\frac{\sqrt[4]{3}}{\sqrt[4]{2x^2}} \bullet \frac{\sqrt[4]{8x^2}}{\sqrt[4]{8x^2}}=\frac{\sqrt[4]{24x^2}}{2x}$

59. Rationalizing the denominator: $\sqrt[4]{\frac{8}{y}}=\frac{\sqrt[4]{8}}{\sqrt[4]{y}} \bullet \frac{\sqrt[4]{y^3}}{\sqrt[4]{y^3}}=\frac{\sqrt[4]{8y^3}}{y}$

61. Rationalizing the denominator: $\sqrt[3]{\frac{4x}{3y}}=\frac{\sqrt[3]{4x}}{\sqrt[3]{3y}} \bullet \frac{\sqrt[3]{9y^2}}{\sqrt[3]{9y^2}}=\frac{\sqrt[3]{36xy^2}}{3y}$

63. Rationalizing the denominator: $\sqrt[3]{\frac{2x}{9y}}=\frac{\sqrt[3]{2x}}{\sqrt[3]{9y}} \bullet \frac{\sqrt[3]{3y^2}}{\sqrt[3]{3y^2}}=\frac{\sqrt[3]{6xy^2}}{3y}$

65. Simplifying: $\sqrt{\frac{27x^3}{5y}} = \frac{\sqrt{27x^3}}{\sqrt{5y}} \bullet \frac{\sqrt{5y}}{\sqrt{5y}} = \frac{\sqrt{135x^3y}}{5y} = \frac{3x\sqrt{15xy}}{5y}$

67. Simplifying: $\sqrt{\frac{75x^3y^2}{2z}} = \frac{\sqrt{75x^3y^2}}{\sqrt{2z}} \bullet \frac{\sqrt{2z}}{\sqrt{2z}} = \frac{\sqrt{150x^3y^2z}}{2z} = \frac{5xy\sqrt{6xz}}{2z}$

69. **a.** Rationalizing the denominator: $\frac{1}{\sqrt{2}} = \frac{1}{\sqrt{2}} \bullet \frac{\sqrt{2}}{\sqrt{2}} = \frac{\sqrt{2}}{2}$

b. Rationalizing the denominator: $\frac{1}{\sqrt[3]{2}} = \frac{1}{\sqrt[3]{2}} \bullet \frac{\sqrt[3]{4}}{\sqrt[3]{4}} = \frac{\sqrt[3]{4}}{2}$

c. Rationalizing the denominator: $\frac{1}{\sqrt[4]{2}} = \frac{1}{\sqrt[4]{2}} \bullet \frac{\sqrt[4]{8}}{\sqrt[4]{8}} = \frac{\sqrt[4]{8}}{2}$

71. Simplifying: $\sqrt{25x^2} = 5|x|$

73. Simplifying: $\sqrt{27x^3y^2} = \sqrt{9x^2y^2 \bullet 3x} = 3|xy|\sqrt{3x}$

75. Simplifying: $\sqrt{x^2 - 10x + 25} = \sqrt{(x-5)^2} = |x-5|$

77. Simplifying: $\sqrt{4x^2 + 12x + 9} = \sqrt{(2x+3)^2} = |2x+3|$

79. Simplifying: $\sqrt{4a^4 + 16a^3 + 16a^2} = \sqrt{4a^2(a^2 + 4a + 4)} = \sqrt{4a^2(a+2)^2} = 2|a(a+2)|$

81. Simplifying: $\sqrt{4x^3 - 8x^2} = \sqrt{4x^2(x-2)} = 2|x|\sqrt{x-2}$

83. Substituting $a = 9$ and $b = 16$:

$\sqrt{a+b} = \sqrt{9+16} = \sqrt{25} = 5$

$\sqrt{a} + \sqrt{b} = \sqrt{9} + \sqrt{16} = 3 + 4 = 7$

Thus $\sqrt{a+b} \neq \sqrt{a} + \sqrt{b}$.

85. Substituting $w = 10$ and $l = 15$:

$d = \sqrt{l^2 + w^2} = \sqrt{15^2 + 10^2} = \sqrt{225 + 100} = \sqrt{325} = \sqrt{25 \bullet 13} = 5\sqrt{13}$ feet

87. **a.** Substituting $k = 1$: $d = \sqrt{8000k + k^2} = \sqrt{8000(1) + (1)^2} = \sqrt{8001} \approx 89.4$ miles

b. Substituting $k = 2$: $d = \sqrt{8000k + k^2} = \sqrt{8000(2) + (2)^2} = \sqrt{16{,}004} \approx 126.5$ miles

c. Substituting $k = 3$: $d = \sqrt{8000k + k^2} = \sqrt{8000(3) + (3)^2} = \sqrt{24{,}009} \approx 154.9$ miles

89. Answers will vary.

91. Finding the first six terms:

$$f(1)=\sqrt{1^2+1}=\sqrt{2}$$

$$f(f(1))=\sqrt{(\sqrt{2})^2+1}=\sqrt{2+1}=\sqrt{3}$$

$$f(f(f(1)))=\sqrt{(\sqrt{3})^2+1}=\sqrt{3+1}=2$$

$$f(f(f(f(1))))=\sqrt{(2)^2+1}=\sqrt{4+1}=\sqrt{5}$$

$$f(f(f(f(f(1)))))=\sqrt{(\sqrt{5})^2+1}=\sqrt{5+1}=\sqrt{6}$$

$$f(f(f(f(f(f(1))))))=\sqrt{(\sqrt{6})^2+1}=\sqrt{6+1}=\sqrt{7}$$

Following this pattern, the 10th term will be $\sqrt{11}$ and the 100th term will be $\sqrt{101}$.

93. Simplifying: $5x-4x+6x=7x$

95. Simplifying: $35xy^2-8xy^2=27xy^2$

97. Simplifying: $\frac{1}{2}x+\frac{1}{3}x=\frac{3}{6}x+\frac{2}{6}x=\frac{5}{6}x$

99. Simplifying: $\sqrt{18}=\sqrt{9\bullet 2}=3\sqrt{2}$

101. Simplifying: $\sqrt{75xy^3}=\sqrt{25y^2\bullet 3xy}=5y\sqrt{3xy}$

103. Simplifying: $\sqrt[3]{8a^4b^2}=\sqrt[3]{8a^3\bullet ab^2}=2a\sqrt[3]{ab^2}$

6.3 Addition and Subtraction of Radical Expressions

1. Combining radicals: $3\sqrt{5}+4\sqrt{5}=7\sqrt{5}$

3. Combining radicals: $3x\sqrt{7}-4x\sqrt{7}=-x\sqrt{7}$

5. Combining radicals: $5\sqrt[3]{10}-4\sqrt[3]{10}=\sqrt[3]{10}$

7. Combining radicals: $8\sqrt[5]{6}-2\sqrt[5]{6}+3\sqrt[5]{6}=9\sqrt[5]{6}$

9. Combining radicals: $3x\sqrt{2}-4x\sqrt{2}+x\sqrt{2}=0$

11. Combining radicals: $\sqrt{20}-\sqrt{80}+\sqrt{45}=2\sqrt{5}-4\sqrt{5}+3\sqrt{5}=\sqrt{5}$

13. Combining radicals: $4\sqrt{8}-2\sqrt{50}-5\sqrt{72}=8\sqrt{2}-10\sqrt{2}-30\sqrt{2}=-32\sqrt{2}$

15. Combining radicals: $5x\sqrt{8}+3\sqrt{32x^2}-5\sqrt{50x^2}=10x\sqrt{2}+12x\sqrt{2}-25x\sqrt{2}=-3x\sqrt{2}$

17. Combining radicals: $5\sqrt[3]{16}-4\sqrt[3]{54}=10\sqrt[3]{2}-12\sqrt[3]{2}=-2\sqrt[3]{2}$

19. Combining radicals: $\sqrt[3]{x^4y^2}+7x\sqrt[3]{xy^2}=x\sqrt[3]{xy^2}+7x\sqrt[3]{xy^2}=8x\sqrt[3]{xy^2}$

21. Combining radicals: $5a^2\sqrt{27ab^3}-6b\sqrt{12a^5b}=15a^2b\sqrt{3ab}-12a^2b\sqrt{3ab}=3a^2b\sqrt{3ab}$

23. Combining radicals: $b\sqrt[3]{24a^5b}+3a\sqrt[3]{81a^2b^4}=2ab\sqrt[3]{3a^2b}+9ab\sqrt[3]{3a^2b}=11ab\sqrt[3]{3a^2b}$

25. Combining radicals:

$$5x\sqrt[4]{3y^5}+y\sqrt[4]{243x^4y}+\sqrt[4]{48x^4y^5}=5xy\sqrt[4]{3y}+3xy\sqrt[4]{3y}+2xy\sqrt[4]{3y}=10xy\sqrt[4]{3y}$$

27. Combining radicals: $\frac{\sqrt{2}}{2}+\frac{1}{\sqrt{2}}=\frac{\sqrt{2}}{2}+\frac{1}{\sqrt{2}}\bullet\frac{\sqrt{2}}{\sqrt{2}}=\frac{\sqrt{2}}{2}+\frac{\sqrt{2}}{2}=\sqrt{2}$

29. Combining radicals: $\frac{\sqrt{5}}{3}+\frac{1}{\sqrt{5}}=\frac{\sqrt{5}}{3}+\frac{1}{\sqrt{5}}\bullet\frac{\sqrt{5}}{\sqrt{5}}=\frac{\sqrt{5}}{3}+\frac{\sqrt{5}}{5}=\frac{5\sqrt{5}}{15}+\frac{3\sqrt{5}}{15}=\frac{8\sqrt{5}}{15}$

31. Combining radicals: $\sqrt{x}-\frac{1}{\sqrt{x}}=\sqrt{x}-\frac{1}{\sqrt{x}}\bullet\frac{\sqrt{x}}{\sqrt{x}}=\sqrt{x}-\frac{\sqrt{x}}{x}=\frac{x\sqrt{x}}{x}-\frac{\sqrt{x}}{x}=\frac{(x-1)\sqrt{x}}{x}$

33. Combining radicals: $\frac{\sqrt{18}}{6}+\sqrt{\frac{1}{2}}+\frac{\sqrt{2}}{2}=\frac{3\sqrt{2}}{6}+\frac{1}{\sqrt{2}}\bullet\frac{\sqrt{2}}{\sqrt{2}}+\frac{\sqrt{2}}{2}=\frac{\sqrt{2}}{2}+\frac{\sqrt{2}}{2}+\frac{\sqrt{2}}{2}=\frac{3\sqrt{2}}{2}$

35. Combining radicals:

$$\sqrt{6}-\sqrt{\frac{2}{3}}+\sqrt{\frac{1}{6}}=\sqrt{6}-\frac{\sqrt{2}}{\sqrt{3}}\bullet\frac{\sqrt{3}}{\sqrt{3}}+\frac{1}{\sqrt{6}}\bullet\frac{\sqrt{6}}{\sqrt{6}}=\sqrt{6}-\frac{\sqrt{6}}{3}+\frac{\sqrt{6}}{6}=\frac{6\sqrt{6}}{6}-\frac{2\sqrt{6}}{6}+\frac{\sqrt{6}}{6}=\frac{5\sqrt{6}}{6}$$

37. Combining radicals: $\sqrt[3]{25}+\frac{3}{\sqrt[3]{5}}=\sqrt[3]{25}+\frac{3}{\sqrt[3]{5}}\bullet\frac{\sqrt[3]{25}}{\sqrt[3]{25}}=\sqrt[3]{25}+\frac{3\sqrt[3]{25}}{5}=\frac{5\sqrt[3]{25}}{5}+\frac{3\sqrt[3]{25}}{5}=\frac{8\sqrt[3]{25}}{5}$

39. Using a calculator:

$\sqrt{12}\approx 3.464$ $\quad$ $2\sqrt{3}\approx 3.464$

41. It is equal to the decimal approximation for $\sqrt{50}$:

$\sqrt{8}+\sqrt{18}\approx 7.071\approx\sqrt{50}$ $\quad$ $\sqrt{26}\approx 5.099$

43. The correct statement is: $3\sqrt{2x}+5\sqrt{2x}=8\sqrt{2x}$

45. The correct statement is: $\sqrt{9+16}=\sqrt{25}=5$

47. Answers will vary.

49. Answers will vary.

51. Answers will vary.

53. If the legs have length x, then the hypotenuse has a length of $x\sqrt{2}$.
Therefore the ratio is: $\sqrt{2}:1$ or 1.414:1

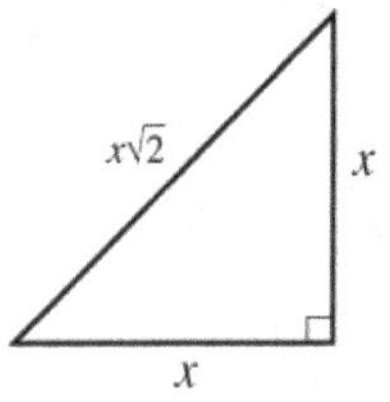

55. **a.** The diagonal of the base is $5\sqrt{2}$, therefore the ratio is: $\frac{5\sqrt{2}}{5}=\frac{\sqrt{2}}{1}$

b. The area of the base is 25, therefore the ratio is: $\frac{25}{5\sqrt{2}}=\frac{5}{\sqrt{2}}$

c. The perimeter of the base is 20, therefore the ratio is: $\frac{25}{20}=\frac{5}{4}$

57. Simplifying: $3\bullet 2=6$

59. Simplifying: $(x+y)(4x-y)=4x^2-xy+4xy-y^2=4x^2+3xy-y^2$

61. Simplifying: $(x+3)^2=x^2+2(3x)+3^2=x^2+6x+9$

63. Simplifying: $(x-2)(x+2)=x^2-2^2=x^2-4$

65. Simplifying: $2\sqrt{18}=2\sqrt{9\bullet 2}=2\bullet 3\sqrt{2}=6\sqrt{2}$

67. Simplifying: $\left(\sqrt{6}\right)^2=6$

69. Simplifying: $\left(3\sqrt{x}\right)^2=9x$

71. Rationalizing the denominator: $\frac{\sqrt{3}}{\sqrt{2}}=\frac{\sqrt{3}}{\sqrt{2}}\bullet\frac{\sqrt{2}}{\sqrt{2}}=\frac{\sqrt{6}}{2}$

6.4 Multiplication and Division of Radical Expressions

1. Multiplying: $\sqrt{6}\sqrt{3} = \sqrt{18} = 3\sqrt{2}$

3. Multiplying: $(2\sqrt{3})(5\sqrt{7}) = 10\sqrt{21}$

5. Multiplying: $(4\sqrt{6})(2\sqrt{15})(3\sqrt{10}) = 24\sqrt{900} = 24 \bullet 30 = 720$

7. Multiplying: $(3\sqrt[3]{3})(6\sqrt[3]{9}) = 18\sqrt[3]{27} = 18 \bullet 3 = 54$

9. Multiplying: $\sqrt{3}(\sqrt{2} - 3\sqrt{3}) = \sqrt{6} - 3\sqrt{9} = \sqrt{6} - 9$

11. Multiplying: $6\sqrt[3]{4}(2\sqrt[3]{2} + 1) = 12\sqrt[3]{8} + 6\sqrt[3]{4} = 24 + 6\sqrt[3]{4}$

13. Multiplying: $(\sqrt{3} + \sqrt{2})(3\sqrt{3} - \sqrt{2}) = 3\sqrt{9} - \sqrt{6} + 3\sqrt{6} - \sqrt{4} = 9 + 2\sqrt{6} - 2 = 7 + 2\sqrt{6}$

15. Multiplying: $(\sqrt{x} + 5)(\sqrt{x} - 3) = x - 3\sqrt{x} + 5\sqrt{x} - 15 = x + 2\sqrt{x} - 15$

17. Multiplying:

$$\begin{aligned}(3\sqrt{6} + 4\sqrt{2})(\sqrt{6} + 2\sqrt{2}) &= 3\sqrt{36} + 4\sqrt{12} + 6\sqrt{12} + 8\sqrt{4} \\ &= 18 + 8\sqrt{3} + 12\sqrt{3} + 16 \\ &= 34 + 20\sqrt{3}\end{aligned}$$

19. Multiplying: $(\sqrt{3} + 4)^2 = (\sqrt{3} + 4)(\sqrt{3} + 4) = \sqrt{9} + 4\sqrt{3} + 4\sqrt{3} + 16 = 19 + 8\sqrt{3}$

21. Multiplying: $(\sqrt{x} - 3)^2 = (\sqrt{x} - 3)(\sqrt{x} - 3) = x - 3\sqrt{x} - 3\sqrt{x} + 9 = x - 6\sqrt{x} + 9$

23. Multiplying:

$$(2\sqrt{a} - 3\sqrt{b})^2 = (2\sqrt{a} - 3\sqrt{b})(2\sqrt{a} - 3\sqrt{b}) = 4a - 6\sqrt{ab} - 6\sqrt{ab} + 9b = 4a - 12\sqrt{ab} + 9b$$

25. Multiplying:

$$(\sqrt{x-4} + 2)^2 = (\sqrt{x-4} + 2)(\sqrt{x-4} + 2) = x - 4 + 2\sqrt{x-4} + 2\sqrt{x-4} + 4 = x + 4\sqrt{x-4}$$

27. Multiplying:

$$(\sqrt{x-5} - 3)^2 = (\sqrt{x-5} - 3)(\sqrt{x-5} - 3) = x - 5 - 3\sqrt{x-5} - 3\sqrt{x-5} + 9 = x + 4 - 6\sqrt{x-5}$$

29. Multiplying: $(\sqrt{3} - \sqrt{2})(\sqrt{3} + \sqrt{2}) = (\sqrt{3})^2 - (\sqrt{2})^2 = 3 - 2 = 1$

31. Multiplying: $(\sqrt{a} + 7)(\sqrt{a} - 7) = (\sqrt{a})^2 - (7)^2 = a - 49$

33. Multiplying: $(5 - \sqrt{x})(5 + \sqrt{x}) = (5)^2 - (\sqrt{x})^2 = 25 - x$

35. Multiplying: $(\sqrt{x-4} + 2)(\sqrt{x-4} - 2) = (\sqrt{x-4})^2 - (2)^2 = x - 4 - 4 = x - 8$

37. Multiplying:

$$(\sqrt{3} + 1)^3 = (\sqrt{3} + 1)(3 + 2\sqrt{3} + 1) = (\sqrt{3} + 1)(4 + 2\sqrt{3}) = 4\sqrt{3} + 4 + 6 + 2\sqrt{3} = 10 + 6\sqrt{3}$$

39. Rationalizing the denominator: $\dfrac{\sqrt{2}}{\sqrt{6} - \sqrt{2}} = \dfrac{\sqrt{2}}{\sqrt{6} - \sqrt{2}} \bullet \dfrac{\sqrt{6} + \sqrt{2}}{\sqrt{6} + \sqrt{2}} = \dfrac{\sqrt{12} + 2}{6 - 2} = \dfrac{2\sqrt{3} + 2}{4} = \dfrac{1 + \sqrt{3}}{2}$

41. Rationalizing the denominator: $\frac{\sqrt{5}}{\sqrt{5}+1}=\frac{\sqrt{5}}{\sqrt{5}+1}\bullet\frac{\sqrt{5}-1}{\sqrt{5}-1}=\frac{5-\sqrt{5}}{5-1}=\frac{5-\sqrt{5}}{4}$

43. Rationalizing the denominator: $\frac{\sqrt{x}}{\sqrt{x}-3}=\frac{\sqrt{x}}{\sqrt{x}-3}\bullet\frac{\sqrt{x}+3}{\sqrt{x}+3}=\frac{x+3\sqrt{x}}{x-9}$

45. Rationalizing the denominator: $\frac{\sqrt{5}}{2\sqrt{5}-3}=\frac{\sqrt{5}}{2\sqrt{5}-3}\bullet\frac{2\sqrt{5}+3}{2\sqrt{5}+3}=\frac{2\sqrt{25}+3\sqrt{5}}{20-9}=\frac{10+3\sqrt{5}}{11}$

47. Rationalizing the denominator: $\frac{3}{\sqrt{x}-\sqrt{y}}=\frac{3}{\sqrt{x}-\sqrt{y}}\bullet\frac{\sqrt{x}+\sqrt{y}}{\sqrt{x}+\sqrt{y}}=\frac{3\sqrt{x}+3\sqrt{y}}{x-y}$

49. Rationalizing the denominator:

$$\frac{\sqrt{6}+\sqrt{2}}{\sqrt{6}-\sqrt{2}}=\frac{\sqrt{6}+\sqrt{2}}{\sqrt{6}-\sqrt{2}}\bullet\frac{\sqrt{6}+\sqrt{2}}{\sqrt{6}+\sqrt{2}}=\frac{6+2\sqrt{12}+2}{6-2}=\frac{8+4\sqrt{3}}{4}=2+\sqrt{3}$$

51. Rationalizing the denominator: $\frac{\sqrt{7}-2}{\sqrt{7}+2}=\frac{\sqrt{7}-2}{\sqrt{7}+2}\bullet\frac{\sqrt{7}-2}{\sqrt{7}-2}=\frac{7-4\sqrt{7}+4}{7-4}=\frac{11-4\sqrt{7}}{3}$

53. **a.** Adding: $(\sqrt{x}+2)+(\sqrt{x}-2)=\sqrt{x}+2+\sqrt{x}-2=2\sqrt{x}$

b. Multiplying: $(\sqrt{x}+2)(\sqrt{x}-2)=x+2\sqrt{x}-2\sqrt{x}-4=x-4$

c. Squaring: $(\sqrt{x}+2)^2=(\sqrt{x}+2)(\sqrt{x}+2)=x+2\sqrt{x}+2\sqrt{x}+4=x+4\sqrt{x}+4$

d. Dividing: $\frac{\sqrt{x}+2}{\sqrt{x}-2}=\frac{\sqrt{x}+2}{\sqrt{x}-2}\bullet\frac{\sqrt{x}+2}{\sqrt{x}+2}=\frac{x+4\sqrt{x}+4}{x-4}$

55. **a.** Adding: $(5+\sqrt{2})+(5-\sqrt{2})=5+\sqrt{2}+5-\sqrt{2}=10$

b. Multiplying: $(5+\sqrt{2})(5-\sqrt{2})=25+5\sqrt{2}-5\sqrt{2}-2=23$

c. Squaring: $(5+\sqrt{2})^2=(5+\sqrt{2})(5+\sqrt{2})=25+5\sqrt{2}+5\sqrt{2}+2=27+10\sqrt{2}$

d. Dividing: $\frac{5+\sqrt{2}}{5-\sqrt{2}}=\frac{5+\sqrt{2}}{5-\sqrt{2}}\bullet\frac{5+\sqrt{2}}{5+\sqrt{2}}=\frac{25+5\sqrt{2}+5\sqrt{2}+2}{25-2}=\frac{27+10\sqrt{2}}{23}$

57. **a.** Adding: $\sqrt{2}+(\sqrt{6}+\sqrt{2})=\sqrt{2}+\sqrt{6}+\sqrt{2}=\sqrt{6}+2\sqrt{2}$

b. Multiplying: $\sqrt{2}(\sqrt{6}+\sqrt{2})=\sqrt{12}+\sqrt{4}=2+2\sqrt{3}$

c. Dividing: $\frac{\sqrt{6}+\sqrt{2}}{\sqrt{2}}=\frac{\sqrt{6}+\sqrt{2}}{\sqrt{2}}\bullet\frac{\sqrt{2}}{\sqrt{2}}=\frac{\sqrt{12}+2}{2}=\frac{2+2\sqrt{3}}{2}=1+\sqrt{3}$

d. Dividing: $\frac{\sqrt{2}}{\sqrt{6}+\sqrt{2}}=\frac{\sqrt{2}}{\sqrt{6}+\sqrt{2}}\bullet\frac{\sqrt{6}-\sqrt{2}}{\sqrt{6}-\sqrt{2}}=\frac{\sqrt{12}-2}{6-2}=\frac{-2+2\sqrt{3}}{4}=\frac{-1+\sqrt{3}}{2}$

59. **a.** Adding: $\left(\frac{1+\sqrt{5}}{2}\right)+\left(\frac{1-\sqrt{5}}{2}\right)=\frac{1+\sqrt{5}}{2}+\frac{1-\sqrt{5}}{2}=\frac{2}{2}=1$

b. Multiplying: $\left(\frac{1+\sqrt{5}}{2}\right)\left(\frac{1-\sqrt{5}}{2}\right)=\frac{1-5}{4}=\frac{-4}{4}=-1$

61. Simplifying the product:

$$\left(\sqrt[3]{2}+\sqrt[3]{3}\right)\left(\sqrt[3]{4}-\sqrt[3]{6}+\sqrt[3]{9}\right)=\sqrt[3]{8}-\sqrt[3]{12}+\sqrt[3]{18}+\sqrt[3]{12}-\sqrt[3]{18}+\sqrt[3]{27}=2+3=5$$

63. The correct statement is: $5\left(2\sqrt{3}\right)=10\sqrt{3}$

65. The correct statement is: $\left(\sqrt{x}+3\right)^2=\left(\sqrt{x}+3\right)\left(\sqrt{x}+3\right)=x+6\sqrt{x}+9$

67. The correct statement is: $\left(5\sqrt{3}\right)^2=\left(5\sqrt{3}\right)\left(5\sqrt{3}\right)=25\bullet 3=75$

69. Substituting $h=50$: $t=\dfrac{\sqrt{100-50}}{4}=\dfrac{\sqrt{50}}{4}=\dfrac{5\sqrt{2}}{4}$ seconds;

Substituting $h=0$: $t=\dfrac{\sqrt{100-0}}{4}=\dfrac{\sqrt{100}}{4}=\dfrac{10}{4}=\dfrac{5}{2}$ seconds

71. Since the large rectangle is a golden rectangle and $AC=6$, then $CE=6\left(\dfrac{1+\sqrt{5}}{2}\right)=3+3\sqrt{5}$.

Since $CD=6$, then $DE=3+3\sqrt{5}-6=3\sqrt{5}-3$. Now computing the ratio:

$$\frac{EF}{DE}=\frac{6}{3\sqrt{5}-3}\bullet\frac{3\sqrt{5}+3}{3\sqrt{5}+3}=\frac{18\left(\sqrt{5}+1\right)}{45-9}=\frac{18\left(\sqrt{5}+1\right)}{36}=\frac{1+\sqrt{5}}{2}$$

Therefore the smaller rectangle *BDEF* is also a golden rectangle.

73. Since the large rectangle is a golden rectangle and $AC=2x$, then $CE=2x\left(\dfrac{1+\sqrt{5}}{2}\right)=x\left(1+\sqrt{5}\right)$.

Since $CD=2x$, then $DE=x\left(1+\sqrt{5}\right)-2x=x\left(-1+\sqrt{5}\right)$. Now computing the ratio:

$$\frac{EF}{DE}=\frac{2x}{x\left(-1+\sqrt{5}\right)}=\frac{2}{-1+\sqrt{5}}\bullet\frac{-1-\sqrt{5}}{-1-\sqrt{5}}=\frac{-2\left(\sqrt{5}+1\right)}{1-5}=\frac{-2\left(\sqrt{5}+1\right)}{-4}=\frac{1+\sqrt{5}}{2}$$

Therefore the smaller rectangle *BDEF* is also a golden rectangle.

75. Simplifying: $(t+5)^2=t^2+2(5t)+5^2=t^2+10t+25$

77. Simplifying: $\sqrt{x}\bullet\sqrt{x}=\sqrt{x^2}=x$

79. Solving the equation:

$$\begin{aligned}3x+4&=5^2\\3x+4&=25\\3x&=21\\x&=7\end{aligned}$$

81. Solving the equation:

$$\begin{aligned}t^2+7t+12&=0\\(t+4)(t+3)&=0\\t&=-4,-3\end{aligned}$$

83. Solving the equation:

$$\begin{aligned}t^2+10t+25&=t+7\\t^2+9t+18&=0\\(t+6)(t+3)&=0\\t&=-6,-3\end{aligned}$$

85. Solving the equation:

$$\begin{aligned}(x+4)^2&=x+6\\x^2+8x+16&=x+6\\x^2+7x+10&=0\\(x+5)(x+2)&=0\\x&=-5,-2\end{aligned}$$

87. Substituting $x = 7$: $\sqrt{3(7)+4} = \sqrt{21+4} = \sqrt{25} = 5$

Yes, $x = 7$ is a solution to the equation.

89. Substituting $t = -6$:

$$-6+5=-1 \qquad \sqrt{-6+7}=\sqrt{1}=1$$

No, $t = -6$ is not a solution to the equation.

6.5 Equations Involving Radicals

1. Solving the equation:

$$\begin{aligned}\sqrt{2x+1} &= 3\\ \left(\sqrt{2x+1}\right)^2 &= 3^2\\ 2x+1 &= 9\\ 2x &= 8\\ x &= 4\end{aligned}$$

3. Solving the equation:

$$\begin{aligned}\sqrt{4x+1} &= -5\\ \left(\sqrt{4x+1}\right)^2 &= (-5)^2\\ 4x+1 &= 25\\ 4x &= 24\\ x &= 6\end{aligned}$$

There is no solution (6 does not check).

5. Solving the equation:

$$\begin{aligned}\sqrt{2y-1} &= 3\\ \left(\sqrt{2y-1}\right)^2 &= 3^2\\ 2y-1 &= 9\\ 2y &= 10\\ y &= 5\end{aligned}$$

7. Solving the equation:

$$\begin{aligned}\sqrt{5x-7} &= -1\\ \left(\sqrt{5x-7}\right)^2 &= (-1)^2\\ 5x-7 &= 1\\ 5x &= 8\\ x &= \frac{8}{5}\end{aligned}$$

There is no solution ($\frac{8}{5}$ does not check).

9. Solving the equation:

$$\begin{aligned}\sqrt{2x-3}-2 &= 4\\ \sqrt{2x-3} &= 6\\ \left(\sqrt{2x-3}\right)^2 &= 6^2\\ 2x-3 &= 36\\ 2x &= 39\\ x &= \frac{39}{2}\end{aligned}$$

11. Solving the equation:

$$\begin{aligned}\sqrt{4a+1}+3 &= 2\\ \sqrt{4a+1} &= -1\\ \left(\sqrt{4a+1}\right)^2 &= (-1)^2\\ 4a+1 &= 1\\ 4a &= 0\\ a &= 0\end{aligned}$$

There is no solution (0 does not check).

13. Solving the equation:

$$\begin{aligned}\sqrt[4]{3x+1}&=2\\ \left(\sqrt[4]{3x+1}\right)^4&=2^4\\ 3x+1&=16\\ 3x&=15\\ x&=5\end{aligned}$$

15. Solving the equation:

$$\begin{aligned}\sqrt[3]{2x-5}&=1\\ \left(\sqrt[3]{2x-5}\right)^3&=1^3\\ 2x-5&=1\\ 2x&=6\\ x&=3\end{aligned}$$

17. Solving the equation:

$$\begin{aligned}\sqrt[3]{3a+5}&=-3\\ \left(\sqrt[3]{3a+5}\right)^3&=(-3)^3\\ 3a+5&=-27\\ 3a&=-32\\ a&=-\frac{32}{3}\end{aligned}$$

19. Solving the equation:

$$\begin{aligned}\sqrt{y-3}&=y-3\\ \left(\sqrt{y-3}\right)^2&=(y-3)^2\\ y-3&=y^2-6y+9\\ 0&=y^2-7y+12\\ 0&=(y-3)(y-4)\\ y&=3,4\end{aligned}$$

21. Solving the equation:

$$\begin{aligned}\sqrt{a+2}&=a+2\\ \left(\sqrt{a+2}\right)^2&=(a+2)^2\\ a+2&=a^2+4a+4\\ 0&=a^2+3a+2\\ 0&=(a+2)(a+1)\\ a&=-2,-1\end{aligned}$$

23. Solving the equation:

$$\begin{aligned}\sqrt{2x+4}&=\sqrt{1-x}\\ \left(\sqrt{2x+4}\right)^2&=\left(\sqrt{1-x}\right)^2\\ 2x+4&=1-x\\ 3x&=-3\\ x&=-1\end{aligned}$$

25. Solving the equation:

$$\begin{aligned}\sqrt{4a+7}&=-\sqrt{a+2}\\ \left(\sqrt{4a+7}\right)^2&=\left(-\sqrt{a+2}\right)^2\\ 4a+7&=a+2\\ 3a&=-5\\ a&=-\frac{5}{3}\end{aligned}$$

There is no solution ($-\frac{5}{3}$ does not check).

27. Solving the equation:

$$\begin{aligned}\sqrt[4]{5x-8}&=\sqrt[4]{4x-1}\\ \left(\sqrt[4]{5x-8}\right)^4&=\left(\sqrt[4]{4x-1}\right)^4\\ 5x-8&=4x-1\\ x&=7\end{aligned}$$

29. Solving the equation:

$$\begin{aligned} x+1 &= \sqrt{5x+1} \\ (x+1)^2 &= \left(\sqrt{5x+1}\right)^2 \\ x^2+2x+1 &= 5x+1 \\ x^2-3x &= 0 \\ x(x-3) &= 0 \\ x &= 0,3 \end{aligned}$$

31. Solving the equation:

$$\begin{aligned} t+5 &= \sqrt{2t+9} \\ (t+5)^2 &= \left(\sqrt{2t+9}\right)^2 \\ t^2+10t+25 &= 2t+9 \\ t^2+8t+16 &= 0 \\ (t+4)^2 &= 0 \\ t &= -4 \end{aligned}$$

33. Solving the equation:

$$\begin{aligned} \sqrt{y-8} &= \sqrt{8-y} \\ \left(\sqrt{y-8}\right)^2 &= \left(\sqrt{8-y}\right)^2 \\ y-8 &= 8-y \\ 2y &= 16 \\ y &= 8 \end{aligned}$$

35. Solving the equation:

$$\begin{aligned} \sqrt[3]{3x+5} &= \sqrt[3]{5-2x} \\ \left(\sqrt[3]{3x+5}\right)^3 &= \left(\sqrt[3]{5-2x}\right)^3 \\ 3x+5 &= 5-2x \\ 5x &= 0 \\ x &= 0 \end{aligned}$$

37. Solving the equation:

$$\begin{aligned} \sqrt{x-8} &= \sqrt{x}-2 \\ \left(\sqrt{x-8}\right)^2 &= \left(\sqrt{x}-2\right)^2 \\ x-8 &= x-4\sqrt{x}+4 \\ -12 &= -4\sqrt{x} \\ \sqrt{x} &= 3 \\ x &= 9 \end{aligned}$$

39. Solving the equation:

$$\begin{aligned} \sqrt{x+1} &= \sqrt{x}+1 \\ \left(\sqrt{x+1}\right)^2 &= \left(\sqrt{x}+1\right)^2 \\ x+1 &= x+2\sqrt{x}+1 \\ 0 &= 2\sqrt{x} \\ \sqrt{x} &= 0 \\ x &= 0 \end{aligned}$$

41. Solving the equation:

$$\begin{aligned} \sqrt{x+8} &= \sqrt{x-4}+2 \\ \left(\sqrt{x+8}\right)^2 &= \left(\sqrt{x-4}+2\right)^2 \\ x+8 &= x-4+4\sqrt{x-4}+4 \\ 8 &= 4\sqrt{x-4} \\ \sqrt{x-4} &= 2 \\ x-4 &= 4 \\ x &= 8 \end{aligned}$$

43. Solving the equation:

$$\begin{aligned} \sqrt{x-5}-3 &= \sqrt{x-8} \\ \left(\sqrt{x-5}-3\right)^2 &= \left(\sqrt{x-8}\right)^2 \\ x-5-6\sqrt{x-5}+9 &= x-8 \\ -6\sqrt{x-5} &= -12 \\ \sqrt{x-5} &= 2 \\ x-5 &= 4 \\ x &= 9 \end{aligned}$$

There is no solution (9 does not check).

45. **a.** Solving the equation:

$$\sqrt{y}-4=6$$
$$\sqrt{y}=10$$
$$\left(\sqrt{y}\right)^2=10^2$$
$$y=100$$

b. Solving the equation:

$$\sqrt{y-4}=6$$
$$\left(\sqrt{y-4}\right)^2=6^2$$
$$y-4=36$$
$$y=40$$

c. Solving the equation:

$$\sqrt{y-4}=-6$$
$$\left(\sqrt{y-4}\right)^2=(-6)^2$$
$$y-4=36$$
$$y=40$$

There is no solution (40 does not check).

d. Solving the equation:

$$\sqrt{y-4}=y-6$$
$$\left(\sqrt{y-4}\right)^2=(y-6)^2$$
$$y-4=y^2-12y+36$$
$$0=y^2-13y+40$$
$$0=(y-5)(y-8)$$
$$y=5,8$$

The solution is 8 (5 does not check).

47. **a.** Solving the equation:

$$x-3=0$$
$$x=3$$

b. Solving the equation:

$$\sqrt{x}-3=0$$
$$\left(\sqrt{x}\right)^2=3^2$$
$$x=9$$

c. Solving the equation:

$$\sqrt{x-3}=0$$
$$\left(\sqrt{x-3}\right)^2=0^2$$
$$x-3=0$$
$$x=3$$

d. Solving the equation:

$$\sqrt{x}+3=0$$
$$\sqrt{x}=-3$$
$$\left(\sqrt{x}\right)^2=(-3)^2$$
$$x=9$$

There is no solution (9 does not check).

e. Solving the equation:

$$\sqrt{x}+3=5$$
$$\sqrt{x}=2$$
$$\left(\sqrt{x}\right)^2=2^2$$
$$x=4$$

f. Solving the equation:

$$\sqrt{x}+3=-5$$
$$\sqrt{x}=-8$$
$$\left(\sqrt{x}\right)^2=(-8)^2$$
$$x=64$$

There is no solution (64 does not check).

g. Solving the equation:

$$\begin{aligned} x-3 &= \sqrt{5-x} \\ (x-3)^2 &= \left(\sqrt{5-x}\right)^2 \\ x^2-6x+9 &= 5-x \\ x^2-5x+4 &= 0 \\ (x-1)(x-4) &= 0 \\ x &= 1,4 \end{aligned}$$

The solution is 4 (1 does not check).

49. Solving for h:

$$\begin{aligned} t &= \frac{\sqrt{100-h}}{4} \\ 4t &= \sqrt{100-h} \\ 16t^2 &= 100-h \\ h &= 100-16t^2 \end{aligned}$$

51. Solving for L:

$$\begin{aligned} 2 &= 2\left(\frac{22}{7}\right)\sqrt{\frac{L}{32}} \\ \frac{7}{22} &= \sqrt{\frac{L}{32}} \\ \left(\frac{7}{22}\right)^2 &= \frac{L}{32} \\ L &= 32\left(\frac{7}{22}\right)^2 \approx 3.24 \text{ feet} \end{aligned}$$

53. Graphing the equation:

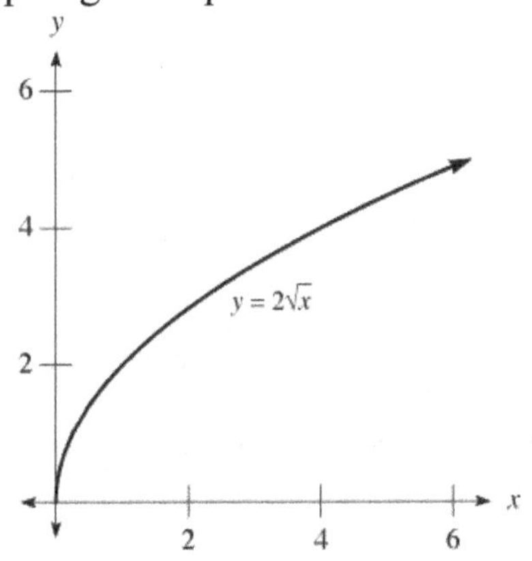

55. Graphing the equation:

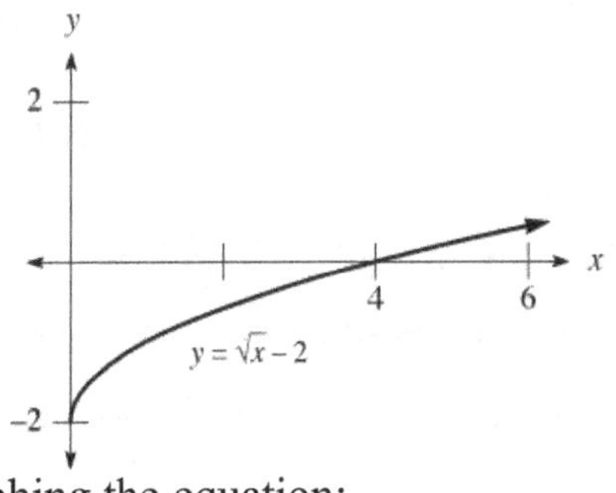

57. Graphing the equation:

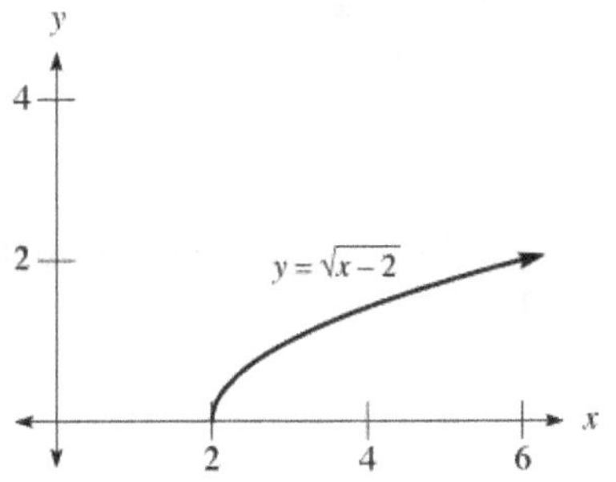

59. Graphing the equation:

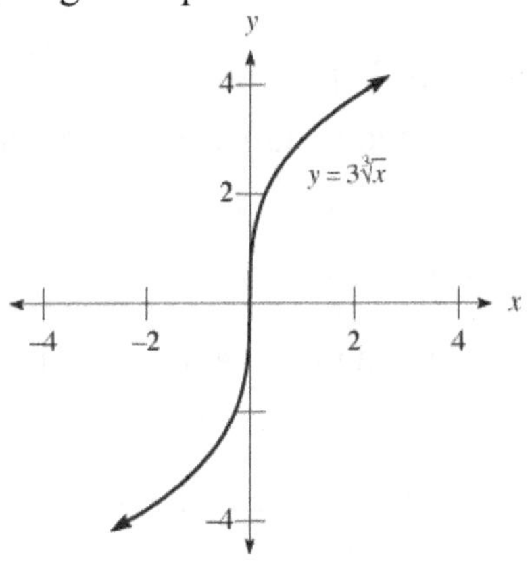

61. Graphing the equation:

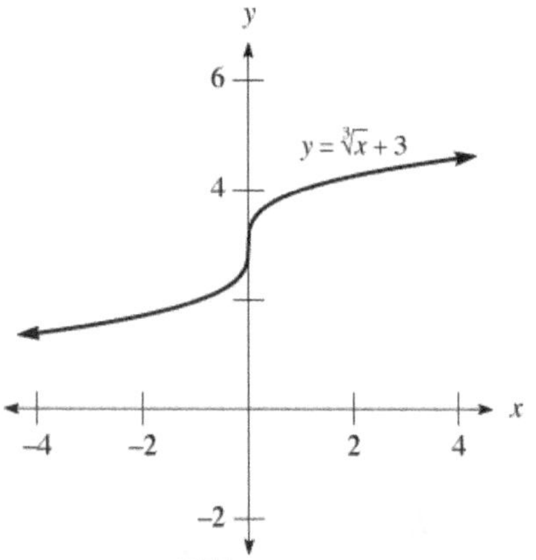

63. Graphing the equation:

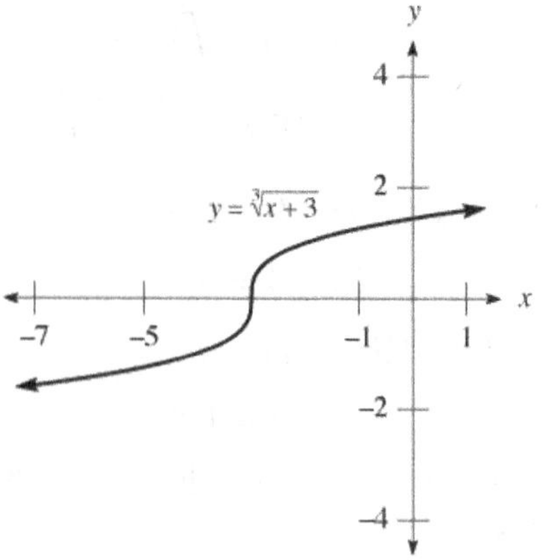

65. Simplifying: $\sqrt{25} = 5$

67. Simplifying: $\sqrt{12} = \sqrt{4 \cdot 3} = 2\sqrt{3}$

69. Simplifying: $(-1)^{15} = -1$

71. Simplifying: $(-1)^{50} = 1$

73. Solving the equation:

$$3x = 12$$
$$x = 4$$

75. Solving the equation:

$$4x - 3 = 5$$
$$4x = 8$$
$$x = 2$$

77. Performing the operations: $(3+4x)+(7-6x) = 10 - 2x$

79. Performing the operations: $(7+3x)-(5+6x) = 7+3x-5-6x = 2-3x$

81. Performing the operations: $(3-4x)(2+5x) = 6+15x-8x-20x^2 = 6+7x-20x^2$

83. Performing the operations: $2x(4-6x) = 8x-12x^2$

85. Performing the operations: $(2+3x)^2 = 2^2 + 2(2)(3x)+(3x)^2 = 4+12x+9x^2$

87. Performing the operations: $(2-3x)(2+3x) = 2^2 - (3x)^2 = 4-9x^2$

6.6 Complex Numbers

1. Writing in terms of i: $\sqrt{-36} = 6i$

3. Writing in terms of i: $-\sqrt{-25} = -5i$

5. Writing in terms of i: $\sqrt{-72} = 6i\sqrt{2}$

7. Writing in terms of i: $-\sqrt{-12} = -2i\sqrt{3}$

9. Rewriting the expression: $i^{28} = \left(i^4\right)^7 = (1)^7 = 1$

11. Rewriting the expression: $i^{26} = i^{24}i^2 = \left(i^4\right)^6 i^2 = (1)^6(-1) = -1$

13. Rewriting the expression: $i^{75} = i^{72}i^3 = \left(i^4\right)^{18} i^2 i = (1)^{18}(-1)i = -i$

15. Setting real and imaginary parts equal:

$$2x = 6 \qquad 3y = -3$$
$$x = 3 \qquad y = -1$$

17. Setting real and imaginary parts equal:

$$-x = 2 \qquad 10y = -5$$
$$x = -2 \qquad y = -\frac{1}{2}$$

19. Setting real and imaginary parts equal:

$$2x = -16 \qquad -2y = 10$$
$$x = -8 \qquad y = -5$$

21. Setting real and imaginary parts equal:

$$2x - 4 = 10 \qquad -6y = -3$$
$$2x = 14 \qquad y = \frac{1}{2}$$
$$x = 7$$

23. Setting real and imaginary parts equal:

$$\begin{aligned} 7x-1&=2 & 5y+2&=4 \\ 7x&=3 & 5y&=2 \\ x&=\frac{3}{7} & y&=\frac{2}{5} \end{aligned}$$

25. Combining the numbers: $(2+3i)+(3+6i)=5+9i$

27. Combining the numbers: $(3-5i)+(2+4i)=5-i$

29. Combining the numbers: $(5+2i)-(3+6i)=5+2i-3-6i=2-4i$

31. Combining the numbers: $(3-5i)-(2+i)=3-5i-2-i=1-6i$

33. Combining the numbers: $[(3+2i)-(6+i)]+(5+i)=3+2i-6-i+5+i=2+2i$

35. Combining the numbers: $[(7-i)-(2+4i)]-(6+2i)=7-i-2-4i-6-2i=-1-7i$

37. Combining the numbers:

$$\begin{aligned} (3+2i)-[(3-4i)-(6+2i)]&=(3+2i)-(3-4i-6-2i) \\ &=(3+2i)-(-3-6i) \\ &=3+2i+3+6i \\ &=6+8i \end{aligned}$$

39. Combining the numbers:

$$(4-9i)+[(2-7i)-(4+8i)]=(4-9i)+(2-7i-4-8i)=(4-9i)+(-2-15i)=2-24i$$

41. Finding the product: $3i(4+5i)=12i+15i^2=-15+12i$

43. Finding the product: $6i(4-3i)=24i-18i^2=18+24i$

45. Finding the product: $(3+2i)(4+i)=12+8i+3i+2i^2=12+11i-2=10+11i$

47. Finding the product: $(4+9i)(3-i)=12+27i-4i-9i^2=12+23i+9=21+23i$

49. Finding the product: $(1+i)^3=(1+i)(1+i)^2=(1+i)(1+2i-1)=(1+i)(2i)=-2+2i$

51. Finding the product:

$$(2-i)^3=(2-i)(2-i)^2=(2-i)(4-4i-1)=(2-i)(3-4i)=6-11i-4=2-11i$$

53. Finding the product: $(2+5i)^2=(2+5i)(2+5i)=4+10i+10i-25=-21+20i$

55. Finding the product: $(1-i)^2=(1-i)(1-i)=1-i-i-1=-2i$

57. Finding the product: $(3-4i)^2=(3-4i)(3-4i)=9-12i-12i-16=-7-24i$

59. Finding the product: $(2+i)(2-i)=4-i^2=4+1=5$

61. Finding the product: $(6-2i)(6+2i)=36-4i^2=36+4=40$

63. Finding the product: $(2+3i)(2-3i)=4-9i^2=4+9=13$

65. Finding the product: $(10+8i)(10-8i)=100-64i^2=100+64=164$

67. Finding the quotient: $\dfrac{2-3i}{i}=\dfrac{2-3i}{i}\bullet\dfrac{i}{i}=\dfrac{2i+3}{-1}=-3-2i$

69. Finding the quotient: $\dfrac{5+2i}{-i}=\dfrac{5+2i}{-i}\bullet\dfrac{i}{i}=\dfrac{5i-2}{1}=-2+5i$

71. Finding the quotient: $\dfrac{4}{2-3i}=\dfrac{4}{2-3i}\bullet\dfrac{2+3i}{2+3i}=\dfrac{8+12i}{4+9}=\dfrac{8+12i}{13}=\dfrac{8}{13}+\dfrac{12}{13}i$

73. Finding the quotient: $\dfrac{6}{-3+2i}=\dfrac{6}{-3+2i}\bullet\dfrac{-3-2i}{-3-2i}=\dfrac{-18-12i}{9+4}=\dfrac{-18-12i}{13}=-\dfrac{18}{13}-\dfrac{12}{13}i$

75. Finding the quotient: $\dfrac{2+3i}{2-3i}=\dfrac{2+3i}{2-3i}\bullet\dfrac{2+3i}{2+3i}=\dfrac{4+12i-9}{4+9}=\dfrac{-5+12i}{13}=-\dfrac{5}{13}+\dfrac{12}{13}i$

77. Finding the quotient: $\dfrac{5+4i}{3+6i}=\dfrac{5+4i}{3+6i}\bullet\dfrac{3-6i}{3-6i}=\dfrac{15-18i+24}{9+36}=\dfrac{39-18i}{45}=\dfrac{13}{15}-\dfrac{2}{5}i$

79. Dividing to find R:

$$R=\frac{80+20i}{-6+2i}=\frac{80+20i}{-6+2i}\bullet\frac{-6-2i}{-6-2i}=\frac{-480-280i+40}{36+4}=\frac{-440-280i}{40}=(-11-7i)\text{ ohms}$$

81. Solving the equation:

$$\frac{t}{3}-\frac{1}{2}=-1$$
$$6\left(\frac{t}{3}-\frac{1}{2}\right)=6(-1)$$
$$2t-3=-6$$
$$2t=-3$$
$$t=-\frac{3}{2}$$

83. Solving the equation:

$$2+\frac{5}{y}=\frac{3}{y^2}$$
$$y^2\left(2+\frac{5}{y}\right)=y^2\left(\frac{3}{y^2}\right)$$
$$2y^2+5y=3$$
$$2y^2+5y-3=0$$
$$(2y-1)(y+3)=0$$
$$y=-3,\frac{1}{2}$$

85. Let x represent the number. The equation is:

$$x+\frac{1}{x}=\frac{41}{20}$$
$$20x\left(x+\frac{1}{x}\right)=20x\left(\frac{41}{20}\right)$$
$$20x^2+20=41x$$
$$20x^2-41x+20=0$$
$$(5x-4)(4x-5)=0$$
$$x=\frac{4}{5},\frac{5}{4}$$

The number is either $\dfrac{5}{4}$ or $\dfrac{4}{5}$.

Chapter 6 Test

See www.mathtv.com for video solutions to all problems in this chapter test.

Chapter 7
Quadratic Equations

7.1 Completing the Square

1. Solving the equation:

$$x^2 = 25$$
$$x = \pm\sqrt{25} = \pm 5$$

3. Solving the equation:

$$a^2 = -9$$
$$a = \pm\sqrt{-9} = \pm 3i$$

5. Solving the equation:

$$y^2 = \frac{3}{4}$$
$$y = \pm\sqrt{\frac{3}{4}} = \pm\frac{\sqrt{3}}{2}$$

7. Solving the equation:

$$x^2 + 12 = 0$$
$$x^2 = -12$$
$$x = \pm\sqrt{-12} = \pm 2i\sqrt{3}$$

9. Solving the equation:

$$4a^2 - 45 = 0$$
$$4a^2 = 45$$
$$a^2 = \frac{45}{4}$$
$$a = \pm\sqrt{\frac{45}{4}} = \pm\frac{3\sqrt{5}}{2}$$

11. Solving the equation:

$$(2y-1)^2 = 25$$
$$2y - 1 = \pm\sqrt{25} = \pm 5$$
$$2y - 1 = -5, 5$$
$$2y = -4, 6$$
$$y = -2, 3$$

13. Solving the equation:

$$(2a+3)^2 = -9$$
$$2a + 3 = \pm\sqrt{-9} = \pm 3i$$
$$2a = -3 \pm 3i$$
$$a = \frac{-3 \pm 3i}{2}$$

15. Solving the equation:

$$(5x+2)^2 = -8$$
$$5x + 2 = \pm\sqrt{-8} = \pm 2i\sqrt{2}$$
$$5x = -2 \pm 2i\sqrt{2}$$
$$x = \frac{-2 \pm 2i\sqrt{2}}{5}$$

17. Solving the equation:

$$x^2 + 8x + 16 = -27$$
$$(x+4)^2 = -27$$
$$x + 4 = \pm\sqrt{-27} = \pm 3i\sqrt{3}$$
$$x = -4 \pm 3i\sqrt{3}$$

19. Solving the equation:

$$4a^2 - 12a + 9 = -4$$
$$(2a-3)^2 = -4$$
$$2a - 3 = \pm\sqrt{-4} = \pm 2i$$
$$2a = 3 \pm 2i$$
$$a = \frac{3 \pm 2i}{2}$$

21. Completing the square: $x^2+12x+36=(x+6)^2$

23. Completing the square: $x^2-4x+4=(x-2)^2$

25. Completing the square: $a^2-10a+25=(a-5)^2$

27. Completing the square: $x^2+5x+\frac{25}{4}=\left(x+\frac{5}{2}\right)^2$

29. Completing the square: $y^2-7y+\frac{49}{4}=\left(y-\frac{7}{2}\right)^2$

31. Completing the square: $x^2+\frac{1}{2}x+\frac{1}{16}=\left(x+\frac{1}{4}\right)^2$

33. Completing the square: $x^2+\frac{2}{3}x+\frac{1}{9}=\left(x+\frac{1}{3}\right)^2$

35. Solving the equation:

$$\begin{aligned} x^2+4x&=12\\ x^2+4x+4&=12+4\\ (x+2)^2&=16\\ x+2&=\pm\sqrt{16}=\pm4\\ x+2&=-4,4\\ x&=-6,2 \end{aligned}$$

37. Solving the equation:

$$\begin{aligned} x^2+12x&=-27\\ x^2+12x+36&=-27+36\\ (x+6)^2&=9\\ x+6&=\pm\sqrt{9}=\pm3\\ x+6&=-3,3\\ x&=-9,-3 \end{aligned}$$

39. Solving the equation:

$$\begin{aligned} a^2-2a+5&=0\\ a^2-2a+1&=-5+1\\ (a-1)^2&=-4\\ a-1&=\pm\sqrt{-4}=\pm2i\\ a&=1\pm2i \end{aligned}$$

41. Solving the equation:

$$\begin{aligned} y^2-8y+1&=0\\ y^2-8y+16&=-1+16\\ (y-4)^2&=15\\ y-4&=\pm\sqrt{15}\\ y&=4\pm\sqrt{15} \end{aligned}$$

43. Solving the equation:

$$\begin{aligned} x^2-5x-3&=0\\ x^2-5x+\frac{25}{4}&=3+\frac{25}{4}\\ \left(x-\frac{5}{2}\right)^2&=\frac{37}{4}\\ x-\frac{5}{2}&=\pm\frac{\sqrt{37}}{2}\\ x&=\frac{5\pm\sqrt{37}}{2} \end{aligned}$$

45. Solving the equation:

$$\begin{aligned} 2x^2-4x-8&=0\\ x^2-2x-4&=0\\ x^2-2x+1&=4+1\\ (x-1)^2&=5\\ x-1&=\pm\sqrt{5}\\ x&=1\pm\sqrt{5} \end{aligned}$$

47. Solving the equation:

$$3t^2 - 8t + 1 = 0$$
$$t^2 - \frac{8}{3}t + \frac{1}{3} = 0$$
$$t^2 - \frac{8}{3}t + \frac{16}{9} = -\frac{1}{3} + \frac{16}{9}$$
$$\left(t - \frac{4}{3}\right)^2 = \frac{13}{9}$$
$$t - \frac{4}{3} = \pm\sqrt{\frac{13}{9}} = \pm\frac{\sqrt{13}}{3}$$
$$t = \frac{4 \pm \sqrt{13}}{3}$$

49. Solving the equation:

$$4x^2 - 3x + 5 = 0$$
$$x^2 - \frac{3}{4}x + \frac{5}{4} = 0$$
$$x^2 - \frac{3}{4}x + \frac{9}{64} = -\frac{5}{4} + \frac{9}{64}$$
$$\left(x - \frac{3}{8}\right)^2 = -\frac{71}{64}$$
$$x - \frac{3}{8} = \pm\sqrt{-\frac{71}{64}} = \pm\frac{i\sqrt{71}}{8}$$
$$x = \frac{3 \pm i\sqrt{71}}{8}$$

51. Solving the equation:

$$3x^2 + 4x - 1 = 0$$
$$x^2 + \frac{4}{3}x - \frac{1}{3} = 0$$
$$x^2 + \frac{4}{3}x + \frac{4}{9} = \frac{1}{3} + \frac{4}{9}$$
$$\left(x + \frac{2}{3}\right)^2 = \frac{7}{9}$$
$$x + \frac{2}{3} = \pm\sqrt{\frac{7}{9}} = \pm\frac{\sqrt{7}}{3}$$
$$x = \frac{-2 \pm \sqrt{7}}{3}$$

53. Solving the equation:

$$2x^2 - 10x = 11$$
$$x^2 - 5x = \frac{11}{2}$$
$$x^2 - 5x + \frac{25}{4} = \frac{11}{2} + \frac{25}{4}$$
$$\left(x - \frac{5}{2}\right)^2 = \frac{47}{4}$$
$$x - \frac{5}{2} = \pm\sqrt{\frac{47}{4}} = \pm\frac{\sqrt{47}}{2}$$
$$x = \frac{5 \pm \sqrt{47}}{2}$$

55. Solving the equation:

$$4x^2 - 10x + 11 = 0$$
$$x^2 - \frac{5}{2}x = -\frac{11}{4}$$
$$x^2 - \frac{5}{2}x + \frac{25}{16} = -\frac{11}{4} + \frac{25}{16}$$
$$\left(x - \frac{5}{4}\right)^2 = -\frac{19}{16}$$
$$x - \frac{5}{4} = \pm\sqrt{-\frac{19}{16}} = \pm\frac{i\sqrt{19}}{4}$$
$$x = \frac{5 \pm i\sqrt{19}}{4}$$

57. **a.** No, it cannot be solved by factoring.

b. Solving the equation:

$$x^2 = -9$$
$$x = \pm\sqrt{-9}$$
$$x = \pm 3i$$

59. **a.** Solving by factoring:

$$x^2 - 6x = 0$$
$$x(x-6) = 0$$
$$x = 0, 6$$

b. Solving by completing the square:

$$x^2 - 6x = 0$$
$$x^2 - 6x + 9 = 0 + 9$$
$$(x-3)^2 = 9$$
$$x - 3 = \pm\sqrt{9}$$
$$x - 3 = -3, 3$$
$$x = 0, 6$$

61. **a.** Solving by factoring:

$$x^2 + 2x = 35$$
$$x^2 + 2x - 35 = 0$$
$$(x+7)(x-5) = 0$$
$$x = -7, 5$$

b. Solving by completing the square:

$$x^2 + 2x = 35$$
$$x^2 + 2x + 1 = 35 + 1$$
$$(x+1)^2 = 36$$
$$x + 1 = \pm\sqrt{36}$$
$$x + 1 = -6, 6$$
$$x = -7, 5$$

63. Substituting:

$$x^2 - 6x - 7 = \left(-3+\sqrt{2}\right)^2 - 6\left(-3+\sqrt{2}\right) - 7 = 9 - 6\sqrt{2} + 2 + 18 - 6\sqrt{2} - 7 = 22 - 12\sqrt{2}$$

No, $x = -3+\sqrt{2}$ is not a solution to the equation.

65. **a.** Solving the equation:

$$5x - 7 = 0$$
$$5x = 7$$
$$x = \frac{7}{5}$$

b. Solving the equation:

$$5x - 7 = 8$$
$$5x = 15$$
$$x = 3$$

c. Solving the equation:

$$(5x-7)^2 = 8$$
$$5x - 7 = \pm\sqrt{8}$$
$$5x - 7 = \pm 2\sqrt{2}$$
$$5x = 7 \pm 2\sqrt{2}$$
$$x = \frac{7 \pm 2\sqrt{2}}{5}$$

d. Solving the equation:

$$\sqrt{5x-7} = 8$$
$$\left(\sqrt{5x-7}\right)^2 = (8)^2$$
$$5x - 7 = 64$$
$$5x = 71$$
$$x = \frac{71}{5}$$

e. Solving the equation:

$$\frac{5}{2} - \frac{7}{2x} = \frac{4}{x}$$
$$2x\left(\frac{5}{2} - \frac{7}{2x}\right) = 2x\left(\frac{4}{x}\right)$$
$$5x - 7 = 8$$
$$5x = 15$$
$$x = 3$$

67. The other two sides are $\frac{\sqrt{3}}{2}$ inch, 1 inch.

69. The hypotenuse is $\sqrt{2}$ inches.

71. Let x represent the horizontal distance. Using the Pythagorean theorem:

$$\begin{aligned} x^2+120^2&=790^2\\ x^2+14400&=624100\\ x^2&=609700\\ x&=\sqrt{609700}\approx 781 \text{ feet}\end{aligned}$$

73. Solving for r:

$$\begin{aligned} 3456&=3000(1+r)^2\\ (1+r)^2&=1.152\\ 1+r&=\sqrt{1.152}\\ r&=\sqrt{1.152}-1\approx 0.073\end{aligned}$$

The annual interest rate is 7.3%.

75. Its length is $20\sqrt{2}\approx 28$ feet.

77. Simplifying: $49-4(6)(-5)=49+120=169$

79. Simplifying: $(-27)^2-4(0.1)(1,700)=729-680=49$

81. Simplifying: $-7+\frac{169}{12}=-\frac{84}{12}+\frac{169}{12}=\frac{85}{12}$

83. Factoring: $27t^3-8=(3t-2)(9t^2+6t+4)$

7.2 The Quadratic Formula

1. Solving the equation:

$$\begin{aligned} x^2+5x+6&=0\\ (x+3)(x+2)&=0\\ x&=-3,-2\end{aligned}$$

3. Solving the equation:

$$a^2-4a+1=0$$

$$a=\frac{4\pm\sqrt{16-4}}{2}=\frac{4\pm\sqrt{12}}{2}=\frac{4\pm2\sqrt{3}}{2}=2\pm\sqrt{3}$$

5. Solving the equation:

$$\begin{aligned} \frac{1}{6}x^2-\frac{1}{2}x+\frac{1}{3}&=0\\ x^2-3x+2&=0\\ (x-1)(x-2)&=0\\ x&=1,2\end{aligned}$$

7. Solving the equation:

$$\frac{x^2}{2}+1=\frac{2x}{3}$$
$$3x^2+6=4x$$
$$3x^2-4x+6=0$$
$$x=\frac{4\pm\sqrt{16-72}}{6}=\frac{4\pm\sqrt{-56}}{6}=\frac{4\pm 2i\sqrt{14}}{6}=\frac{2\pm i\sqrt{14}}{3}$$

9. Solving the equation:

$$y^2-5y=0$$
$$y(y-5)=0$$
$$y=0,5$$

11. Solving the equation:

$$30x^2+40x=0$$
$$10x(3x+4)=0$$
$$x=-\frac{4}{3},0$$

13. Solving the equation:

$$\frac{2t^2}{3}-t=-\frac{1}{6}$$
$$4t^2-6t=-1$$
$$4t^2-6t+1=0$$
$$t=\frac{6\pm\sqrt{36-16}}{8}=\frac{6\pm\sqrt{20}}{8}=\frac{6\pm 2\sqrt{5}}{8}=\frac{3\pm\sqrt{5}}{4}$$

15. Solving the equation:

$$0.01x^2+0.06x-0.08=0$$
$$x^2+6x-8=0$$
$$x=\frac{-6\pm\sqrt{36+32}}{2}=\frac{-6\pm\sqrt{68}}{2}=\frac{-6\pm 2\sqrt{17}}{2}=-3\pm\sqrt{17}$$

17. Solving the equation:

$$2x+3=-2x^2$$
$$2x^2+2x+3=0$$
$$x=\frac{-2\pm\sqrt{4-24}}{4}=\frac{-2\pm\sqrt{-20}}{4}=\frac{-2\pm 2i\sqrt{5}}{4}=\frac{-1\pm i\sqrt{5}}{2}$$

19. Solving the equation:

$$100x^2-200x+100=0$$
$$100\left(x^2-2x+1\right)=0$$
$$100(x-1)^2=0$$
$$x=1$$

21. Solving the equation:

$$\frac{1}{2}r^2 = \frac{1}{6}r - \frac{2}{3}$$
$$3r^2 = r - 4$$
$$3r^2 - r + 4 = 0$$
$$r = \frac{1 \pm \sqrt{1-48}}{6} = \frac{1 \pm \sqrt{-47}}{6} = \frac{1 \pm i\sqrt{47}}{6}$$

23. Solving the equation:

$$(x-3)(x-5) = 1$$
$$x^2 - 8x + 15 = 1$$
$$x^2 - 8x + 14 = 0$$
$$x = \frac{8 \pm \sqrt{64-56}}{2} = \frac{8 \pm \sqrt{8}}{2} = \frac{8 \pm 2\sqrt{2}}{2} = 4 \pm \sqrt{2}$$

25. Solving the equation:

$$(x+3)^2 + (x-8)(x-1) = 16$$
$$x^2 + 6x + 9 + x^2 - 9x + 8 = 16$$
$$2x^2 - 3x + 1 = 0$$
$$(2x-1)(x-1) = 0$$
$$x = \frac{1}{2}, 1$$

27. Solving the equation:

$$\frac{x^2}{3} - \frac{5x}{6} = \frac{1}{2}$$
$$2x^2 - 5x = 3$$
$$2x^2 - 5x - 3 = 0$$
$$(2x+1)(x-3) = 0$$
$$x = -\frac{1}{2}, 3$$

29. Solving the equation:

$$\frac{1}{x+1} - \frac{1}{x} = \frac{1}{2}$$
$$2x(x+1)\left(\frac{1}{x+1} - \frac{1}{x}\right) = 2x(x+1) \bullet \frac{1}{2}$$
$$2x - (2x+2) = x^2 + x$$
$$2x - 2x - 2 = x^2 + x$$
$$x^2 + x + 2 = 0$$
$$x = \frac{-1 \pm \sqrt{1-8}}{2} = \frac{-1 \pm \sqrt{-7}}{2} = \frac{-1 \pm i\sqrt{7}}{2}$$

31. Solving the equation:

$$\frac{1}{y-1}+\frac{1}{y+1}=1$$

$$(y+1)(y-1)\left(\frac{1}{y-1}+\frac{1}{y+1}\right)=(y+1)(y-1)\bullet 1$$

$$y+1+y-1=y^2-1$$

$$2y=y^2-1$$

$$y^2-2y-1=0$$

$$y=\frac{2\pm\sqrt{4+4}}{2}=\frac{2\pm\sqrt{8}}{2}=\frac{2\pm 2\sqrt{2}}{2}=1\pm\sqrt{2}$$

33. Solving the equation:

$$\frac{1}{x+2}+\frac{1}{x+3}=1$$

$$(x+2)(x+3)\left(\frac{1}{x+2}+\frac{1}{x+3}\right)=(x+2)(x+3)\bullet 1$$

$$x+3+x+2=x^2+5x+6$$

$$2x+5=x^2+5x+6$$

$$x^2+3x+1=0$$

$$x=\frac{-3\pm\sqrt{9-4}}{2}=\frac{-3\pm\sqrt{5}}{2}$$

35. Solving the equation:

$$\frac{6}{r^2-1}-\frac{1}{2}=\frac{1}{r+1}$$

$$2(r+1)(r-1)\left(\frac{6}{(r+1)(r-1)}-\frac{1}{2}\right)=2(r+1)(r-1)\bullet\frac{1}{r+1}$$

$$12-(r^2-1)=2r-2$$

$$12-r^2+1=2r-2$$

$$r^2+2r-15=0$$

$$(r+5)(r-3)=0$$

$$r=-5,3$$

37. Solving the equation:

$$x^3-8=0$$

$$(x-2)(x^2+2x+4)=0$$

$$x=2 \quad \text{or} \quad x=\frac{-2\pm\sqrt{4-16}}{2}=\frac{-2\pm\sqrt{-12}}{2}=\frac{-2\pm 2i\sqrt{3}}{2}=-1\pm i\sqrt{3}$$

$$x=2,-1\pm i\sqrt{3}$$

39. Solving the equation:

$$8a^3+27=0$$
$$(2a+3)(4a^2-6a+9)=0$$
$$a=-\frac{3}{2} \quad \text{or} \quad a=\frac{6\pm\sqrt{36-144}}{8}=\frac{6\pm\sqrt{-108}}{8}=\frac{6\pm 6i\sqrt{3}}{8}=\frac{3\pm 3i\sqrt{3}}{4}$$
$$a=-\frac{3}{2},\frac{3\pm 3i\sqrt{3}}{4}$$

41. Solving the equation:

$$125t^3-1=0$$
$$(5t-1)(25t^2+5t+1)=0$$
$$t=\frac{1}{5} \quad \text{or} \quad t=\frac{-5\pm\sqrt{25-100}}{50}=\frac{-5\pm\sqrt{-75}}{50}=\frac{-5\pm 5i\sqrt{3}}{50}=\frac{-1\pm i\sqrt{3}}{10}$$
$$t=\frac{1}{5},\frac{-1\pm i\sqrt{3}}{10}$$

43. Solving the equation:

$$2x^3+2x^2+3x=0$$
$$x(2x^2+2x+3)=0$$
$$x=0 \quad \text{or} \quad x=\frac{-2\pm\sqrt{4-24}}{4}=\frac{-2\pm\sqrt{-20}}{4}=\frac{-2\pm 2i\sqrt{5}}{4}=\frac{-1\pm i\sqrt{5}}{2}$$
$$x=0,\frac{-1\pm i\sqrt{5}}{2}$$

45. Solving the equation:

$$3y^4=6y^3-6y^2$$
$$3y^4-6y^3+6y^2=0$$
$$3y^2(y^2-2y+2)=0$$
$$y=0 \quad \text{or} \quad y=\frac{2\pm\sqrt{4-8}}{2}=\frac{2\pm\sqrt{-4}}{2}=\frac{2\pm 2i}{2}=1\pm i$$
$$y=0,1\pm i$$

47. Solving the equation:

$$6t^5+4t^4=-2t^3$$
$$6t^5+4t^4+2t^3=0$$
$$2t^3(3t^2+2t+1)=0$$
$$t=0 \quad \text{or} \quad t=\frac{-2\pm\sqrt{4-12}}{6}=\frac{-2\pm\sqrt{-8}}{6}=\frac{-2\pm 2i\sqrt{2}}{6}=\frac{-1\pm i\sqrt{2}}{3}$$
$$t=0,\frac{-1\pm i\sqrt{2}}{3}$$

49. The expressions from **a** and **b** are equivalent, since: $\frac{6+2\sqrt{3}}{4}=\frac{2\left(3+\sqrt{3}\right)}{4}=\frac{3+\sqrt{3}}{2}$

51. **a.** Solving by factoring:

$$\begin{aligned}3x^2-5x&=0\\ x(3x-5)&=0\\ x&=0,\frac{5}{3}\end{aligned}$$

b. Using the quadratic formula: $x=\frac{5\pm\sqrt{(-5)^2-4(3)(0)}}{2(3)}=\frac{5\pm\sqrt{25-0}}{6}=\frac{5\pm5}{6}=0,\frac{5}{3}$

53. No, it cannot be solved by factoring. Using the quadratic formula:

$$x=\frac{4\pm\sqrt{(-4)^2-4(1)(7)}}{2(1)}=\frac{4\pm\sqrt{16-28}}{2}=\frac{4\pm\sqrt{-12}}{2}=\frac{4\pm2i\sqrt{3}}{2}=2\pm i\sqrt{3}$$

55. Substituting: $x^2+2x=(-1+i)^2+2(-1+i)=1-2i+i^2-2+2i=1-2i-1-2+2i=-2$

Yes, $x=-1+i$ is a solution to the equation.

57. Substituting $s=74$:

$$\begin{aligned}5t+16t^2&=74\\ 16t^2+5t-74&=0\\ (16t+37)(t-2)&=0\\ t&=2 \quad \left(t=-\frac{37}{16}\text{ is impossible}\right)\end{aligned}$$

It will take 2 seconds for the object to fall 74 feet.

59. Since profit is revenue minus the cost, the equation is:

$$\begin{aligned}100x-0.5x^2-(60x+300)&=300\\ 100x-0.5x^2-60x-300&=300\\ -0.5x^2+40x-600&=0\\ x^2-80x+1{,}200&=0\\ (x-20)(x-60)&=0\\ x&=20,60\end{aligned}$$

The weekly profit is \$300 if 20 items or 60 items are sold.

61. Evaluating b^2-4ac: $b^2-4ac=(-3)^2-4(1)(-40)=9+160=169$

63. Evaluating b^2-4ac: $b^2-4ac=12^2-4(4)(9)=144-144=0$

65. Solving the equation:

$$\begin{aligned}k^2-144&=0\\ (k+12)(k-12)&=0\\ k&=-12,12\end{aligned}$$

67. Multiplying: $(x-3)(x+2)=x^2+2x-3x-6=x^2-x-6$

69. Multiplying:

$$\begin{aligned}(x-3)(x-3)(x+2) &= (x^2-6x+9)(x+2)\\ &= x^3-6x^2+9x+2x^2-12x+18\\ &= x^3-4x^2-3x+18\end{aligned}$$

7.3 Additional Items Involving Solutions to Equations

1. Computing the discriminant: $D=(-6)^2-4(1)(5)=36-20=16$.
The equation will have two rational solutions.
3. First write the equation as $4x^2-4x+1=0$. Computing the discriminant:
$$D=(-4)^2-4(4)(1)=16-16=0$$
The equation will have one rational solution.
5. Computing the discriminant: $D=1^2-4(1)(-1)=1+4=5$.
The equation will have two irrational solutions.
7. First write the equation as $2y^2-3y-1=0$. Computing the discriminant:
$$D=(-3)^2-4(2)(-1)=9+8=17$$
The equation will have two irrational solutions.
9. Computing the discriminant: $D=0^2-4(1)(-9)=36$.
The equation will have two rational solutions.
11. First write the equation as $5a^2-4a-5=0$. Computing the discriminant:
$$D=(-4)^2-4(5)(-5)=16+100=116$$
The equation will have two irrational solutions.
13. Setting the discriminant equal to 0:
$$\begin{aligned}(-k)^2-4(1)(25)&=0\\ k^2-100&=0\\ k^2&=100\\ k&=\pm 10\end{aligned}$$
15. First write the equation as $x^2-kx+36=0$. Setting the discriminant equal to 0:
$$\begin{aligned}(-k)^2-4(1)(36)&=0\\ k^2-144&=0\\ k^2&=144\\ k&=\pm 12\end{aligned}$$
17. Setting the discriminant equal to 0:
$$\begin{aligned}(-12)^2-4(4)(k)&=0\\ 144-16k&=0\\ 16k&=144\\ k&=9\end{aligned}$$

19. First write the equation as $kx^2 - 40x - 25 = 0$. Setting the discriminant equal to 0:

$$(-40)^2 - 4(k)(-25) = 0$$
$$1600 + 100k = 0$$
$$100k = -1600$$
$$k = -16$$

21. Setting the discriminant equal to 0:

$$(-k)^2 - 4(3)(2) = 0$$
$$k^2 - 24 = 0$$
$$k^2 = 24$$
$$k = \pm\sqrt{24} = \pm 2\sqrt{6}$$

23. Writing the equation:

$$(x-5)(x-2) = 0$$
$$x^2 - 7x + 10 = 0$$

25. Writing the equation:

$$(t+3)(t-6) = 0$$
$$t^2 - 3t - 18 = 0$$

27. Writing the equation:

$$(y-2)(y+2)(y-4) = 0$$
$$(y^2 - 4)(y-4) = 0$$
$$y^3 - 4y^2 - 4y + 16 = 0$$

29. Writing the equation:

$$(2x-1)(x-3) = 0$$
$$2x^2 - 7x + 3 = 0$$

31. Writing the equation:

$$(4t+3)(t-3) = 0$$
$$4t^2 - 9t - 9 = 0$$

33. Writing the equation:

$$(x-3)(x+3)(6x-5) = 0$$
$$(x^2 - 9)(6x-5) = 0$$
$$6x^3 - 5x^2 - 54x + 45 = 0$$

35. Writing the equation:

$$(2a+1)(5a-3) = 0$$
$$10a^2 - a - 3 = 0$$

37. Writing the equation:

$$(3x+2)(3x-2)(x-1) = 0$$
$$(9x^2 - 4)(x-1) = 0$$
$$9x^3 - 9x^2 - 4x + 4 = 0$$

39. Writing the equation:

$$(x-2)(x+2)(x-3)(x+3) = 0$$
$$(x^2 - 4)(x^2 - 9) = 0$$
$$x^4 - 13x^2 + 36 = 0$$

41. Writing the equation:

$$\left(x - \sqrt{7}\right)\left(x + \sqrt{7}\right) = 0$$
$$x^2 - 7 = 0$$

43. Writing the equation:

$$(x-5i)(x+5i) = 0$$
$$x^2 - 25i^2 = 0$$
$$x^2 + 25 = 0$$

45. Writing the equation:

$$(x-1-i)(x-1+i) = 0$$
$$(x-1)^2 - i^2 = 0$$
$$x^2 - 2x + 1 + 1 = 0$$
$$x^2 - 2x + 2 = 0$$

47. Writing the equation:

$$(x+2+3i)(x+2-3i) = 0$$
$$(x+2)^2 - 9i^2 = 0$$
$$x^2 + 4x + 4 + 9 = 0$$
$$x^2 + 4x + 13 = 0$$

49. Writing the equation:

$$(x-3)(x+5)^2=0$$
$$(x-3)(x^2+10x+25)=0$$
$$x^3+7x^2-5x-75=0$$

51. Writing the equation:

$$(x-3)^2(x+3)^2=0$$
$$(x^2-6x+9)(x^2+6x+9)=0$$
$$x^4-18x^2+81=0$$

53. First use long division:

$$\begin{array}{r} x^2+3x+2 \\ x+3\overline{)x^3+6x^2+11x+6} \\ \underline{x^3+3x^2} \qquad\qquad \\ 3x^2+11x \qquad \\ \underline{3x^2+\ 9x} \qquad \\ 2x+6 \\ \underline{2x+6} \\ 0 \end{array}$$

Thus $x^3+6x^2+11x+6=(x+3)(x^2+3x+2)=(x+3)(x+2)(x+1)$. The solutions are –3,–2,–1.

55. First use long division:

$$\begin{array}{r} y^2+2y-8 \\ y+3\overline{)y^3+5y^2-2y-24} \\ \underline{y^3+3y^2} \qquad\qquad \\ 2y^2-2y \qquad \\ \underline{2y^2+6y} \qquad \\ -8y-24 \\ \underline{-8y-24} \\ 0 \end{array}$$

Thus $y^3+5y^2-2y-24=(y+3)(y^2+2y-8)=(y+3)(y+4)(y-2)$. The solutions are –4,–3,2.

57. Write the equation as $x^3-5x^2+8x-6=0$. Using long division:

$$\begin{array}{r} x^2-2x+2 \\ x-3\overline{)x^3-5x^2+8x-6} \\ \underline{x^3-3x^2} \qquad\qquad \\ -2x^2+8x \qquad \\ \underline{-2x^2+6x} \qquad \\ 2x-6 \\ \underline{2x-6} \\ 0 \end{array}$$

Now solving the equation:

$$x^2 - 2x + 2 = 0$$

$$x = \frac{2 \pm \sqrt{4-8}}{2} = \frac{2 \pm \sqrt{-4}}{2} = \frac{2 \pm 2i}{2} = 1 \pm i$$

The other solutions are $1 \pm i$.

59. Write the equation as $t^3 - 13t^2 + 65t - 125 = 0$. Using long division:

$$\begin{array}{r} t^2 - 8t + 25 \\ t-5 \overline{)\, t^3 - 13t^2 + 65t - 125} \\ \underline{t^3 - 5t^2} \qquad\qquad\quad \\ -8t^2 + 65t \qquad\quad \\ \underline{-8t^2 + 40t} \qquad\quad \\ 25t - 125 \\ \underline{25t - 125} \\ 0 \end{array}$$

Now solving the equation:

$$t^2 - 8t + 25 = 0$$

$$t = \frac{8 \pm \sqrt{64-100}}{2} = \frac{8 \pm \sqrt{-36}}{2} = \frac{8 \pm 6i}{2} = 4 \pm 3i$$

The solutions are $5, 4 \pm 3i$.

61. Simplifying: $(x+3)^2 - 2(x+3) - 8 = x^2 + 6x + 9 - 2x - 6 - 8 = x^2 + 4x - 5$

63. Simplifying: $(2a-3)^2 - 9(2a-3) + 20 = 4a^2 - 12a + 9 - 18a + 27 + 20 = 4a^2 - 30a + 56$

65. Simplifying:

$$\begin{aligned} 2(4a+2)^2 - 3(4a+2) - 20 &= 2(16a^2 + 16a + 4) - 3(4a+2) - 20 \\ &= 32a^2 + 32a + 8 - 12a - 6 - 20 \\ &= 32a^2 + 20a - 18 \end{aligned}$$

67. Solving the equation:

$$x^2 = \frac{1}{4}$$

$$x = \pm\sqrt{\frac{1}{4}} = \pm\frac{1}{2}$$

69. Since $\sqrt{x} \geq 0$, this equation has no solution.

71. Solving the equation:

$$x + 3 = 4$$

$$x = 1$$

73. Solving the equation:

$$y^2 - 2y - 8 = 0$$

$$(y+2)(y-4) = 0$$

$$y = -2, 4$$

75. Solving the equation:

$$4y^2 + 7y - 2 = 0$$

$$(4y-1)(y+2) = 0$$

$$y = -2, \frac{1}{4}$$

7.4 More Equations

1. Solving the equation:

$$\begin{aligned}(x-3)^2+3(x-3)+2&=0\\(x-3+2)(x-3+1)&=0\\(x-1)(x-2)&=0\\x&=1,2\end{aligned}$$

3. Solving the equation:

$$\begin{aligned}2(x+4)^2+5(x+4)-12&=0\\\left[2(x+4)-3\right]\left[(x+4)+4\right]&=0\\(2x+5)(x+8)&=0\\x&=-8,-\frac{5}{2}\end{aligned}$$

5. Solving the equation:

$$\begin{aligned}x^4-6x^2-27&=0\\\left(x^2-9\right)\left(x^2+3\right)&=0\\x^2&=9,-3\\x&=\pm 3,\pm i\sqrt{3}\end{aligned}$$

7. Solving the equation:

$$\begin{aligned}x^4+9x^2&=-20\\x^4+9x^2+20&=0\\\left(x^2+4\right)\left(x^2+5\right)&=0\\x^2&=-4,-5\\x&=\pm 2i,\pm i\sqrt{5}\end{aligned}$$

9. Solving the equation:

$$\begin{aligned}(2a-3)^2-9(2a-3)&=-20\\(2a-3)^2-9(2a-3)+20&=0\\(2a-3-4)(2a-3-5)&=0\\(2a-7)(2a-8)&=0\\a&=\frac{7}{2},4\end{aligned}$$

11. Solving the equation:

$$\begin{aligned}2(4a+2)^2&=3(4a+2)+20\\2(4a+2)^2-3(4a+2)-20&=0\\\left[2(4a+2)+5\right]\left[(4a+2)-4\right]&=0\\(8a+9)(4a-2)&=0\\a&=-\frac{9}{8},\frac{1}{2}\end{aligned}$$

13. Solving the equation:

$$\begin{aligned} 6t^4 &= -t^2+5 \\ 6t^4+t^2-5 &= 0 \\ (6t^2-5)(t^2+1) &= 0 \\ t^2 &= \frac{5}{6},-1 \\ t &= \pm\sqrt{\frac{5}{6}} = \pm\frac{\sqrt{30}}{6},\pm i \end{aligned}$$

15. Solving the equation:

$$\begin{aligned} 9x^4-49 &= 0 \\ (3x^2-7)(3x^2+7) &= 0 \\ x^2 &= \frac{7}{3},-\frac{7}{3} \\ x &= \pm\sqrt{\frac{7}{3}},\pm\sqrt{-\frac{7}{3}} \\ t &= \pm\frac{\sqrt{21}}{3},\pm\frac{i\sqrt{21}}{3} \end{aligned}$$

17. Solving the equation:

$$\begin{aligned} x-7\sqrt{x}+10 &= 0 \\ (\sqrt{x}-5)(\sqrt{x}-2) &= 0 \\ \sqrt{x} &= 2,5 \\ x &= 4,25 \end{aligned}$$

Both values check in the original equation.

19. Solving the equation:

$$\begin{aligned} t-2\sqrt{t}-15 &= 0 \\ (\sqrt{t}-5)(\sqrt{t}+3) &= 0 \\ \sqrt{t} &= -3,5 \\ t &= 9,25 \end{aligned}$$

Only $t = 25$ checks in the original equation.

21. Solving the equation:

$$\begin{aligned} 6x+11\sqrt{x} &= 35 \\ 6x+11\sqrt{x}-35 &= 0 \\ (3\sqrt{x}-5)(2\sqrt{x}+7) &= 0 \\ \sqrt{x} &= \frac{5}{3},-\frac{7}{2} \\ x &= \frac{25}{9},\frac{49}{4} \end{aligned}$$

Only $x = \frac{25}{9}$ checks in the original equation.

23. Solving the equation:

$$\begin{aligned} (a-2)-11\sqrt{a-2}+30 &= 0 \\ (\sqrt{a-2}-6)(\sqrt{a-2}-5) &= 0 \\ \sqrt{a-2} &= 5,6 \\ a-2 &= 25,36 \\ a &= 27,38 \end{aligned}$$

25. Solving the equation:

$$\begin{aligned} (2x+1)-8\sqrt{2x+1}+15 &= 0 \\ (\sqrt{2x+1}-3)(\sqrt{2x+1}-5) &= 0 \\ \sqrt{2x+1} &= 3,5 \\ 2x+1 &= 9,25 \\ 2x &= 8,24 \\ x &= 4,12 \end{aligned}$$

27. Solving for t:

$$16t^2-vt-h=0$$

$$t = \frac{v\pm\sqrt{v^2-4(16)(-h)}}{32} = \frac{v\pm\sqrt{v^2+64h}}{32}$$

29. Solving for x:

$$kx^2+8x+4=0$$

$$x=\frac{-8\pm\sqrt{64-16k}}{2k}=\frac{-8\pm4\sqrt{4-k}}{2k}=\frac{-4\pm2\sqrt{4-k}}{k}$$

31. Solving for x:

$$x^2+2xy+y^2=0$$

$$x=\frac{-2y\pm\sqrt{4y^2-4y^2}}{2}=\frac{-2y}{2}=-y$$

33. Solving for t (note that $t>0$):

$$16t^2-8t-h=0$$

$$t=\frac{8+\sqrt{64+64h}}{32}=\frac{8+8\sqrt{1+h}}{32}=\frac{1+\sqrt{1+h}}{4}$$

35. **a.** Sketching the graph:

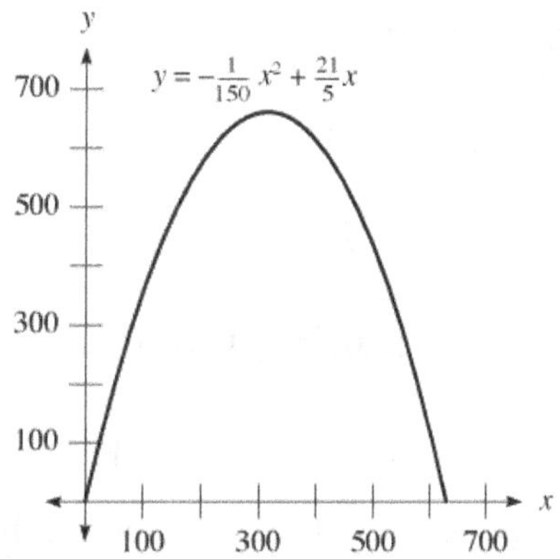

b. Finding the x-intercepts:

$$-\frac{1}{150}x^2+\frac{21}{5}x=0$$

$$-\frac{1}{150}x(x-630)=0$$

$$x=0,630$$

The width is 630 feet.

37. **a.** The equation is $l+2w=160$. **b.** The area is $A=-2w^2+160w$.

c. Completing the table:

w	0	10	20	30	40	50	60	70	80
A	0	1400	2400	3000	3200	3000	2400	1400	0

d. The maximum area is 3,200 square yards.

39. Evaluating when $x=1$: $y=3(1)^2-6(1)+1=3-6+1=-2$

41. Evaluating: $P(135)=-0.1(135)^2+27(135)-500=-1,822.5+3,645-500=1,322.5$

43. Solving the equation:

$$\begin{aligned} 0 &= a(80)^2 + 70 \\ 0 &= 6400a + 70 \\ 6400a &= -70 \\ a &= -\frac{7}{640} \end{aligned}$$

45. Solving the equation:

$$\begin{aligned} x^2 - 6x + 5 &= 0 \\ (x-1)(x-5) &= 0 \\ x &= 1,5 \end{aligned}$$

47. Solving the equation:

$$\begin{aligned} -x^2 - 2x + 3 &= 0 \\ x^2 + 2x - 3 &= 0 \\ (x+3)(x-1) &= 0 \\ x &= -3,1 \end{aligned}$$

49. Solving the equation:

$$2x^2 - 6x + 5 = 0$$

$$x = \frac{6 \pm \sqrt{(-6)^2 - 4(2)(5)}}{2(2)} = \frac{6 \pm \sqrt{36-40}}{4} = \frac{6 \pm \sqrt{-4}}{4} = \frac{6 \pm 2i}{4} = \frac{3 \pm i}{2} = \frac{3}{2} \pm \frac{1}{2}i$$

51. Completing the square: $x^2 - 6x + 9 = (x-3)^2$

53. Completing the square: $y^2 + 2y + 1 = (y+1)^2$

7.5 Graphing Parabolas

1. First complete the square: $y = x^2 + 2x - 3 = (x^2 + 2x + 1) - 1 - 3 = (x+1)^2 - 4$

The x-intercepts are –3,1 and the vertex is (–1,–4). Graphing the parabola:

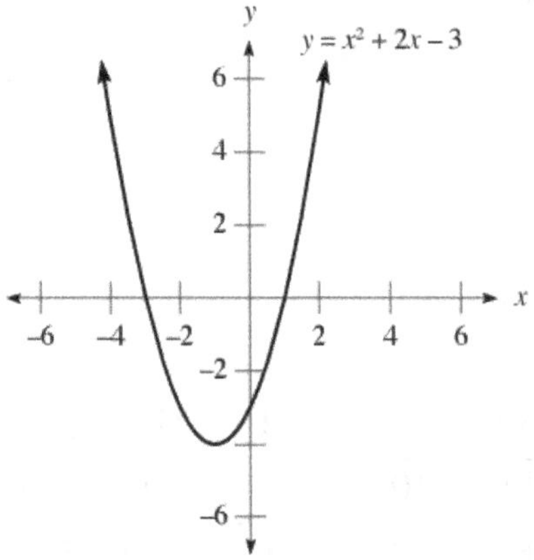

3. First complete the square: $y=-x^2-4x+5=-(x^2+4x+4)+4+5=-(x+2)^2+9$
The x-intercepts are –5,1 and the vertex is (–2,9). Graphing the parabola:

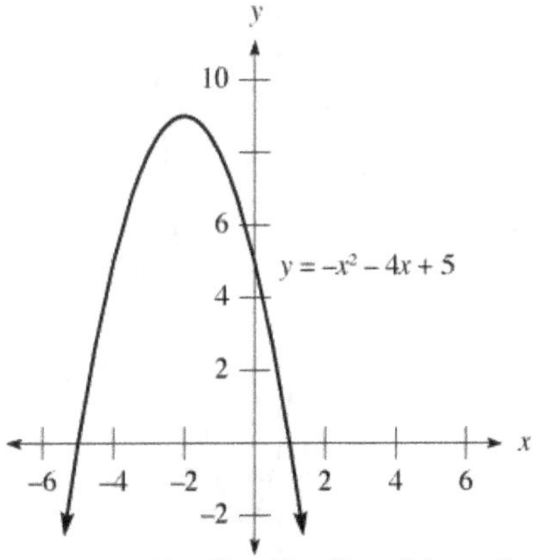

5. The x-intercepts are –1,1 and the vertex is (0,–1). Graphing the parabola:

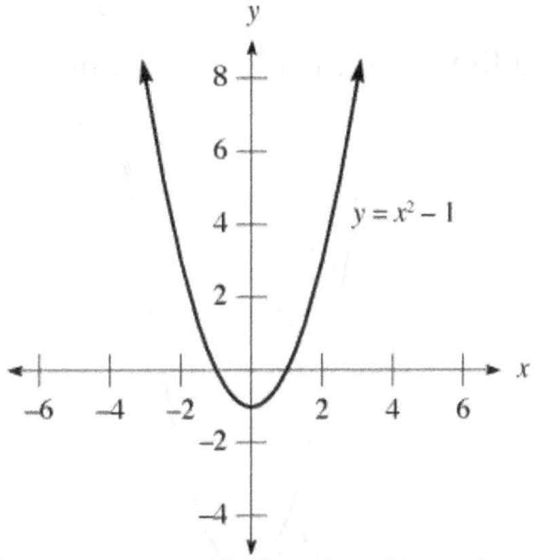

7. The x-intercepts are –3,3 and the vertex is (0,9). Graphing the parabola:

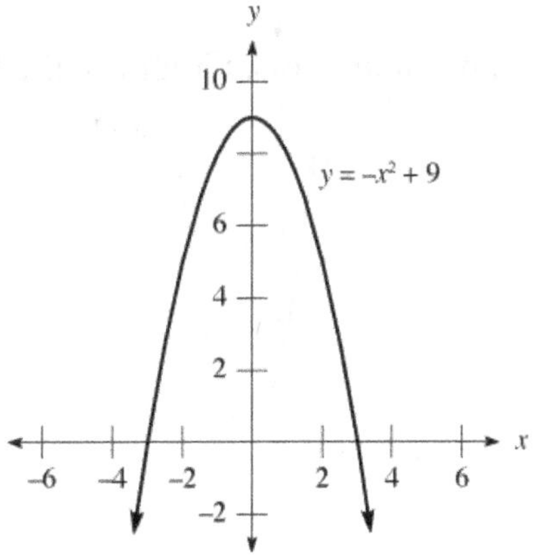

9. First complete the square: $y = 2x^2 - 4x - 6 = 2(x^2 - 2x + 1) - 2 - 6 = 2(x-1)^2 - 8$
The x-intercepts are $-1,3$ and the vertex is $(1,-8)$. Graphing the parabola:

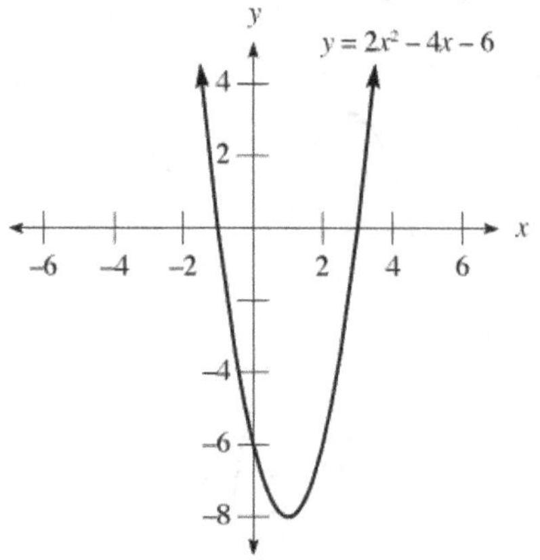

11. First complete the square: $y = x^2 - 2x - 4 = (x^2 - 2x + 1) - 1 - 4 = (x-1)^2 - 5$
The x-intercepts are $1 \pm \sqrt{5}$ and the vertex is $(1,-5)$. Graphing the parabola:

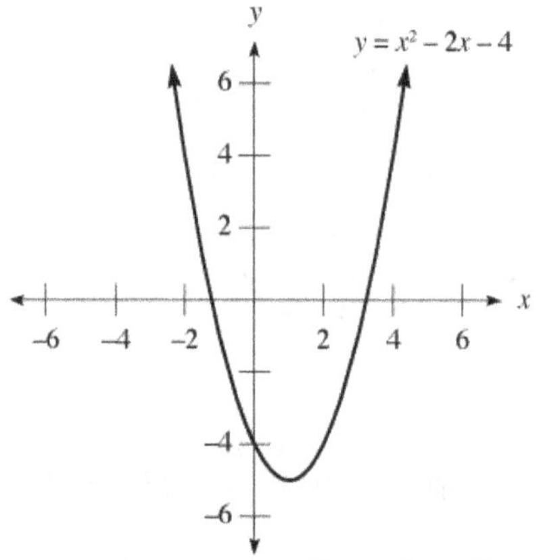

13. The vertex is $(1,3)$ and there are no x-intercepts. Graphing the parabola:

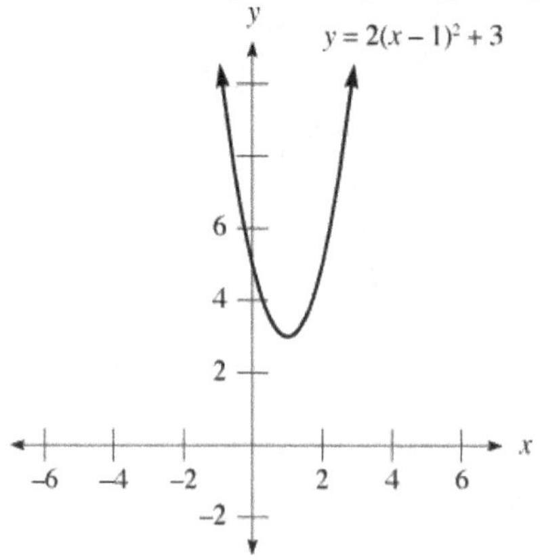

15. The vertex is (–2,4) and the x-intercepts are –4,0. Graphing the parabola:

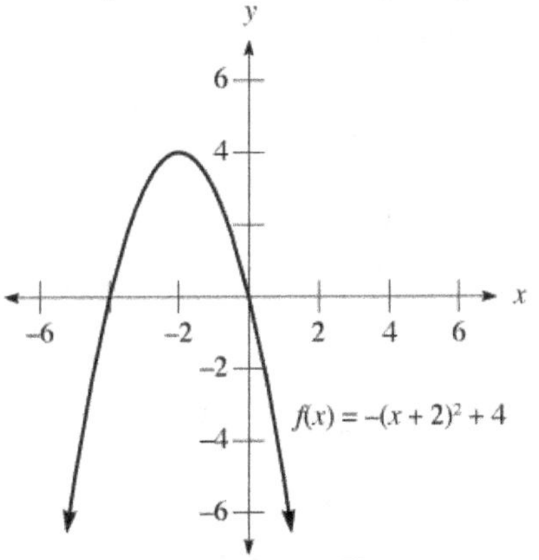

17. The vertex is (2,–4) and the x-intercepts are $2 \pm 2\sqrt{2}$. Graphing the parabola:

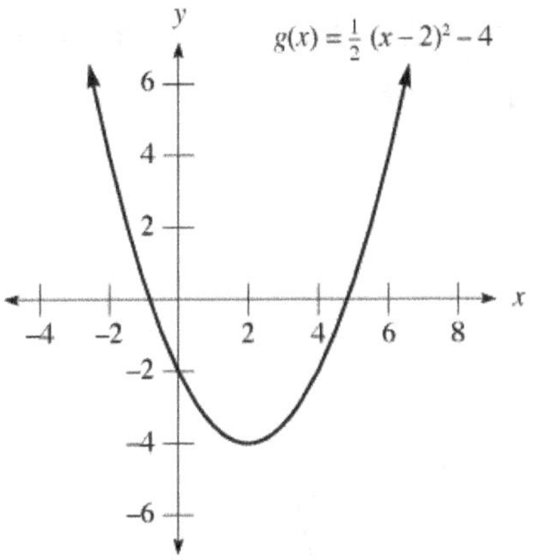

19. The vertex is (4,–1) and there are no x-intercepts. Graphing the parabola:

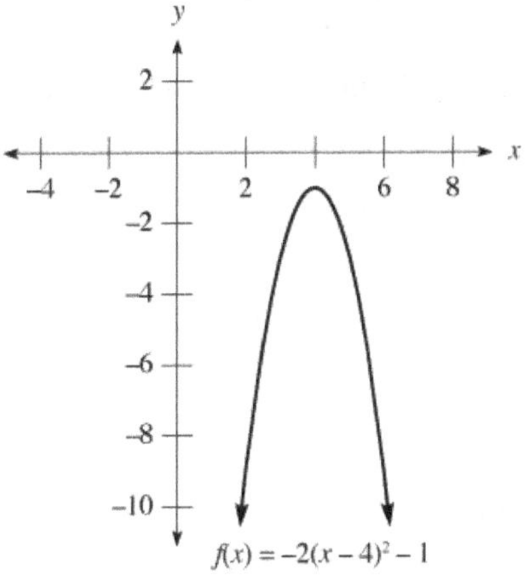

21. First complete the square: $y = x^2 - 4x - 4 = (x^2 - 4x + 4) - 4 - 4 = (x-2)^2 - 8$
The vertex is (2,–8). Graphing the parabola:

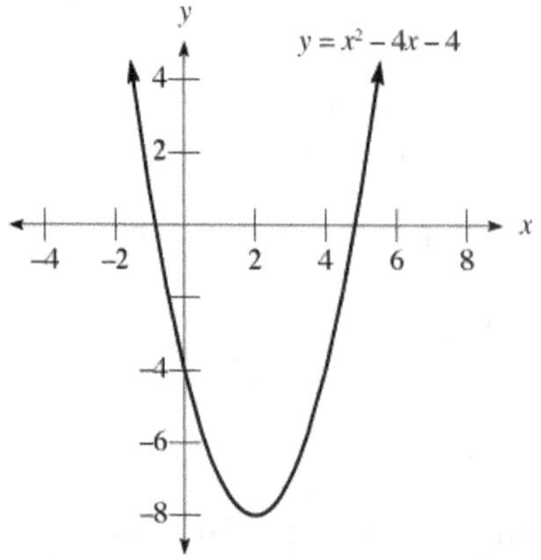

23. First complete the square: $y = -x^2 + 2x - 5 = -(x^2 - 2x + 1) + 1 - 5 = -(x-1)^2 - 4$
The vertex is (1,–4). Graphing the parabola:

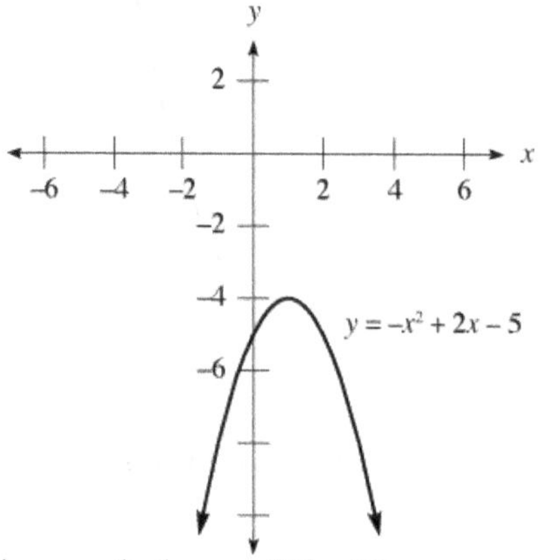

25. The vertex is (0,1). Graphing the parabola:

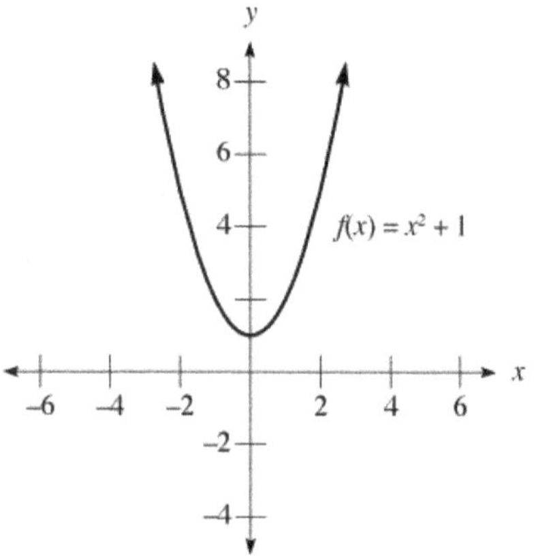

27. The vertex is (0,–3). Graphing the parabola:

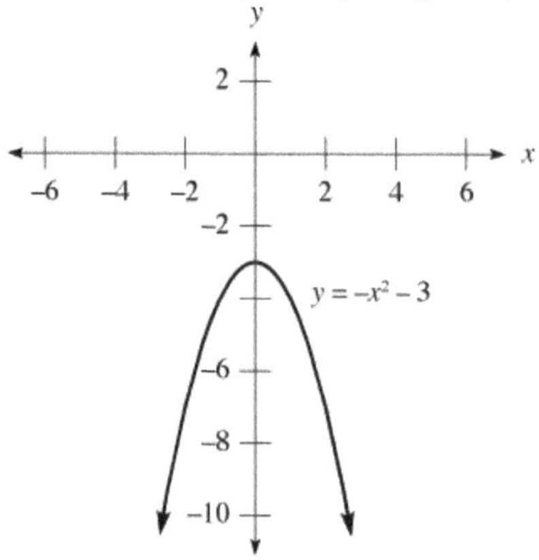

29. First complete the square: $g(x) = 3x^2 + 4x + 1 = 3\left(x^2 + \frac{4}{3}x + \frac{4}{9}\right) - \frac{4}{3} + 1 = 3\left(x + \frac{2}{3}\right)^2 - \frac{1}{3}$

The vertex is $\left(-\frac{2}{3}, -\frac{1}{3}\right)$. Graphing the parabola:

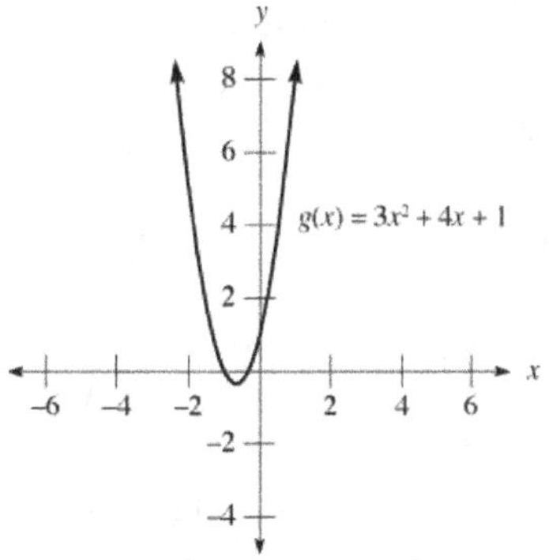

31. Completing the square: $y = x^2 - 6x + 5 = (x^2 - 6x + 9) - 9 + 5 = (x-3)^2 - 4$

The vertex is (3,–4), which is the lowest point on the graph.

33. Completing the square: $y = -x^2 + 2x + 8 = -(x^2 - 2x + 1) + 1 + 8 = -(x-1)^2 + 9$

The vertex is (1,9), which is the highest point on the graph.

35. Completing the square: $y = -x^2 + 4x + 12 = -(x^2 - 4x + 4) + 4 + 12 = -(x-2)^2 + 16$

The vertex is (2,16), which is the highest point on the graph.

37. Completing the square: $y = -x^2 - 8x = -(x^2 + 8x + 16) + 16 = -(x+4)^2 + 16$

The vertex is (–4,16), which is the highest point on the graph.

39. First complete the square:

$$\begin{aligned} P(x) &= -0.002x^2 + 3.5x - 800 \\ &= -0.002(x^2 - 1750x + 765{,}625) + 1{,}531.25 - 800 \\ &= -0.002(x - 875)^2 + 731.25 \end{aligned}$$

It must sell 875 patterns to obtain a maximum profit of \$731.25.

41. The ball is in her hand at times 0 sec and 2 sec.

Completing the square: $h(t) = -16t^2 + 32t = -16(t^2 - 2t + 1) + 16 = -16(t-1)^2 + 16$

The maximum height of the ball is 16 feet.

43. Completing the square:

$$R = xp = 1200p - 100p^2 = -100(p^2 - 12p + 36) + 3600 = -100(p-6)^2 + 3600$$

The price is \$6.00 and the maximum revenue is \$3,600. Sketching the graph:

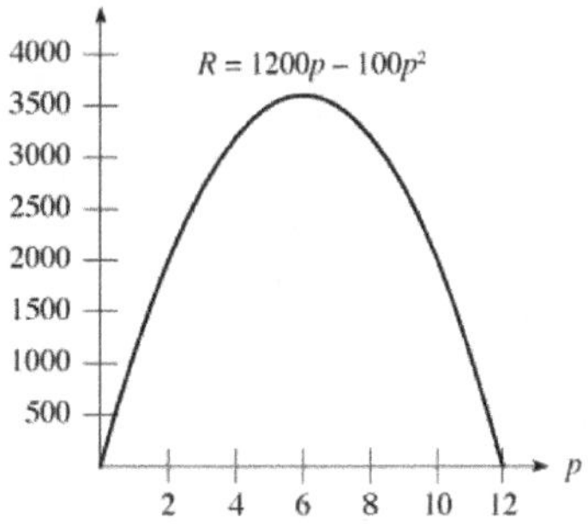

45. Completing the square:

$$R = xp = 1700p - 100p^2 = -100\left(p^2 - 17p + 72.25\right) + 7225 = -100\left(p - 8.5\right)^2 + 7225$$

The price is \$8.50 and the maximum revenue is \$7,225. Sketching the graph:

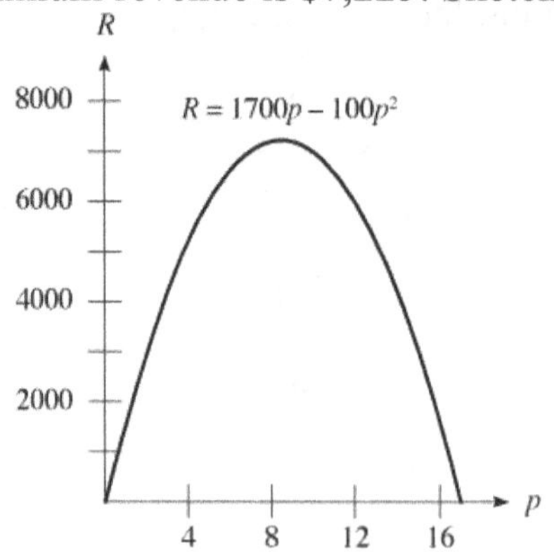

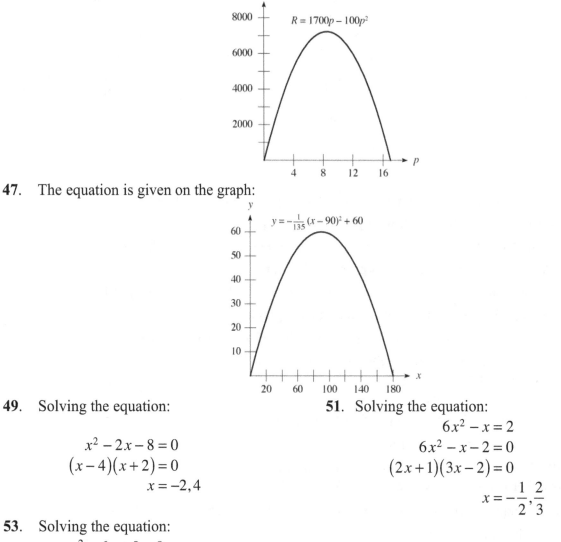

47. The equation is given on the graph:

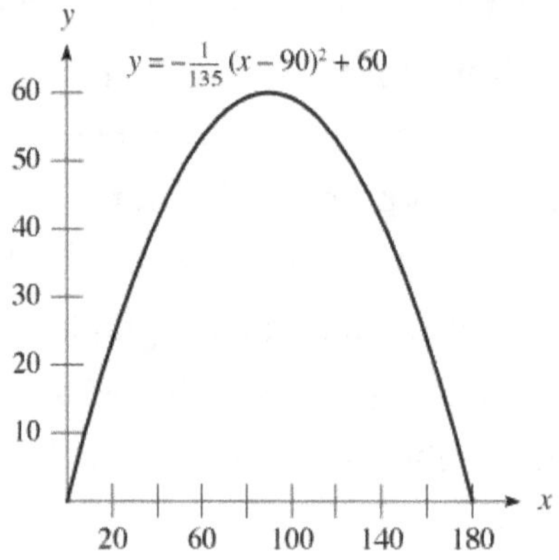

49. Solving the equation:

$$x^2 - 2x - 8 = 0$$
$$(x-4)(x+2) = 0$$
$$x = -2, 4$$

51. Solving the equation:

$$6x^2 - x = 2$$
$$6x^2 - x - 2 = 0$$
$$(2x+1)(3x-2) = 0$$
$$x = -\frac{1}{2}, \frac{2}{3}$$

53. Solving the equation:

$$x^2 - 6x + 9 = 0$$
$$(x-3)^2 = 0$$
$$x = 3$$

7.6 Quadratic Inequalities

1. Factoring the inequality:

$$x^2 + x - 6 > 0$$
$$(x+3)(x-2) > 0$$

Forming the sign chart:

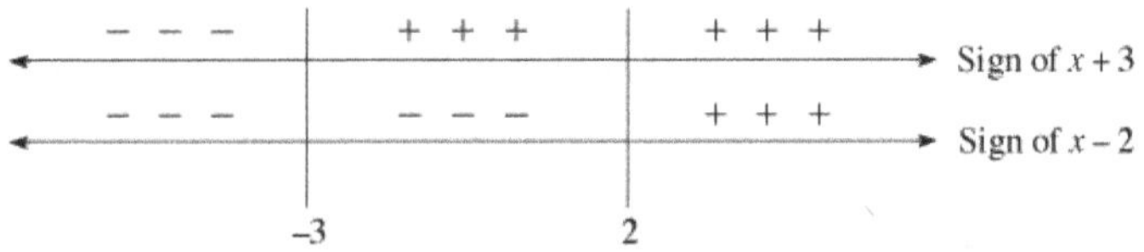

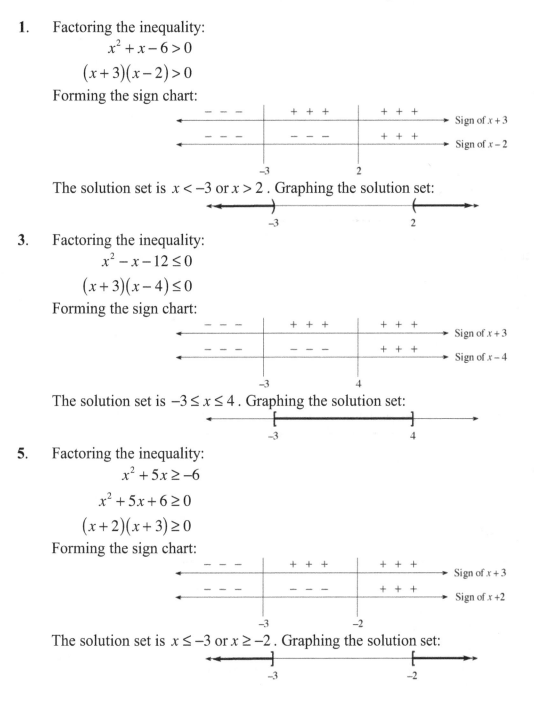

The solution set is $x < -3$ or $x > 2$. Graphing the solution set:

3. Factoring the inequality:

$$x^2 - x - 12 \le 0$$
$$(x+3)(x-4) \le 0$$

Forming the sign chart:

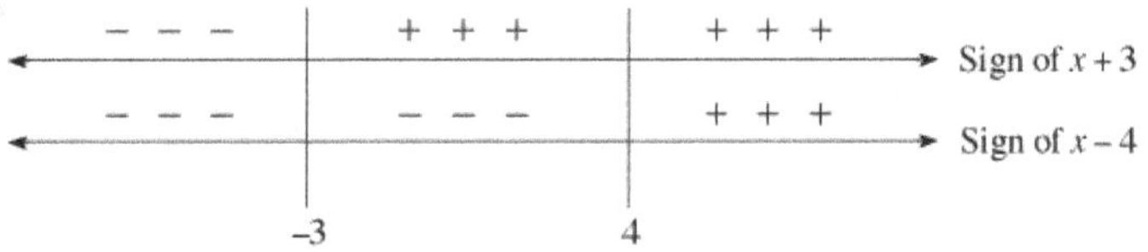

The solution set is $-3 \le x \le 4$. Graphing the solution set:

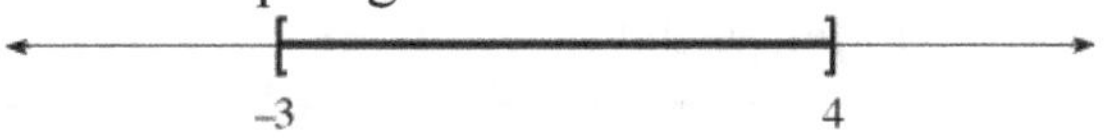

5. Factoring the inequality:

$$x^2 + 5x \ge -6$$
$$x^2 + 5x + 6 \ge 0$$
$$(x+2)(x+3) \ge 0$$

Forming the sign chart:

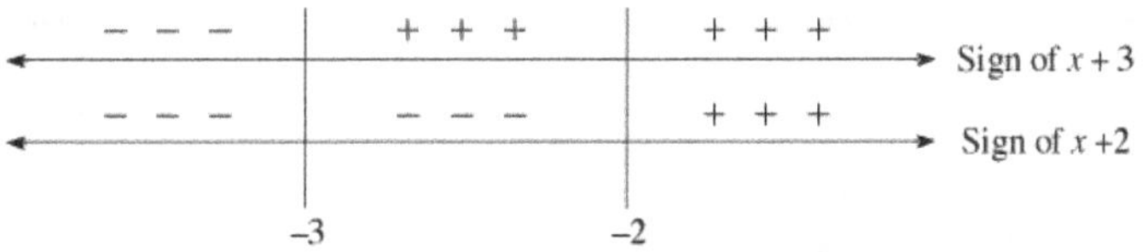

The solution set is $x \le -3$ or $x \ge -2$. Graphing the solution set:

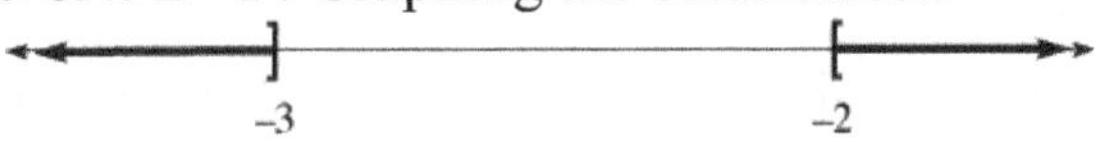

7. Factoring the inequality:

$$6x^2 < 5x - 1$$
$$6x^2 - 5x + 1 < 0$$
$$(3x-1)(2x-1) < 0$$

Forming the sign chart:

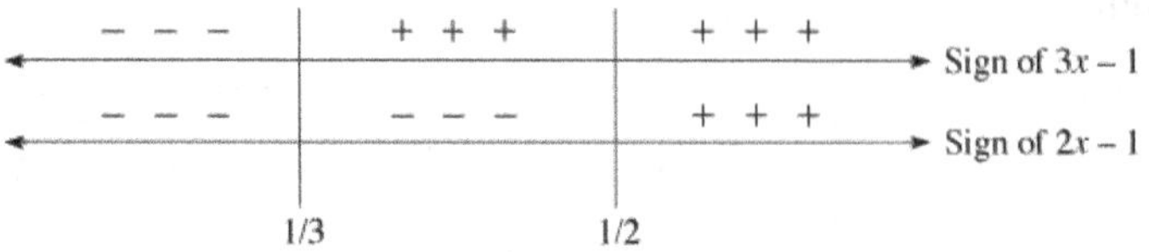

The solution set is $\frac{1}{3} < x < \frac{1}{2}$. Graphing the solution set:

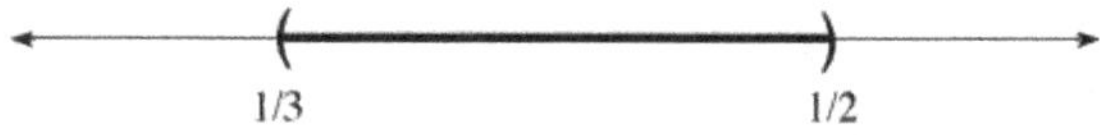

9. Factoring the inequality:

$$x^2 - 9 < 0$$
$$(x+3)(x-3) < 0$$

Forming the sign chart:

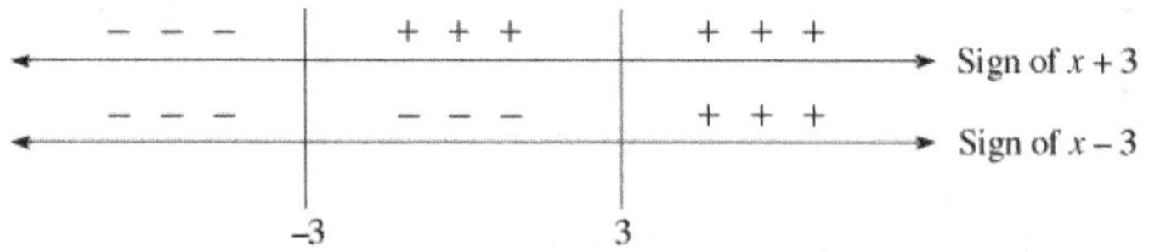

The solution set is $-3 < x < 3$. Graphing the solution set:

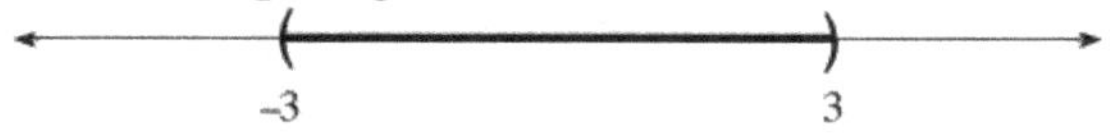

11. Factoring the inequality:

$$4x^2 - 9 \geq 0$$
$$(2x+3)(2x-3) \geq 0$$

Forming the sign chart:

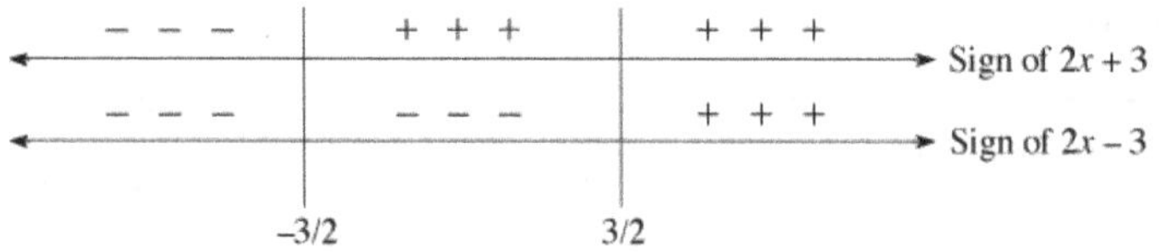

The solution set is $x \leq -\frac{3}{2}$ or $x \geq \frac{3}{2}$. Graphing the solution set:

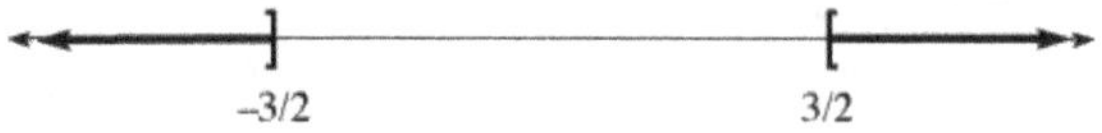

13. Factoring the inequality:

$$2x^2 - x - 3 < 0$$

$$(2x-3)(x+1) < 0$$

Forming the sign chart:

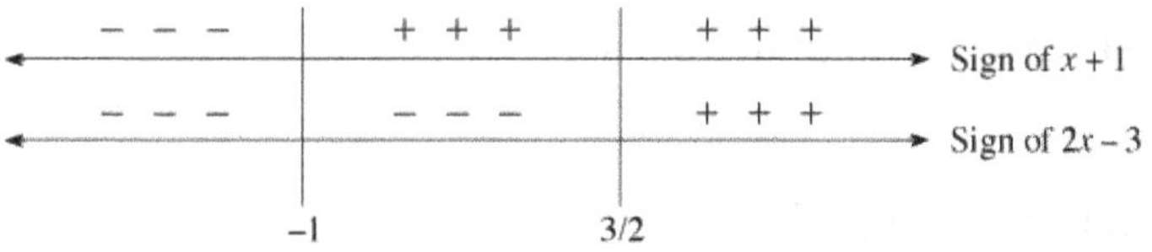

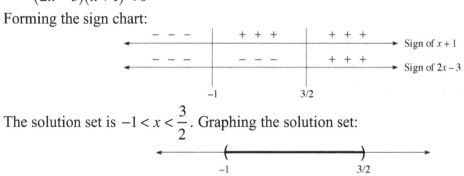

The solution set is $-1 < x < \frac{3}{2}$. Graphing the solution set:

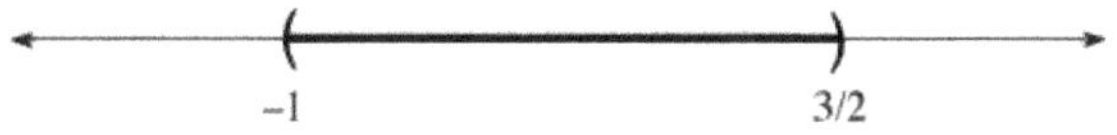

15. Factoring the inequality:

$$x^2 - 4x + 4 \geq 0$$

$$(x-2)^2 \geq 0$$

Since this inequality is always true, the solution set is all real numbers. Graphing the solution set:

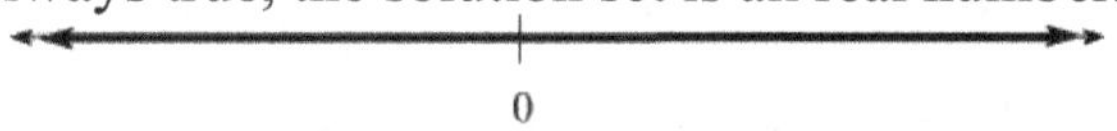

17. Factoring the inequality:

$$x^2 - 10x + 25 < 0$$

$$(x-5)^2 < 0$$

Since this inequality is never true, there is no solution.

19. Forming the sign chart:

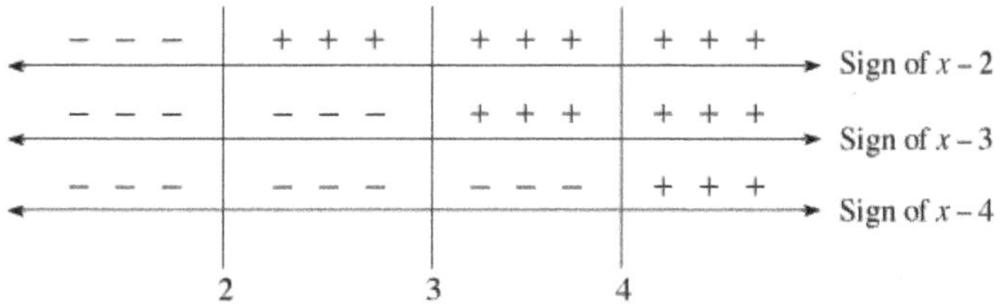

The solution set is $2 < x < 3$ or $x > 4$. Graphing the solution set:

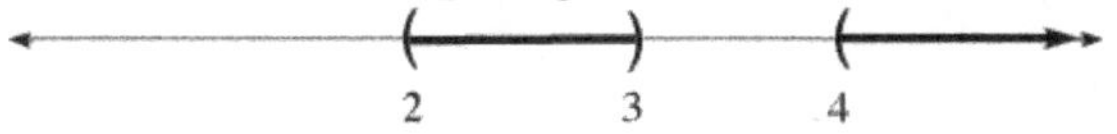

21. Forming the sign chart:

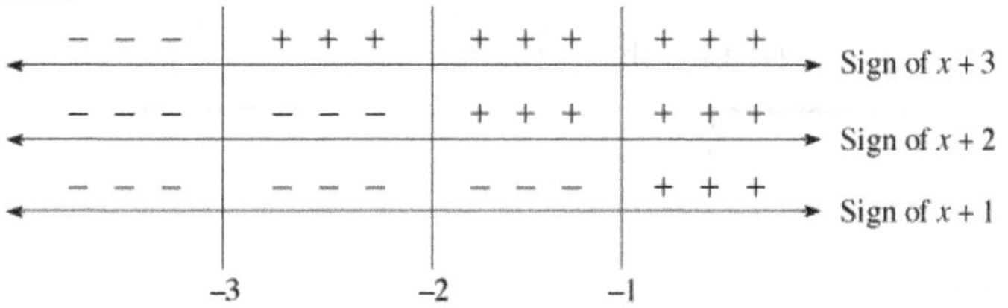

The solution set is $x \leq -3$ or $-2 \leq x \leq -1$. Graphing the solution set:

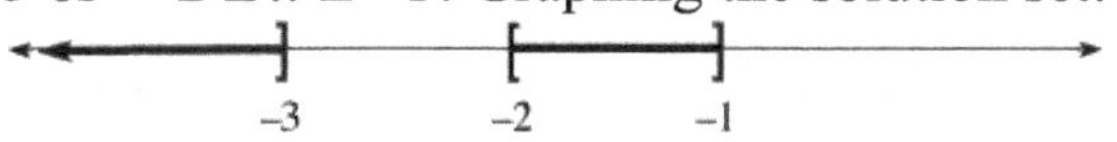

23. Forming the sign chart:

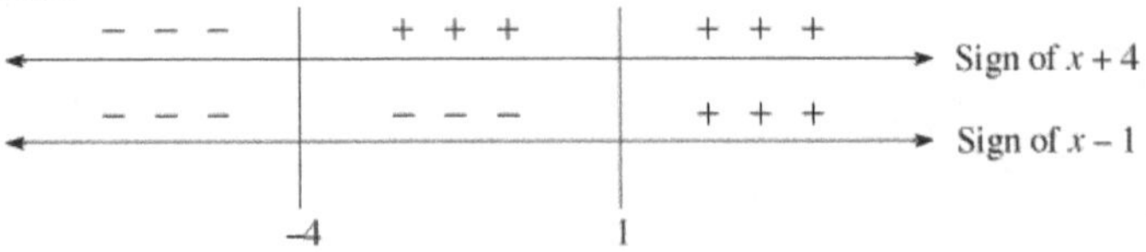

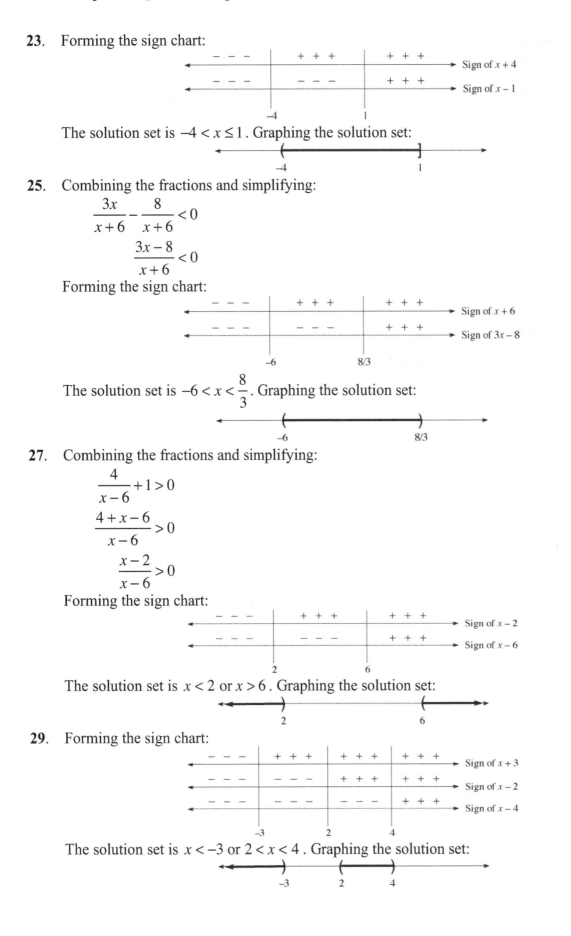

The solution set is $-4 < x \leq 1$. Graphing the solution set:

25. Combining the fractions and simplifying:

$$\frac{3x}{x+6} - \frac{8}{x+6} < 0$$

$$\frac{3x-8}{x+6} < 0$$

Forming the sign chart:

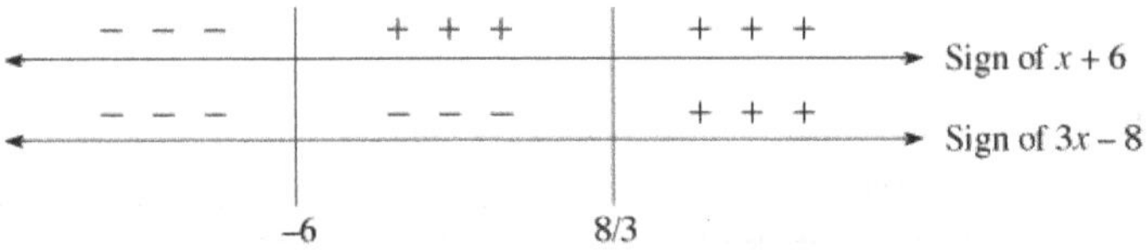

The solution set is $-6 < x < \frac{8}{3}$. Graphing the solution set:

27. Combining the fractions and simplifying:

$$\frac{4}{x-6} + 1 > 0$$

$$\frac{4+x-6}{x-6} > 0$$

$$\frac{x-2}{x-6} > 0$$

Forming the sign chart:

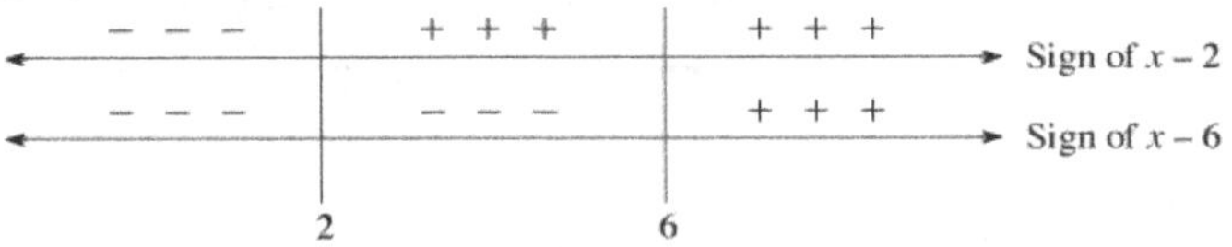

The solution set is $x < 2$ or $x > 6$. Graphing the solution set:

29. Forming the sign chart:

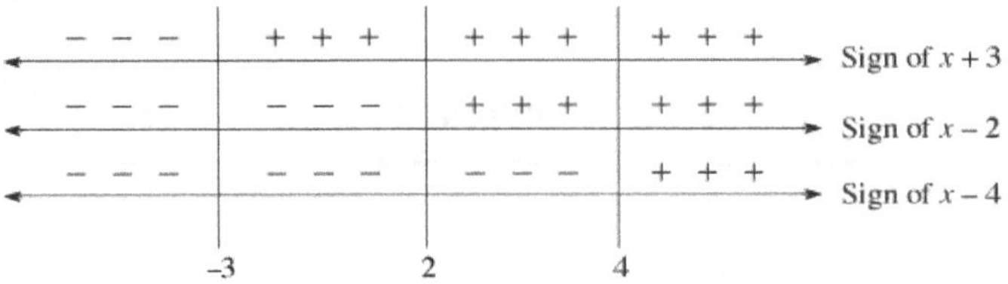

The solution set is $x < -3$ or $2 < x < 4$. Graphing the solution set:

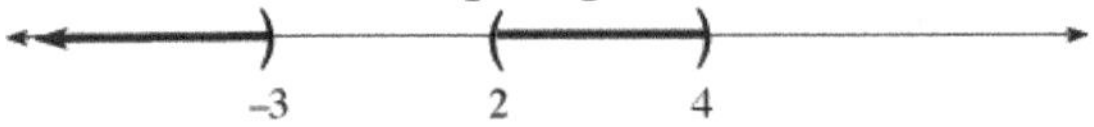

31. Combining the fractions and simplifying:

$$\frac{2}{x-4}-\frac{1}{x-3}>0$$

$$\frac{2x-6-x+4}{(x-4)(x-3)}>0$$

$$\frac{x-2}{(x-4)(x-3)}>0$$

Forming the sign chart:

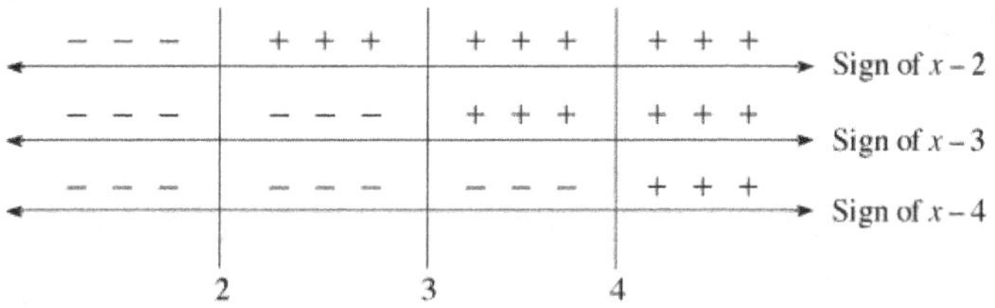

The solution set is $2<x<3$ or $x>4$. Graphing the solution set:

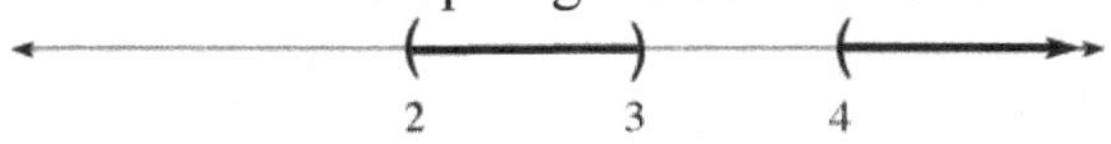

33. Combining the fractions and simplifying:

$$\frac{x+7}{2x+12}+\frac{6}{x^2-36}\le 0$$

$$\frac{x+7}{2(x+6)}+\frac{6}{(x+6)(x-6)}\le 0$$

$$\frac{(x+7)(x-6)+12}{2(x+6)(x-6)}\le 0$$

$$\frac{x^2+x-42+12}{2(x+6)(x-6)}\le 0$$

$$\frac{x^2+x-30}{2(x+6)(x-6)}\le 0$$

$$\frac{(x+6)(x-5)}{2(x+6)(x-6)}\le 0$$

$$\frac{x-5}{2(x-6)}\le 0$$

Forming the sign chart:

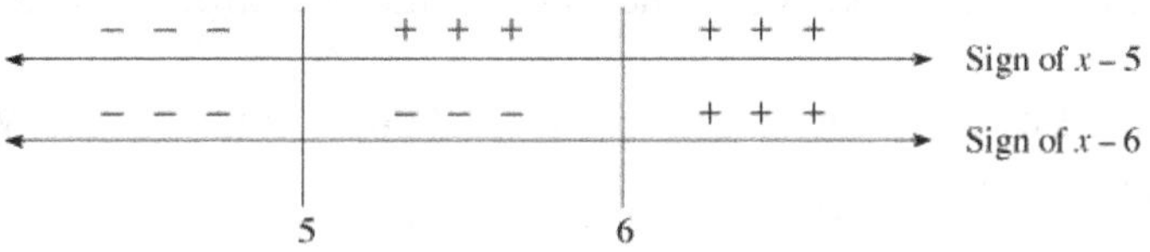

The solution set is $5\le x<6$. Graphing the solution set:

35. a. The solution set is $-2 < x < 2$.
b. The solution set is $x < -2$ or $x > 2$.
c. The solution set is $x = -2, 2$.

37. a. The solution set is $-2 < x < 5$.
b. The solution set is $x < -2$ or $x > 5$.
c. The solution set is $x = -2, 5$.

39. a. The solution set is $x < -1$ or $1 < x < 3$.
b. The solution set is $-1 < x < 1$ or $x > 3$.
c. The solution set is $x = -1, 1, 3$.

41. Let w represent the width and $2w + 3$ represent the length. Using the area formula:

$$w(2w+3) \ge 44$$
$$2w^2 + 3w \ge 44$$
$$2w^2 + 3w - 44 \ge 0$$
$$(2w+11)(w-4) \ge 0$$

Forming the sign chart:

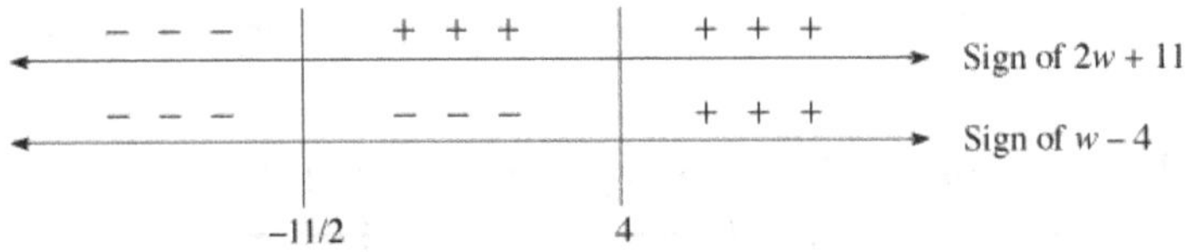

The width is at least 4 inches.

43. Solving the inequality:

$$1300p - 100p^2 \ge 4000$$
$$-100p^2 + 1300p - 4000 \ge 0$$
$$p^2 - 13p + 40 \le 0$$
$$(p-8)(p-5) \le 0$$

Forming the sign chart:

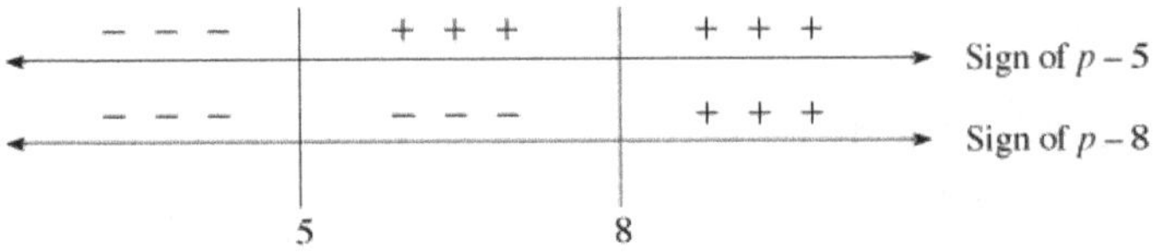

Charge at least \$5 but no more than \$8 per radio.

45. Let x represent the number of \$10 increases in dues. Then the revenue is given by:

$$\begin{aligned} y &= (10000 - 200x)(100 + 10x) \\ &= -2000x^2 + 80000x + 1000000 \\ &= -2000\left(x^2 - 40x + 400\right) + 1000000 + 800000 \\ &= -2000(x-20)^2 + 1{,}800{,}000 \end{aligned}$$

The dues should be increased by \$200, so the dues should be \$300 to result in a maximum income of \$1,800,000.

47. Let x represent the number of \$2 increases in price. Then the income is given by:

$$\begin{aligned} y &= (40 - 2x)(20 + 2x) \\ &= -4x^2 + 40x + 800 \\ &= -4\left(x^2 - 10x + 25\right) + 800 + 100 \\ &= -4(x-5)^2 + 900 \end{aligned}$$

The owner should make five \$2 increases in price, resulting in a price of \$30.

49. Using a calculator: $\frac{50{,}000}{32{,}000}=1.5625$

51. Using a calculator: $\frac{1}{2}\left(\frac{4.5926}{1.3876}-2\right)\approx 0.6549$

53. Solving the equation:

$$\begin{aligned}\sqrt{3t-1}&=2\\ \left(\sqrt{3t-1}\right)^2&=(2)^2\\ 3t-1&=4\\ 3t&=5\\ t&=\frac{5}{3}\end{aligned}$$

The solution is $\frac{5}{3}$.

55. Solving the equation:

$$\begin{aligned}\sqrt{x+3}&=x-3\\ \left(\sqrt{x+3}\right)^2&=(x-3)^2\\ x+3&=x^2-6x+9\\ 0&=x^2-7x+6\\ 0&=(x-6)(x-1)\\ x&=1,6 \qquad (x=1 \text{ does not check})\end{aligned}$$

The solution is 6.

57. Graphing the equation:

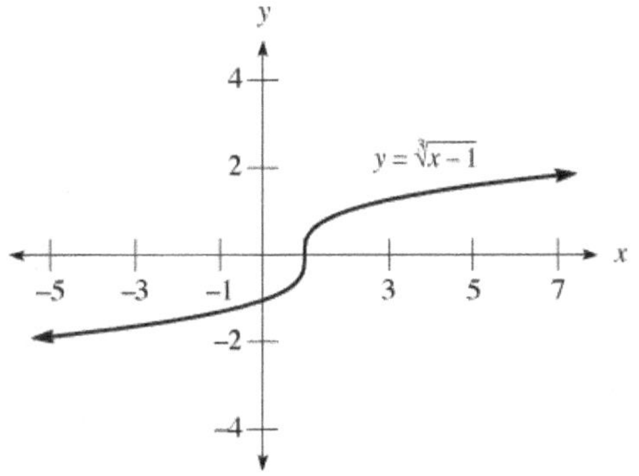

Chapter 7 Test

See www.mathtv.com for video solutions to all problems in this chapter test.

Chapter 8
Exponential and Logarithmic Functions

8.1 Exponential Functions

1. Evaluating: $g(0)=\left(\frac{1}{2}\right)^0=1$

3. Evaluating: $g(-1)=\left(\frac{1}{2}\right)^{-1}=2$

5. Evaluating: $f(-3)=3^{-3}=\frac{1}{27}$

7. Evaluating: $f(2)+g(-2)=3^2+\left(\frac{1}{2}\right)^{-2}=9+4=13$

9. Evaluating: $f(-1)+g(1)=4^{-1}+\left(\frac{1}{3}\right)^1=\frac{1}{4}+\frac{1}{3}=\frac{3}{12}+\frac{4}{12}=\frac{7}{12}$

11. Evaluating: $\frac{f(-2)}{g(1)}=\frac{4^{-2}}{\left(\frac{1}{3}\right)^1}=\frac{1}{16}\bullet 3=\frac{3}{16}$

13. Graphing the function:

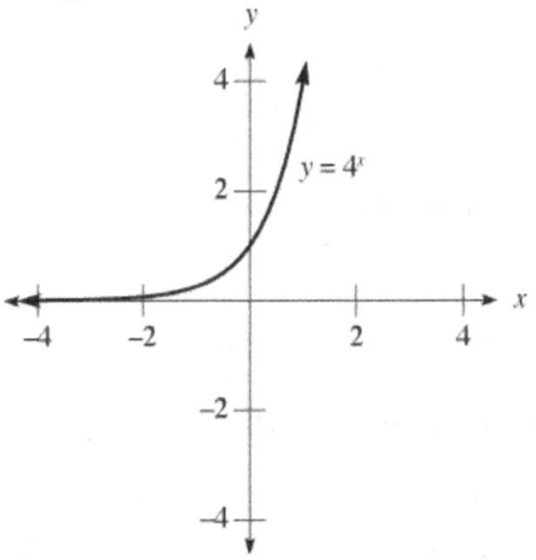

15. Graphing the function:

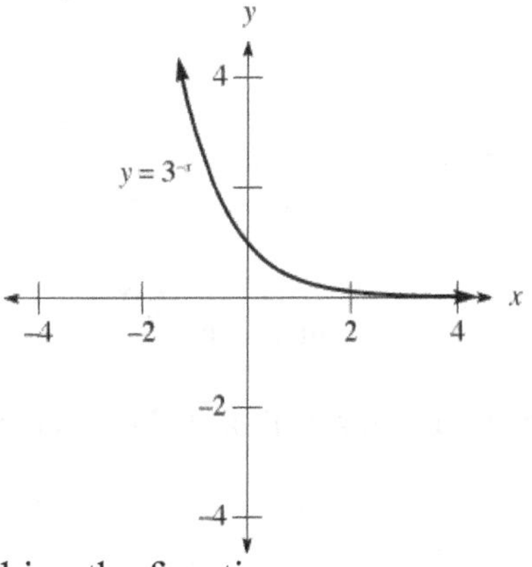

17. Graphing the function:

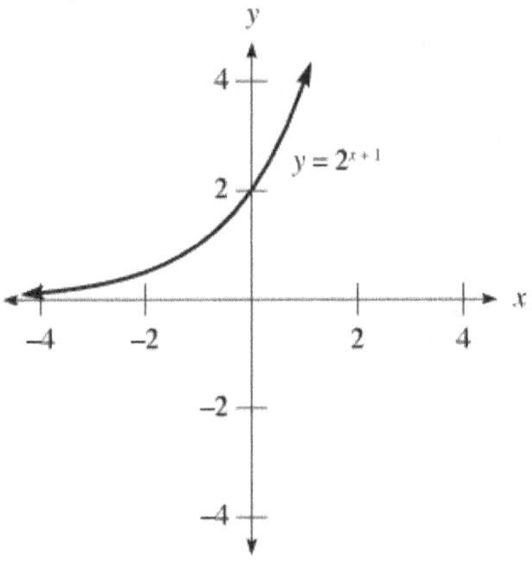

19. Graphing the function:

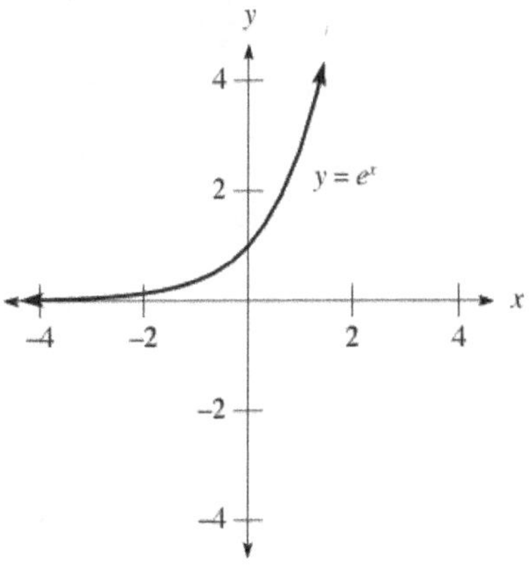

21. Graphing the function:

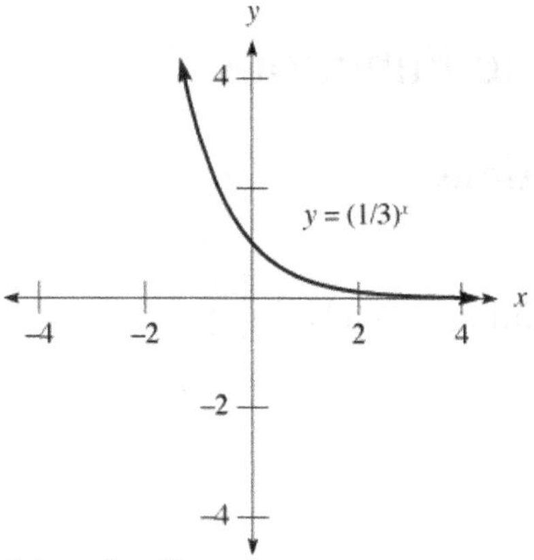

23. Graphing the function:

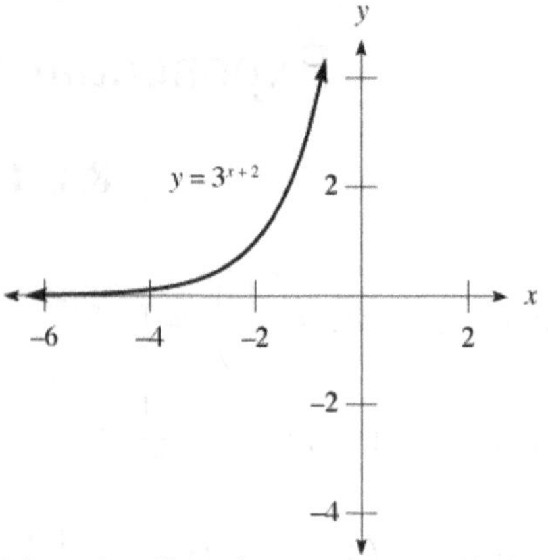

25. Graphing the functions:

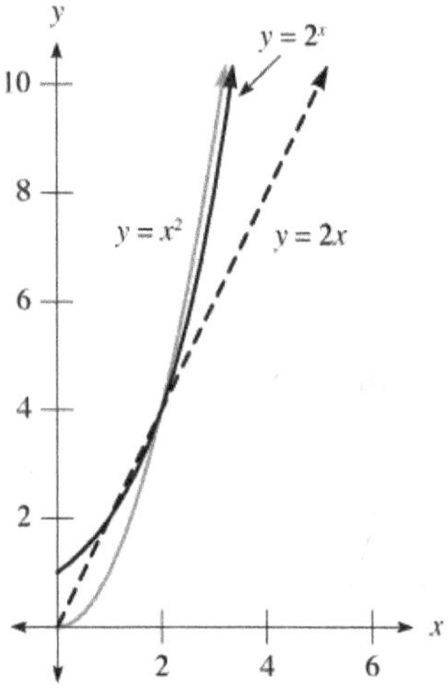

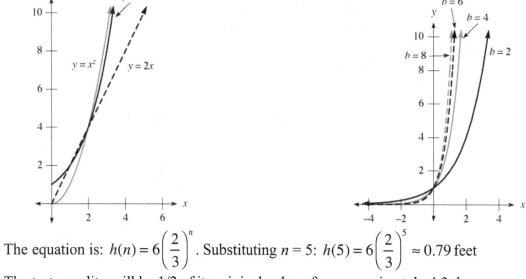

27. Graphing the functions:

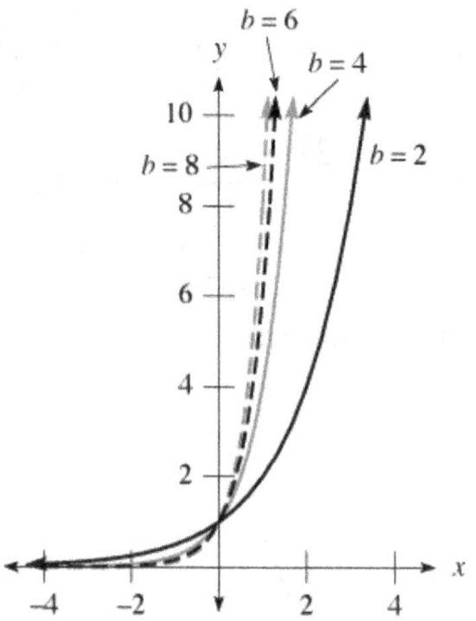

29. The equation is: $h(n)=6\left(\frac{2}{3}\right)^n$. Substituting $n=5$: $h(5)=6\left(\frac{2}{3}\right)^5\approx 0.79$ feet

31. The taste quality will be 1/2 of its original value after approximately 4.3 days.

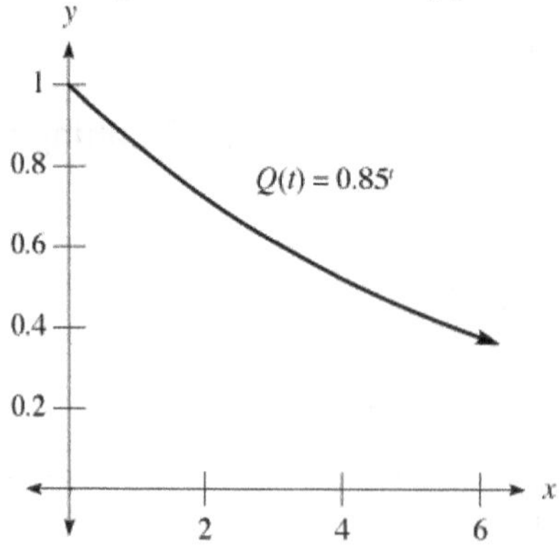

33. a. The equation is $A(t)=1200\left(1+\dfrac{0.06}{4}\right)^{4t}$.

b. Substitute $t=8$: $A(8)=1200\left(1+\dfrac{0.06}{4}\right)^{32}\approx \$1,932.39$

c. Using a graphing calculator, the time is approximately 11.6 years.

d. Substitute $t=8$ into the compound interest formula: $A(8)=1200e^{0.06\times 8}\approx \$1,939.29$

35. a. Substitute $t=3.5$: $V(5)=450,000(1-0.30)^5\approx \$129,138.48$

b. The domain is $\{t \mid 0\le t\le 6\}$.

c. Sketching the graph:

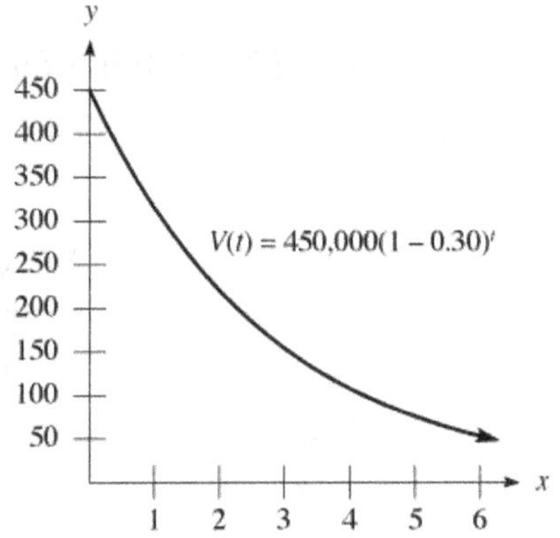

d. The range is $\{V(t) \mid 52,942.05\le V(t)\le 450,000\}$.

e. From the graph, the crane will be worth \$85,000 after approximately 4.7 years, or 4 years 8 months.

37. For 1 day, substitute $x=1$: $f(1)=50\bullet 4^1=200$ bacteria

For 2 days, substitute $x=2$: $f(2)=50\bullet 4^2=800$ bacteria

For 3 days, substitute $x=3$: $f(3)=50\bullet 4^3=3200$ bacteria

39. Graphing the function:

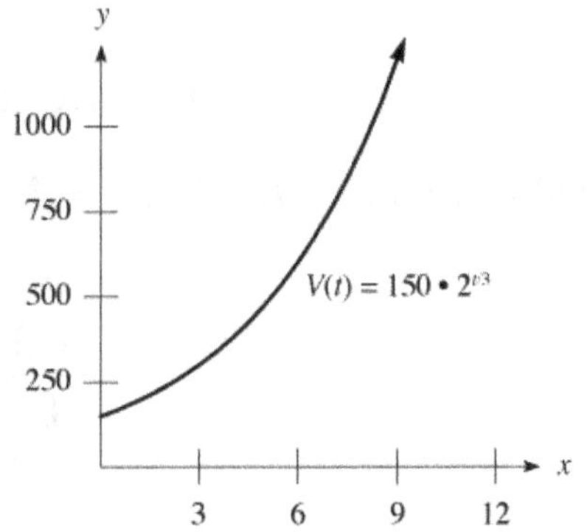

41. a. Substitute $t=25$: $C(25)=0.10e^{0.0576\times 25}\approx \$0.42$

b. Substitute $t=40$: $C(40)=0.10e^{0.0576\times 40}\approx \$1.00$

c. Substitute $t=50$: $C(50)=0.10e^{0.0576\times 50}\approx \$1.78$

d. Substitute $t=90$: $C(90)=0.10e^{0.0576\times 90}\approx \$17.84$

43. Substitute $t = 3$: $B(3) = 0.798 \bullet 1.164^3 \approx 1{,}258{,}525$ bankruptcies
This is 58,474 less than the actual number of bankruptcies filed.

45. **a.** Substitute $t = 5$: $A(5) = 5{,}000{,}000e^{-0.598\times 5} \approx 251{,}437$ cells

b. Substitute $t = 10$: $A(10) = 5{,}000{,}000e^{-0.598\times 10} \approx 12{,}644$ cells

c. Substitute $t = 20$: $A(20) = 5{,}000{,}000e^{-0.598\times 20} \approx 32$ cells

47. Solving for y:

$$\begin{aligned} x &= 2y - 3 \\ 2y &= x + 3 \\ y &= \frac{x+3}{2} \end{aligned}$$

49. Solving for y:

$$\begin{aligned} x &= y^2 - 2 \\ y^2 &= x + 2 \\ y &= \pm\sqrt{x+2} \end{aligned}$$

51. Solving for y:

$$\begin{aligned} x &= \frac{y-4}{y-2} \\ x(y-2) &= y - 4 \\ xy - 2x &= y - 4 \\ xy - y &= 2x - 4 \\ y(x-1) &= 2x - 4 \\ y &= \frac{2x-4}{x-1} \end{aligned}$$

53. Solving for y:

$$\begin{aligned} x &= \sqrt{y-3} \\ x^2 &= y - 3 \\ y &= x^2 + 3 \end{aligned}$$

8.2 The Inverse of a Function

1. Let $y = f(x)$. Switch x and y and solve for y:

$$\begin{aligned} 3y - 1 &= x \\ 3y &= x + 1 \\ y &= \frac{x+1}{3} \end{aligned}$$

The inverse is $f^{-1}(x) = \frac{x+1}{3}$.

3. Let $y = f(x)$. Switch x and y and solve for y:

$$\begin{aligned} y^3 &= x \\ y &= \sqrt[3]{x} \end{aligned}$$

The inverse is $f^{-1}(x) = \sqrt[3]{x}$.

5. Let $y = f(x)$. Switch x and y and solve for y:

$$\begin{aligned} \frac{y-3}{y-1} &= x \\ y - 3 &= xy - x \\ y - xy &= 3 - x \\ y(1-x) &= 3 - x \\ y &= \frac{3-x}{1-x} = \frac{x-3}{x-1} \end{aligned}$$

The inverse is $f^{-1}(x) = \frac{x-3}{x-1}$.

7. Let $y = f(x)$. Switch x and y and solve for y:

$$\begin{aligned} \frac{y-3}{4} &= x \\ y - 3 &= 4x \\ y &= 4x + 3 \end{aligned}$$

The inverse is $f^{-1}(x) = 4x + 3$.

9. Let $y = f(x)$. Switch x and y and solve for y:

$$\frac{1}{2}y - 3 = x$$
$$y - 6 = 2x$$
$$y = 2x + 6$$

The inverse is $f^{-1}(x) = 2x + 6$.

11. Let $y = f(x)$. Switch x and y and solve for y:

$$\frac{2}{3}y - 3 = x$$
$$2y - 9 = 3x$$
$$2y = 3x + 9$$
$$y = \frac{3}{2}x + \frac{9}{2}$$

The inverse is $f^{-1}(x) = \frac{3}{2}x + \frac{9}{2}$.

13. Let $y = f(x)$. Switch x and y and solve for y:

$$y^3 - 4 = x$$
$$y^3 = x + 4$$
$$y = \sqrt[3]{x + 4}$$

The inverse is $f^{-1}(x) = \sqrt[3]{x + 4}$.

15. Let $y = f(x)$. Switch x and y and solve for y:

$$\frac{4y - 3}{2y + 1} = x$$
$$4y - 3 = 2xy + x$$
$$4y - 2xy = x + 3$$
$$y(4 - 2x) = x + 3$$
$$y = \frac{x + 3}{4 - 2x}$$

The inverse is $f^{-1}(x) = \frac{x + 3}{4 - 2x}$.

17. Let $y = f(x)$. Switch x and y and solve for y:

$$\frac{2y + 1}{3y + 1} = x$$
$$2y + 1 = 3xy + x$$
$$2y - 3xy = x - 1$$
$$y(2 - 3x) = x - 1$$
$$y = \frac{x - 1}{2 - 3x} = \frac{1 - x}{3x - 2}$$

The inverse is $f^{-1}(x) = \frac{1 - x}{3x - 2}$.

19. Finding the inverse:

$$2y - 1 = x$$
$$2y = x + 1$$
$$y = \frac{x + 1}{2}$$

21. Finding the inverse:

$$y^2 - 3 = x$$
$$y^2 = x + 3$$
$$y = \pm\sqrt{x + 3}$$

The inverse is $y^{-1} = \dfrac{x+1}{2}$. Graphing:

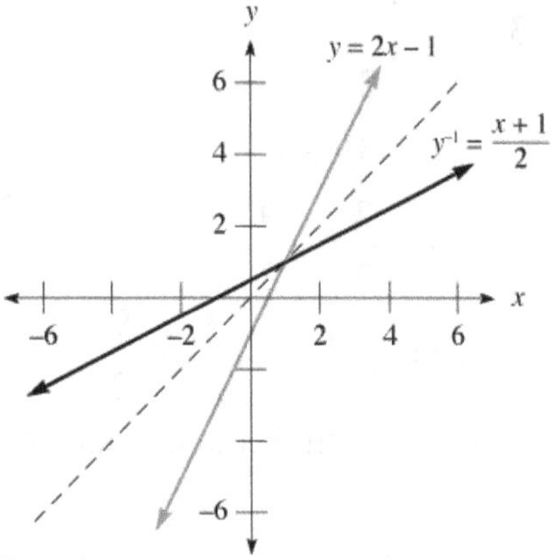

The inverse is $y^{-1} = \pm\sqrt{x+3}$. Graphing:

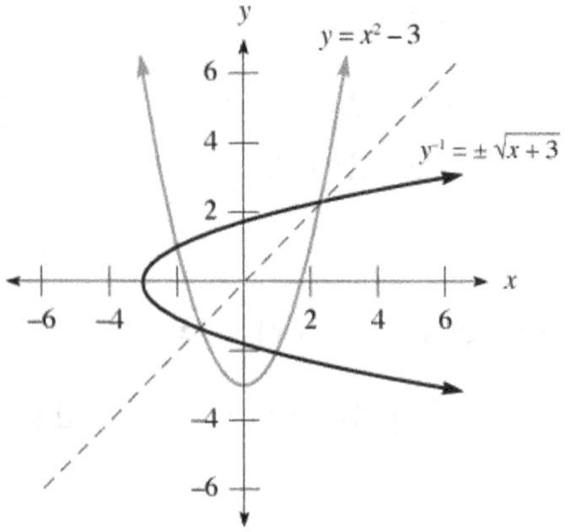

23. Finding the inverse:

$$y^2 - 2y - 3 = x$$
$$y^2 - 2y + 1 = x + 3 + 1$$
$$(y-1)^2 = x + 4$$
$$y - 1 = \pm\sqrt{x+4}$$
$$y = 1 \pm \sqrt{x+4}$$

The inverse is $y^{-1} = 1 \pm \sqrt{x+4}$. Graphing:

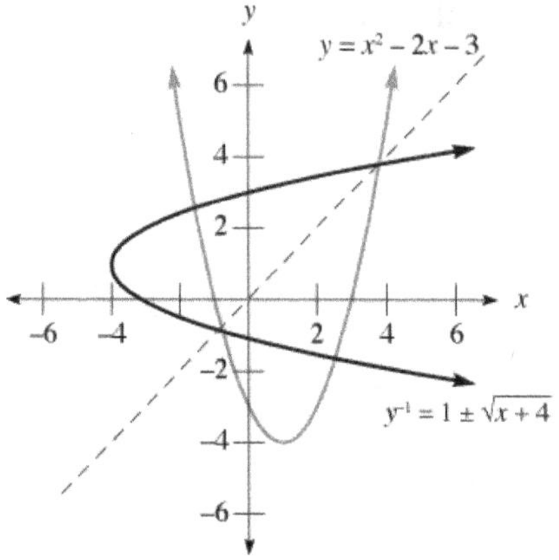

25. The inverse is $x = 3^y$. Graphing each curve:

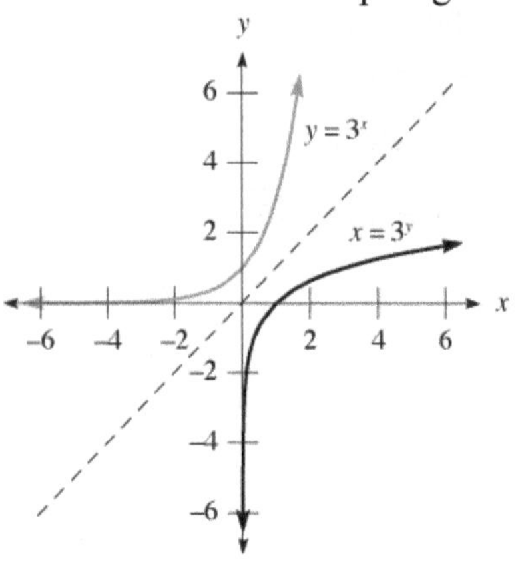

27. The inverse is $x = 4$. Graphing each curve:

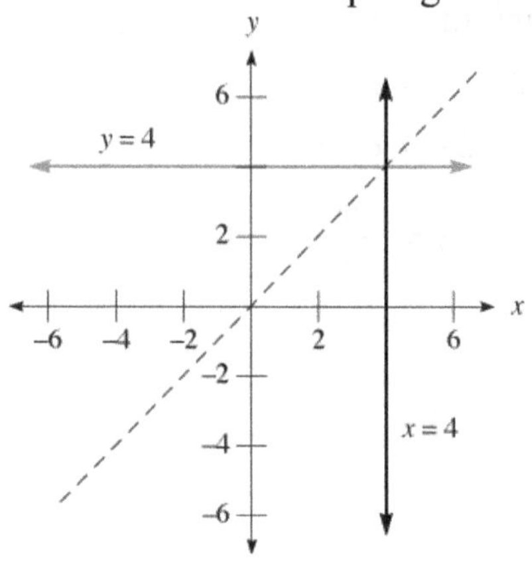

29. Finding the inverse:

$$\frac{1}{2}y^3 = x$$
$$y^3 = 2x$$
$$y = \sqrt[3]{2x}$$

The inverse is $y^{-1} = \sqrt[3]{2x}$. Graphing:

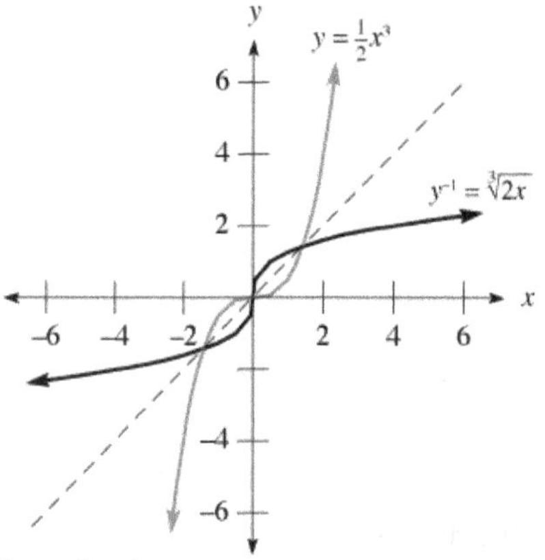

31. Finding the inverse:

$$\frac{1}{2}y + 2 = x$$
$$y + 4 = 2x$$
$$y = 2x - 4$$

The inverse is $y^{-1} = 2x - 4$. Graphing:

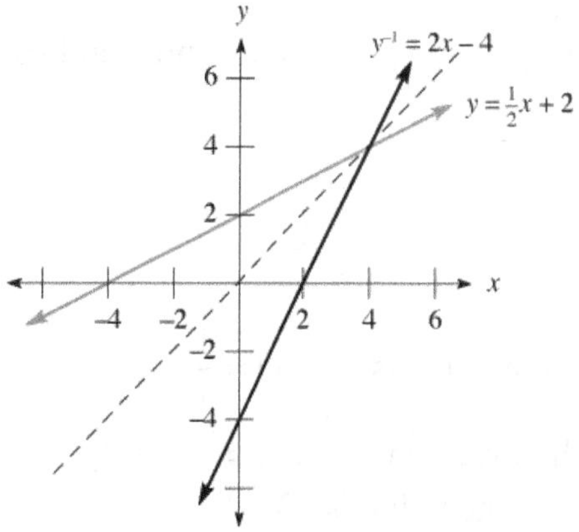

33. Finding the inverse:

$$\sqrt{y+2} = x$$
$$y + 2 = x^2$$
$$y = x^2 - 2$$

The inverse is $y^{-1} = x^2 - 2, x \geq 0$. Graphing each curve:

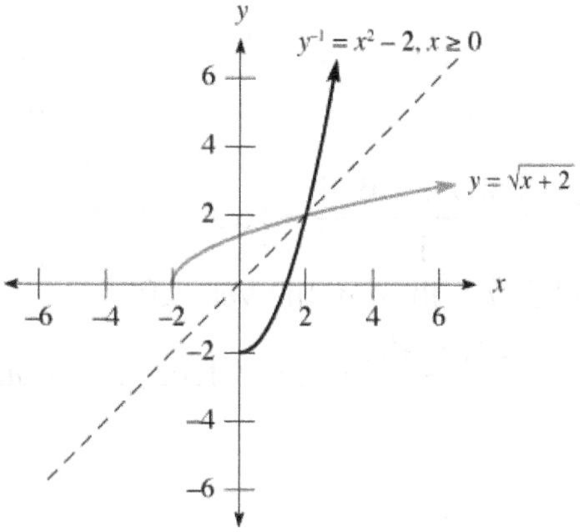

35. **a.** Yes, this function is one-to-one.
b. No, this function is not one-to-one.
c. Yes, this function is one-to-one.

37. **a.** Evaluating the function: $f(2)=3(2)-2=6-2=4$

b. Evaluating the function: $f^{-1}(2)=\dfrac{2+2}{3}=\dfrac{4}{3}$

c. Evaluating the function: $f\left[f^{-1}(2)\right]=f\left(\dfrac{4}{3}\right)=3\left(\dfrac{4}{3}\right)-2=4-2=2$

d. Evaluating the function: $f^{-1}\left[f(2)\right]=f^{-1}(4)=\dfrac{4+2}{3}=\dfrac{6}{3}=2$

39. Let $y=f(x)$. Switch x and y and solve for y:

$$\frac{1}{y}=x$$

$$y=\frac{1}{x}$$

The inverse is $f^{-1}(x)=\dfrac{1}{x}$.

41. The inverse is $f^{-1}(x)=7(x+2)$.

43. **a.** The value is –3. **b.** The value is –6.
c. The value is 2. **d.** The value is 3.
e. The value is –2. **f.** The value is 3.
g. Each is an inverse of the other.

45. **a.** Substituting $t=15$: $s(15)=16(15)+249.4=\$489.4$ billion

b. Finding the inverse:

$$s=16t+249.4$$

$$16t=s-249.4$$

$$t(s)=\frac{s-249.4}{16}$$

c. Substitute $s=507$: $t(507)=\dfrac{507-249.4}{16}\approx 16$.

The payments will reach $507 billion in the year 2006.

47. **a.** Substituting $m=4520$: $f=\dfrac{22(4520)}{15}\approx 6629$ feet per second

b. Finding the inverse:

$$f=\frac{22m}{15}$$

$$15f=22m$$

$$m(f)=\frac{15f}{22}$$

c. Substituting $f=2$: $m(2)=\dfrac{15(2)}{22}\approx 1.36$ mph

49. Simplifying: $3^{-2}=\dfrac{1}{3^2}=\dfrac{1}{9}$

51. Solving the equation:
$$2 = 3x$$
$$x = \frac{2}{3}$$

53. Solving the equation:
$$4 = x^3$$
$$x = \sqrt[3]{4}$$

55. Completing the statement: $8 = 2^3$

57. Completing the statement: $10{,}000 = 10^4$

59. Completing the statement: $81 = 3^4$

61. Completing the statement: $6 = 6^1$

8.3 Logarithms are Exponents

1. Writing in logarithmic form: $\log_2 16 = 4$

3. Writing in logarithmic form: $\log_5 125 = 3$

5. Writing in logarithmic form: $\log_{10} 0.01 = -2$

7. Writing in logarithmic form: $\log_2 \frac{1}{32} = -5$

9. Writing in logarithmic form: $\log_{1/2} 8 = -3$

11. Writing in logarithmic form: $\log_3 27 = 3$

13. Writing in exponential form: $10^2 = 100$

15. Writing in exponential form: $2^6 = 64$

17. Writing in exponential form: $8^0 = 1$

19. Writing in exponential form: $10^{-3} = 0.001$

21. Writing in exponential form: $6^2 = 36$

23. Writing in exponential form: $5^{-2} = \frac{1}{25}$

25. Solving the equation:
$$\log_3 x = 2$$
$$x = 3^2 = 9$$

27. Solving the equation:
$$\log_5 x = -3$$
$$x = 5^{-3} = \frac{1}{125}$$

29. Solving the equation:
$$\log_2 16 = x$$
$$2^x = 16$$
$$x = 4$$

31. Solving the equation:
$$\log_8 2 = x$$
$$8^x = 2$$
$$x = \frac{1}{3}$$

33. Solving the equation:
$$\log_x 4 = 2$$
$$x^2 = 4$$
$$x = 2$$

35. Solving the equation:
$$\log_x 5 = 3$$
$$x^3 = 5$$
$$x = \sqrt[3]{5}$$

37. Solving the equation:
$$\log_5 25 = x$$
$$5^x = 25$$
$$x = 2$$

39. Solving the equation:
$$\log_x 36 = 2$$
$$x^2 = 36$$
$$x = 6$$

41. Solving the equation:

$$\log_8 4 = x$$
$$8^x = 4$$
$$2^{3x} = 2^2$$
$$3x = 2$$
$$x = \frac{2}{3}$$

43. Solving the equation:

$$\log_9 \frac{1}{3} = x$$
$$9^x = \frac{1}{3}$$
$$3^{2x} = 3^{-1}$$
$$2x = -1$$
$$x = -\frac{1}{2}$$

45. Solving the equation:

$$\log_8 x = -2$$
$$x = 8^{-2} = \frac{1}{64}$$

47. Sketching the graph:

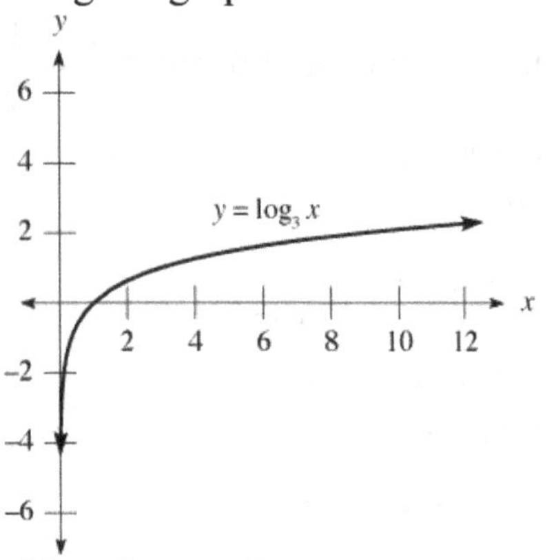

49. Sketching the graph:

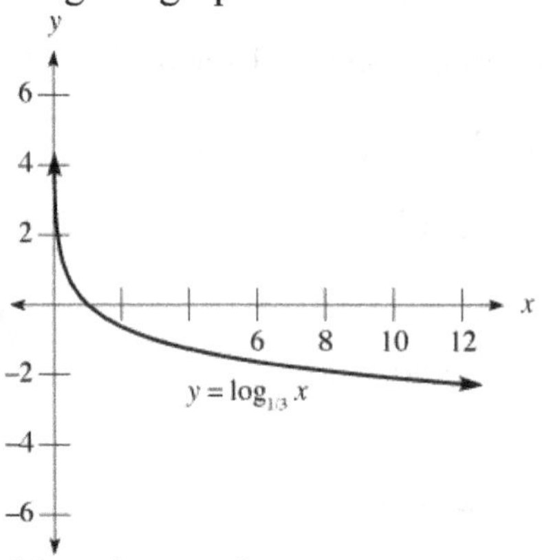

51. Sketching the graph:

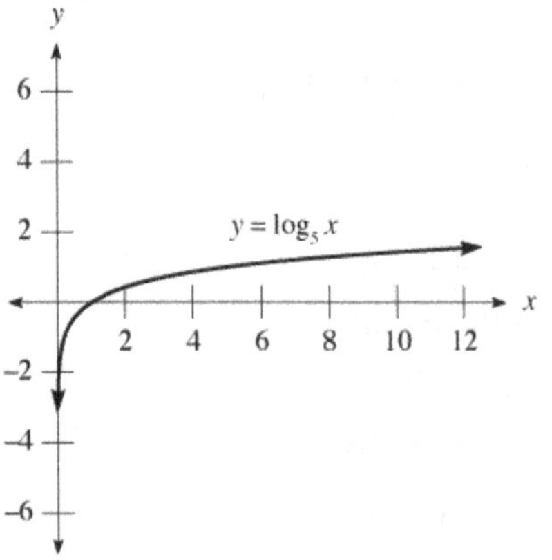

53. Sketching the graph:

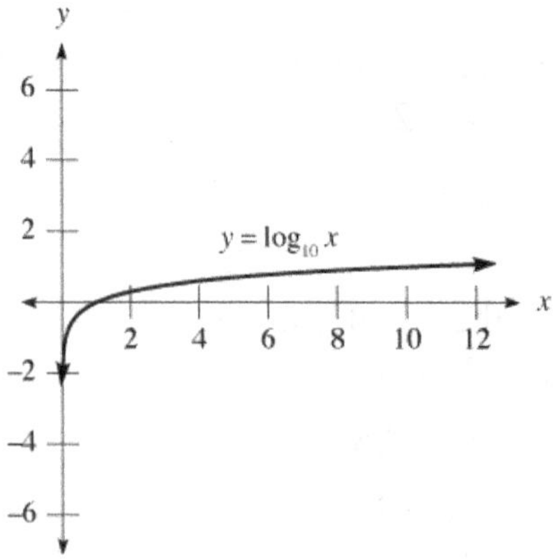

55. The equation is $y = 3^x$.

57. The equation is $y = \log_{1/3} x$.

59. Simplifying the logarithm:

$$x = \log_2 16$$
$$2^x = 16$$
$$x = 4$$

61. Simplifying the logarithm:

$$x = \log_{25} 125$$
$$25^x = 125$$
$$5^{2x} = 5^3$$
$$2x = 3$$
$$x = \frac{3}{2}$$

63. Simplifying the logarithm:

$$x = \log_{10} 1000$$
$$10^x = 1000$$
$$x = 3$$

65. Simplifying the logarithm:

$$x = \log_3 3$$
$$3^x = 3$$
$$x = 1$$

67. Simplifying the logarithm:

$$x = \log_5 1$$
$$5^x = 1$$
$$x = 0$$

69. Simplifying the logarithm:

$$x = \log_{17} 1$$
$$17^x = 1$$
$$x = 0$$

71. Simplifying the logarithm:

$$x = \log_{16} 4$$
$$16^x = 4$$
$$4^{2x} = 4^1$$
$$2x = 1$$
$$x = \frac{1}{2}$$

73. Simplifying the logarithm:

$$x = \log_{100} 1000$$
$$100^x = 1000$$
$$10^{2x} = 10^3$$
$$2x = 3$$
$$x = \frac{3}{2}$$

75. First find $\log_2 8$:

$$x = \log_2 8$$
$$2^x = 8$$
$$x = 3$$

Now find $\log_3 3$:

$$x = \log_3 3$$
$$3^x = 3$$
$$x = 1$$

77. First find $\log_3 81$:

$$x = \log_3 81$$
$$3^x = 81$$
$$x = 4$$

Now find $\log_{1/2} 4$:

$$x = \log_{1/2} 4$$
$$\left(\frac{1}{2}\right)^x = 4$$
$$2^{-x} = 2^2$$
$$x = -2$$

79. First find $\log_6 6$:

$$x = \log_6 6$$
$$6^x = 6$$
$$x = 1$$

Now find $\log_3 1$:

$$x = \log_3 1$$
$$3^x = 1$$
$$x = 0$$

81. First find $\log_2 16$:

$$x = \log_2 16$$
$$2^x = 16$$
$$x = 4$$

Now find $\log_2 4$:

$$x = \log_2 4$$
$$2^x = 4$$
$$x = 2$$

Now find $\log_4 2$:

$$x = \log_4 2$$
$$4^x = 2$$
$$2^{2x} = 2$$
$$2x = 1$$
$$x = \frac{1}{2}$$

83. Completing the table:

Prefix	Multiplying Factor	$\log_{10}(\text{Multiplying Factor})$
Nano	0.000000001	–9
Micro	0.000001	–6
Deci	0.1	–1
Giga	1,000,000,000	9
Peta	1,000,000,000,000,000	15

85. Using the relationship $M = \log_{10} T$:

$$M = \log_{10} 100$$
$$10^M = 100$$
$$M = 2$$

87. It is 10^8 times as large.

89. Since $M = 6$, the average is 120.

91. Simplifying; $8^{2/3} = \left(8^{1/3}\right)^2 = \left(\sqrt[3]{8}\right)^2 = 2^2 = 4$

93. Solving the equation:

$$(x+2)(x) = 2^3$$
$$x^2 + 2x = 8$$
$$x^2 + 2x - 8 = 0$$
$$(x+4)(x-2) = 0$$
$$x = -4, 2$$

95. Solving the equation:

$$\frac{x-2}{x+1} = 9$$
$$x - 2 = 9(x+1)$$
$$x - 2 = 9x + 9$$
$$-8x = 11$$
$$x = -\frac{11}{8}$$

97. Writing in exponential form: $2^3 = (x+2)(x)$

99. Writing in exponential form: $3^4 = \dfrac{x-2}{x+1}$

8.4 Properties of Logarithms

1. Using properties of logarithms: $\log_3 4x = \log_3 4 + \log_3 x$

3. Using properties of logarithms: $\log_6 \frac{5}{x} = \log_6 5 - \log_6 x$

5. Using properties of logarithms: $\log_2 y^5 = 5\log_2 y$

7. Using properties of logarithms: $\log_9 \sqrt[3]{z} = \log_9 z^{1/3} = \frac{1}{3}\log_9 z$

9. Using properties of logarithms: $\log_6 x^2y^4 = \log_6 x^2 + \log_6 y^4 = 2\log_6 x + 4\log_6 y$

11. Using properties of logarithms: $\log_5 \sqrt{x} \bullet y^4 = \log_5 x^{1/2} + \log_5 y^4 = \frac{1}{2}\log_5 x + 4\log_5 y$

13. Using properties of logarithms: $\log_b \frac{xy}{z} = \log_b xy - \log_b z = \log_b x + \log_b y - \log_b z$

15. Using properties of logarithms: $\log_{10} \frac{4}{xy} = \log_{10} 4 - \log_{10} xy = \log_{10} 4 - \log_{10} x - \log_{10} y$

17. Using properties of logarithms:

$$\log_{10} \frac{x^2 y}{\sqrt{z}} = \log_{10} x^2 + \log_{10} y - \log_{10} z^{1/2} = 2\log_{10} x + \log_{10} y - \frac{1}{2}\log_{10} z$$

19. Using properties of logarithms:

$$\log_{10} \frac{x^3\sqrt{y}}{z^4} = \log_{10} x^3 + \log_{10} y^{1/2} - \log_{10} z^4 = 3\log_{10} x + \frac{1}{2}\log_{10} y - 4\log_{10} z$$

21. Using properties of logarithms:

$$\log_b \sqrt[3]{\frac{x^2 y}{z^4}} = \log_b \frac{x^{2/3} y^{1/3}}{z^{4/3}} = \log_b x^{2/3} + \log_b y^{1/3} - \log_b z^{4/3} = \frac{2}{3}\log_b x + \frac{1}{3}\log_b y - \frac{4}{3}\log_b z$$

23. Using properties of logarithms:

$$\log_3 \sqrt[3]{\frac{x^2 y}{z^6}} = \log_3 \frac{x^{2/3} y^{1/3}}{z^2} = \log_3 x^{2/3} + \log_3 y^{1/3} - \log_3 z^2 = \frac{2}{3}\log_3 x + \frac{1}{3}\log_3 y - 2\log_3 z$$

25. Using properties of logarithms:

$$\begin{aligned}\log_a \frac{4x^5}{9a^2} &= \log_a 4x^5 - \log_a 9a^2 \\ &= \log_a 2^2 + \log_a x^5 - \log_a 3^2 - \log_a a^2 \\ &= 2\log_a 2 + 5\log_a x - 2\log_a 3 - 2\end{aligned}$$

27. Writing as a single logarithm: $\log_b x + \log_b z = \log_b xz$

29. Writing as a single logarithm: $2\log_3 x - 3\log_3 y = \log_3 x^2 - \log_3 y^3 = \log_3 \frac{x^2}{y^3}$

31. Writing as a single logarithm: $\frac{1}{2}\log_{10} x + \frac{1}{3}\log_{10} y = \log_{10} x^{1/2} + \log_{10} y^{1/3} = \log_{10} \sqrt{x}\sqrt[3]{y}$

33. Writing as a single logarithm: $3\log_2 x + \frac{1}{2}\log_2 y - \log_2 z = \log_2 x^3 + \log_2 y^{1/2} - \log_2 z = \log_2 \frac{x^3\sqrt{y}}{z}$

35. Writing as a single logarithm:

$$\frac{1}{2}\log_2 x - 3\log_2 y - 4\log_2 z = \log_2 x^{1/2} - \log_2 y^3 - \log_2 z^4 = \log_2 \frac{\sqrt{x}}{y^3 z^4}$$

37. Writing as a single logarithm:

$$\frac{3}{2}\log_{10} x - \frac{3}{4}\log_{10} y - \frac{4}{5}\log_{10} z = \log_{10} x^{3/2} - \log_{10} y^{3/4} - \log_{10} z^{4/5} = \log_{10} \frac{x^{3/2}}{y^{3/4} z^{4/5}}$$

39. Writing as a single logarithm:

$$\frac{1}{2}\log_5 x + \tfrac{2}{3}\log_5 y - 4\log_5 z = \log_5 x^{1/2} + \log_5 y^{2/3} - \log_5 z^4 = \log_5 \frac{\sqrt{x} \bullet \sqrt[3]{y^2}}{z^4}$$

41. Writing as a single logarithm:

$$\log_3\left(x^2 - 16\right) - 2\log_3(x+4) = \log_3\left(x^2 - 16\right) - \log_3(x+4)^2 = \log_3 \frac{(x+4)(x-4)}{(x+4)^2} = \log_3 \frac{x-4}{x+4}$$

43. Solving the equation:

$$\begin{aligned} \log_2 x + \log_2 3 &= 1 \\ \log_2 3x &= 1 \\ 3x &= 2^1 \\ 3x &= 2 \\ x &= \frac{2}{3} \end{aligned}$$

45. Solving the equation:

$$\begin{aligned} \log_3 x - \log_3 2 &= 2 \\ \log_3 \frac{x}{2} &= 2 \\ \frac{x}{2} &= 3^2 \\ \frac{x}{2} &= 9 \\ x &= 18 \end{aligned}$$

47. Solving the equation:

$$\begin{aligned} \log_3 x + \log_3(x-2) &= 1 \\ \log_3\left(x^2 - 2x\right) &= 1 \\ x^2 - 2x &= 3^1 \\ x^2 - 2x - 3 &= 0 \\ (x-3)(x+1) &= 0 \\ x &= 3, -1 \end{aligned}$$

The solution is 3 (–1 does not check).

49. Solving the equation:

$$\begin{aligned} \log_3(x+3) - \log_3(x-1) &= 1 \\ \log_3 \frac{x+3}{x-1} &= 1 \\ \frac{x+3}{x-1} &= 3^1 \\ x+3 &= 3x - 3 \\ -2x &= -6 \\ x &= 3 \end{aligned}$$

51. Solving the equation:

$$\begin{aligned}\log_2 x+\log_2(x-2)&=3\\ \log_2(x^2-2x)&=3\\ x^2-2x&=2^3\\ x^2-2x-8&=0\\ (x-4)(x+2)&=0\\ x&=4,-2\end{aligned}$$

The solution is 4 (–2 does not check).

53. Solving the equation:

$$\begin{aligned}\log_8 x+\log_8(x-3)&=\frac{2}{3}\\ \log_8(x^2-3x)&=\frac{2}{3}\\ x^2-3x&=8^{2/3}\\ x^2-3x-4&=0\\ (x-4)(x+1)&=0\\ x&=4,-1\end{aligned}$$

The solution is 4 (–1 does not check).

55. Solving the equation:

$$\begin{aligned}\log_3(x+2)-\log_3 x&=1\\ \log_3\frac{x+2}{x}&=1\\ \frac{x+2}{x}&=3^1\\ x+2&=3x\\ 2x&=2\\ x&=1\end{aligned}$$

57. Solving the equation:

$$\begin{aligned}\log_2(x+1)+\log_2(x+2)&=1\\ \log_2(x^2+3x+2)&=1\\ x^2+3x+2&=2^1\\ x^2+3x&=0\\ x(x+3)&=0\\ x&=0,-3\end{aligned}$$

The solution is 0 (–3 does not check).

59. Solving the equation:

$$\begin{aligned}\log_9\sqrt{x}+\log_9\sqrt{2x+3}&=\frac{1}{2}\\ \log_9\sqrt{2x^2+3x}&=\frac{1}{2}\\ \sqrt{2x^2+3x}&=9^{1/2}\\ 2x^2+3x&=9\\ 2x^2+3x-9&=0\\ (2x-3)(x+3)&=0\\ x&=\frac{3}{2},-3\end{aligned}$$

The solution is $\frac{3}{2}$ (–3 does not check).

61. Solving the equation:

$$\begin{aligned}4\log_3 x-\log_3 x^2&=6\\ 4\log_3 x-2\log_3 x&=6\\ 2\log_3 x&=6\\ \log_3 x&=3\\ x&=3^3\\ x&=27\end{aligned}$$

63. Solving the equation:

$$\begin{aligned}\log_5 \sqrt{x} + \log_5 \sqrt{6x+5} &= 1 \\ \log_5 \sqrt{6x^2+5x} &= 1 \\ \frac{1}{2}\log_5\left(6x^2+5x\right) &= 1 \\ \log_5\left(6x^2+5x\right) &= 2 \\ 6x^2+5x &= 5^2 \\ 6x^2+5x-25 &= 0 \\ (3x-5)(2x+5) &= 0 \\ x &= \frac{5}{3}, -\frac{5}{2}\end{aligned}$$

The solution is $\frac{5}{3}$ ($-\frac{5}{2}$ does not check).

65. Rewriting the formula:

$$D = 10\log_{10}\left(\frac{I}{I_0}\right)$$

$$D = 10\left(\log_{10} I - \log_{10} I_0\right)$$

67. **a.** Finding the value: $\log_{10} 40 = \log_{10}(8 \bullet 5) = \log_{10} 8 + \log_{10} 5 = 0.903 + 0.699 = 1.602$

b. Finding the value:

$$\log_{10} 320 = \log_{10}\left(8^2 \bullet 5\right) = \log_{10} 8^2 + \log_{10} 5 = 2\log_{10} 8 + \log_{10} 5 = 2(0.903) + 0.699 = 2.505$$

c. Finding the value:

$$\begin{aligned}\log_{10} 1600 &= \log_{10}\left(8^2 \bullet 5^2\right) \\ &= \log_{10} 8^2 + \log_{10} 5^2 \\ &= 2\log_{10} 8 + 2\log_{10} 5 \\ &= 2(0.903) + 2(0.699) \\ &= 3.204\end{aligned}$$

69. Rewriting the expression: $\text{pH} = 6.1 + \log_{10}\left(\frac{x}{y}\right) = 6.1 + \log_{10} x - \log_{10} y$

71. Solving for M: $M = 0.21\log_{10}\frac{1}{10^{-12}} = 0.21\log_{10} 10^{12} = 0.21(12) = 2.52$

73. Simplifying: $5^0 = 1$

75. Simplifying: $\log_3 3 = \log_3 3^1 = 1$

77. Simplifying: $\log_b b^4 = 4$

79. Using a calculator: $10^{-5.6} \approx 2.5 \times 10^{-6}$

81. Using a calculator: $\frac{2.00 \times 10^8}{3.96 \times 10^6} \approx 51$

8.5 Common Logarithms and Natural Logarithms

1. Evaluating the logarithm: $\log 378 \approx 2.5775$

3. Evaluating the logarithm: $\log 37.8 \approx 1.5775$

5. Evaluating the logarithm: $\log 3,780 \approx 3.5775$

7. Evaluating the logarithm: $\log 0.0378 \approx -1.4225$

9. Evaluating the logarithm: $\log 37,800 \approx 4.5775$

11. Evaluating the logarithm: $\log 600 \approx 2.7782$

13. Evaluating the logarithm: $\log 2,010 \approx 3.3032$

15. Evaluating the logarithm: $\log 0.00971 \approx -2.0128$

17. Evaluating the logarithm: $\log 0.0314 \approx -1.5031$

19. Evaluating the logarithm: $\log 0.399 \approx -0.3990$

21. Solving for x:
$$\log x = 2.8802$$
$$x = 10^{2.8802} \approx 759$$

23. Solving for x:
$$\log x = -2.1198$$
$$x = 10^{-2.1198} \approx 0.00759$$

25. Solving for x:
$$\log x = 3.1553$$
$$x = 10^{3.1553} \approx 1,430$$

27. Solving for x:
$$\log x = -5.3497$$
$$x = 10^{-5.3497} \approx 0.00000447$$

29. Solving for x:
$$\log x = -7.0372$$
$$x = 10^{-7.0372} \approx 0.0000000918$$

31. Solving for x:
$$\log x = 10$$
$$x = 10^{10}$$

33. Solving for x:
$$\log x = -10$$
$$x = 10^{-10}$$

35. Solving for x:
$$\log x = 20$$
$$x = 10^{20}$$

37. Solving for x:
$$\log x = -2$$
$$x = 10^{-2} = \frac{1}{100}$$

39. Solving for x:
$$\log x = \log_2 8$$
$$\log x = 3$$
$$x = 10^3 = 1,000$$

41. Solving for x:
$$\ln x = 1$$
$$x = e^{-1} = \frac{1}{e}$$

43. Solving for x:
$$\log x = 2\log 5$$
$$\log x = \log 5^2$$
$$x = 25$$

45. Solving for x:
$$\ln x = -3\ln 2$$
$$\ln x = \ln 2^{-3}$$
$$x = \frac{1}{8}$$

47. Simplifying the logarithm: $\ln e = \ln e^1 = 1$

49. Simplifying the logarithm: $\ln e^5 = 5$

51. Simplifying the logarithm: $\ln e^x = x$

53. Simplifying the logarithm: $\log 10,000 = \log 10^4 = 4$

55. Simplifying the logarithm: $\ln\frac{1}{e^3} = \log e^{-3} = -3$

57. Simplifying the logarithm: $\log\sqrt{1000} = \log 10^{3/2} = \frac{3}{2}$

59. Using properties of logarithms: $\ln 10e^{3t} = \ln 10 + \ln e^{3t} = \ln 10 + 3t$

61. Using properties of logarithms: $\ln Ae^{-2t} = \ln A + \ln e^{-2t} = \ln A - 2t$

63. Using properties of logarithms: $\log\left[100(1.01)^{3t}\right] = \log 10^2 + \log 1.01^{3t} = 2 + 3t\log 1.01$

65. Using properties of logarithms: $\ln\left(Pe^{rt}\right) = \ln P + \ln e^{rt} = \ln P + rt$

67. Using properties of logarithms: $-\log\left(4.2\times 10^{-3}\right) = -\log 4.2 - \log 10^{-3} = 3 - \log 4.2$

69. Evaluating the logarithm: $\ln 15 = \ln(3\bullet 5) = \ln 3 + \ln 5 = 1.0986 + 1.6094 = 2.7080$

71. Evaluating the logarithm: $\ln\frac{1}{3} = \ln 3^{-1} = -\ln 3 = -1.0986$

73. Evaluating the logarithm: $\ln 9 = \ln 3^2 = 2\ln 3 = 2(1.0986) = 2.1972$

75. Evaluating the logarithm: $\ln 16 = \ln 2^4 = 4\ln 2 = 4(0.6931) = 2.7724$

77. For the 1906 earthquake:

$$\log T = 8.3$$
$$T = 10^{8.3} = 1.995\times 10^8$$

For the atomic bomb test:

$$\log T = 5.0$$
$$T = 10^{5.0} = 1\times 10^5$$

Dividing the two values: $\frac{1.995\times 10^8}{1\times 10^5} \approx 2000$. San Francisco earthquake was approximately 2,000 times greater.

79. It appears to approach *e*. Completing the table:

x	$(1+x)^{1/x}$
1	2
0.5	2.25
0.1	2.5937
0.01	2.7048
0.001	2.7169
0.0001	2.7181
0.00001	2.7183

81. Substituting $s = 15$:

$$5\ln x = 15$$
$$\ln x = 3$$
$$x = e^3 \approx 20$$

Approximately 15% of students enrolled are in the age range in the year 2009.

83. Computing the pH: $\text{pH} = -\log\left(6.50\times 10^{-4}\right) \approx 3.19$

85. Finding the concentration:

$$4.75 = -\log\left[H^+\right]$$
$$-4.75 = \log\left[H^+\right]$$
$$\left[H^+\right] = 10^{-4.75} \approx 1.78 \times 10^{-5}$$

87. Finding the magnitude:

$$5.5 = \log T$$
$$T = 10^{5.5} \approx 3.16 \times 10^5$$

89. Finding the magnitude:

$$8.3 = \log T$$
$$T = 10^{8.3} \approx 2.00 \times 10^8$$

91. Completing the table:

Location	Date	Magnitude (M)	Shockwave (T)
Moresby Island	January 23	4.0	1.00×10^4
Vancouver Island	April 30	5.3	1.99×10^5
Quebec City	June 29	3.2	1.58×10^3
Mould Bay	November 13	5.2	1.58×10^5
St. Lawrence	December 14	3.7	5.01×10^3

93. Finding the rate of depreciation:

$$\log(1-r) = \frac{1}{5}\log\frac{4500}{9000}$$
$$\log(1-r) \approx -0.0602$$
$$1-r \approx 10^{-0.0602}$$
$$r = 1 - 10^{-0.0602}$$
$$r \approx 0.129 = 12.9\%$$

95. Finding the rate of depreciation:

$$\log(1-r) = \frac{1}{5}\log\frac{5750}{7550}$$
$$\log(1-r) \approx -0.0237$$
$$1-r \approx 10^{-0.0237}$$
$$r = 1 - 10^{-0.0237}$$
$$r \approx 0.053 = 5.3\%$$

97. Solving the equation:

$$5(2x+1) = 12$$
$$10x + 5 = 12$$
$$10x = 7$$
$$x = \frac{7}{10} = 0.7$$

99. Using a calculator: $\frac{100{,}000}{32{,}000} = 3.125$

101. Using a calculator: $\frac{1}{2}\left(\frac{-0.6931}{1.4289} + 3\right) \approx 1.2575$

103. Rewriting the logarithm: $\log 1.05^t = t \log 1.05$

105. Simplifying: $\ln e^{0.05t} = 0.05t$

8.6 Exponential Equations and Change of Base

1. Solving the equation:

$$\begin{aligned} 3^x &= 5 \\ \ln 3^x &= \ln 5 \\ x\ln 3 &= \ln 5 \\ x &= \frac{\ln 5}{\ln 3} \approx 1.4650 \end{aligned}$$

3. Solving the equation:

$$\begin{aligned} 5^x &= 3 \\ \ln 5^x &= \ln 3 \\ x\ln 5 &= \ln 3 \\ x &= \frac{\ln 3}{\ln 5} \approx 0.6826 \end{aligned}$$

5. Solving the equation:

$$\begin{aligned} 5^{-x} &= 12 \\ \ln 5^{-x} &= \ln 12 \\ -x\ln 5 &= \ln 12 \\ x &= -\frac{\ln 12}{\ln 5} \approx -1.5440 \end{aligned}$$

7. Solving the equation:

$$\begin{aligned} 12^{-x} &= 5 \\ \ln 12^{-x} &= \ln 5 \\ -x\ln 12 &= \ln 5 \\ x &= -\frac{\ln 5}{\ln 12} \approx -0.6477 \end{aligned}$$

9. Solving the equation:

$$\begin{aligned} 8^{x+1} &= 4 \\ 2^{3x+3} &= 2^2 \\ 3x+3 &= 2 \\ 3x &= -1 \\ x &= -\frac{1}{3} \end{aligned}$$

11. Solving the equation:

$$\begin{aligned} 4^{x-1} &= 4 \\ 4^{x-1} &= 4^1 \\ x-1 &= 1 \\ x &= 2 \end{aligned}$$

13. Solving the equation:

$$\begin{aligned} 3^{2x+1} &= 2 \\ \ln 3^{2x+1} &= \ln 2 \\ (2x+1)\ln 3 &= \ln 2 \\ 2x+1 &= \frac{\ln 2}{\ln 3} \\ 2x &= \frac{\ln 2}{\ln 3} - 1 \\ x &= \frac{1}{2}\left(\frac{\ln 2}{\ln 3} - 1\right) \approx -0.1845 \end{aligned}$$

15. Solving the equation:

$$\begin{aligned} 3^{1-2x} &= 2 \\ \ln 3^{1-2x} &= \ln 2 \\ (1-2x)\ln 3 &= \ln 2 \\ 1-2x &= \frac{\ln 2}{\ln 3} \\ -2x &= \frac{\ln 2}{\ln 3} - 1 \\ x &= \frac{1}{2}\left(1 - \frac{\ln 2}{\ln 3}\right) \approx 0.1845 \end{aligned}$$

17. Solving the equation:

$$15^{3x-4} = 10$$
$$\ln 15^{3x-4} = \ln 10$$
$$(3x-4)\ln 15 = \ln 10$$
$$3x-4 = \frac{\ln 10}{\ln 15}$$
$$3x = \frac{\ln 10}{\ln 15} + 4$$
$$x = \frac{1}{3}\left(\frac{\ln 10}{\ln 15} + 4\right) \approx 1.6168$$

19. Solving the equation:

$$6^{5-2x} = 4$$
$$\ln 6^{5-2x} = \ln 4$$
$$(5-2x)\ln 6 = \ln 4$$
$$5-2x = \frac{\ln 4}{\ln 6}$$
$$-2x = \frac{\ln 4}{\ln 6} - 5$$
$$x = \frac{1}{2}\left(5 - \frac{\ln 4}{\ln 6}\right) \approx 2.1131$$

21. Solving the equation:

$$3^{-4x} = 81$$
$$3^{-4x} = 3^4$$
$$-4x = 4$$
$$x = -1$$

23. Solving the equation:

$$5^{3x-2} = 15$$
$$\ln 5^{3x-2} = \ln 15$$
$$(3x-2)\ln 5 = \ln 15$$
$$3x-2 = \frac{\ln 15}{\ln 5}$$
$$3x = \frac{\ln 15}{\ln 5} + 2$$
$$x = \frac{1}{3}\left(\frac{\ln 15}{\ln 5} + 2\right) \approx 1.2275$$

25. Solving the equation:

$$100e^{3t} = 250$$
$$e^{3t} = \frac{5}{2}$$
$$3t = \ln\frac{5}{2}$$
$$t = \frac{1}{3}\ln\frac{5}{2} \approx 0.3054$$

27. Solving the equation:

$$1200\left(1+\frac{0.072}{4}\right)^{4t} = 25000$$

$$\left(1+\frac{0.072}{4}\right)^{4t} = \frac{125}{6}$$

$$\ln\left(1+\frac{0.072}{4}\right)^{4t} = \ln\frac{125}{6}$$

$$4t\ln\left(1+\frac{0.072}{4}\right) = \ln\frac{125}{6}$$

$$t = \frac{\ln\frac{125}{6}}{4\ln\left(1+\frac{0.072}{4}\right)} \approx 42.5528$$

29. Solving the equation:

$$50e^{-0.0742t} = 32$$

$$e^{-0.0742t} = \frac{16}{25}$$

$$-0.0742t = \ln\frac{16}{25}$$

$$t = \frac{\ln\frac{16}{25}}{-0.0742} \approx 6.0147$$

31. Evaluating the logarithm: $\log_8 16 = \frac{\log 16}{\log 8} \approx 1.3333$

33. Evaluating the logarithm: $\log_{16} 8 = \frac{\log 8}{\log 16} = 0.7500$

35. Evaluating the logarithm: $\log_7 15 = \frac{\log 15}{\log 7} \approx 1.3917$

37. Evaluating the logarithm: $\log_{15} 7 = \frac{\log 7}{\log 15} \approx 0.7186$

39. Evaluating the logarithm: $\log_8 240 = \frac{\log 240}{\log 8} \approx 2.6356$

41. Evaluating the logarithm: $\log_4 321 = \frac{\log 321}{\log 4} \approx 4.1632$

43. Evaluating the logarithm: $\ln 345 \approx 5.8435$

45. Evaluating the logarithm: $\ln 0.345 \approx -1.0642$

47. Evaluating the logarithm: $\ln 10 \approx 2.3026$

49. Evaluating the logarithm: $\ln 45{,}000 \approx 10.7144$

51. Using the compound interest formula:

$$500\left(1+\frac{0.06}{2}\right)^{2t}=1000$$
$$\left(1+\frac{0.06}{2}\right)^{2t}=2$$
$$\ln\left(1+\frac{0.06}{2}\right)^{2t}=\ln 2$$
$$2t\ln\left(1+\frac{0.06}{2}\right)=\ln 2$$
$$t=\frac{\ln 2}{2\ln\left(1+\frac{0.06}{2}\right)}\approx 11.72$$

It will take 11.72 years.

53. Using the compound interest formula:

$$1000\left(1+\frac{0.12}{6}\right)^{6t}=3000$$
$$\left(1+\frac{0.12}{6}\right)^{6t}=3$$
$$\ln\left(1+\frac{0.12}{6}\right)^{6t}=\ln 3$$
$$6t\ln\left(1+\frac{0.12}{6}\right)=\ln 3$$
$$t=\frac{\ln 3}{6\ln\left(1+\frac{0.12}{6}\right)}\approx 9.25$$

It will take 9.25 years.

55. Using the compound interest formula:

$$P\left(1+\frac{0.08}{4}\right)^{4t}=2P$$
$$\left(1+\frac{0.08}{4}\right)^{4t}=2$$
$$\ln\left(1+\frac{0.08}{4}\right)^{4t}=\ln 2$$
$$4t\ln\left(1+\frac{0.08}{4}\right)=\ln 2$$
$$t=\frac{\ln 2}{4\ln\left(1+\frac{0.08}{4}\right)}\approx 8.75$$

It will take 8.75 years.

57. Using the compound interest formula:

$$25\left(1+\frac{0.06}{2}\right)^{2t}=75$$
$$\left(1+\frac{0.06}{2}\right)^{2t}=3$$
$$\ln\left(1+\frac{0.06}{2}\right)^{2t}=\ln 3$$
$$2t\ln\left(1+\frac{0.06}{2}\right)=\ln 3$$
$$t=\frac{\ln 3}{2\ln\left(1+\frac{0.06}{2}\right)}\approx 18.58$$

It was invested 18.58 years ago.

59. Using the continuous interest formula:

$$500e^{0.06t}=1000$$
$$e^{0.06t}=2$$
$$0.06t=\ln 2$$
$$t=\frac{\ln 2}{0.06}\approx 11.55$$

It will take 11.55 years.

61. Using the continuous interest formula:

$$500e^{0.06t}=1500$$
$$e^{0.06t}=3$$
$$0.06t=\ln 3$$
$$t=\frac{\ln 3}{0.06}\approx 18.31$$

It will take 18.31 years.

63. Using the continuous interest formula:

$$\begin{aligned} 1000e^{0.08t} &= 2500 \\ e^{0.08t} &= 2.5 \\ 0.08t &= \ln 2.5 \\ t &= \frac{\ln 2.5}{0.08} \approx 11.45 \end{aligned}$$

It will take 11.45 years.

65. Using the population model:

$$\begin{aligned} 32000e^{0.05t} &= 64000 \\ e^{0.05t} &= 2 \\ 0.05t &= \ln 2 \\ t &= \frac{\ln 2}{0.05} \approx 13.9 \end{aligned}$$

The city will reach 64,000 toward the end of the year 2007 (October).

67. Using the exponential model:

$$\begin{aligned} 466 \bullet 1.035^t &= 900 \\ 1.035^t &= \tfrac{900}{466} \\ \ln 1.035^t &= \ln \tfrac{900}{466} \\ t \ln 1.035 &= \ln \tfrac{900}{466} \\ t &= \frac{\ln \frac{900}{466}}{\ln 1.035} \approx 19 \end{aligned}$$

In the year 2009 it is predicted that 900 million passengers will travel by airline.

69. Using the exponential model:

$$\begin{aligned} 78.16(1.11)^t &= 800 \\ 1.11^t &= \frac{800}{78.16} \\ \ln 1.11^t &= \ln \frac{800}{78.16} \\ t \ln 1.11 &= \ln \frac{800}{78.16} \\ t &= \frac{\ln \dfrac{800}{78.16}}{\ln 1.11} \approx 22 \end{aligned}$$

In the year 1992 it was estimated that $800 billion will be spent on health care expenditures.

71. Using the compound interest formula:

$$16552\left(1+\frac{0.07}{2}\right)^{2t}=33104$$
$$\left(1+\frac{0.07}{2}\right)^{2t}=2$$
$$\ln\left(1+\frac{0.07}{2}\right)^{2t}=\ln 2$$
$$2t\ln\left(1+\frac{0.07}{2}\right)=\ln 2$$
$$t=\frac{\ln 2}{2\ln\left(1+\frac{0.07}{2}\right)}\approx 10.07$$

It will take 10.07 years for the money to double.

73. Using the exponential formula:

$$0.10e^{0.0576t}=1.00$$
$$e^{0.0576t}=10$$
$$0.0576t=\ln 10$$
$$t=\frac{\ln 10}{0.0576}\approx 40$$

A Coca Cola will cost \$1.00 in the year 2000.

75. Completing the square: $y=2x^2+8x-15=2\left(x^2+4x+4\right)-8-15=2(x+2)^2-23$.
The lowest point is (–2,–23).

77. Completing the square: $y=12x-4x^2=-4\left(x^2-3x+\frac{9}{4}\right)+9=-4\left(x-\frac{3}{2}\right)^2+9$.

The highest point is $\left(\frac{3}{2},9\right)$.

79. Completing the square: $y=64t-16t^2=-16\left(t^2-4t+4\right)+64=-16(t-2)^2+64$
The object reaches a maximum height after 2 seconds, and the maximum height is 64 feet.

Chapter 8 Test

See www.mathtv.com for video solutions to all problems in this chapter test.

Chapter 9
Sequences and Series

9.1 More About Sequences

1. The first five terms are: $4,7,10,13,16$
3. The first five terms are: $3,7,11,15,19$
5. The first five terms are: $1,2,3,4,5$
7. The first five terms are: $4,7,12,19,28$
9. The first five terms are: $\frac{1}{4},\frac{2}{5},\frac{1}{2},\frac{4}{7},\frac{5}{8}$
11. The first five terms are: $1,\frac{1}{4},\frac{1}{9},\frac{1}{16},\frac{1}{25}$
13. The first five terms are: $2,4,8,16,32$
15. The first five terms are: $2,\frac{3}{2},\frac{4}{3},\frac{5}{4},\frac{6}{5}$
17. The first five terms are: $-2,4,-8,16,-32$
19. The first five terms are: $3,5,3,5,3$
21. The first five terms are: $1,-\frac{2}{3},\frac{3}{5},-\frac{4}{7},\frac{5}{9}$
23. The first five terms are: $\frac{1}{2},1,\frac{9}{8},1,\frac{25}{32}$
25. The first five terms are: $3,-9,27,-81,243$
27. The first five terms are: $1,5,13,29,61$
29. The first five terms are: $2,3,5,9,17$
31. The first five terms are: $5,11,29,83,245$
33. The first five terms are: $4,4,4,4,4$
35. The general term is: $a_n = 4n$
37. The general term is: $a_n = n^2$
39. The general term is: $a_n = 2^{n+1}$
41. The general term is: $a_n = \frac{1}{2^{n+1}}$
43. The general term is: $a_n = 3n+2$
45. The general term is: $a_n = -4n+2$
47. The general term is: $a_n = (-2)^{n-1}$
49. The general term is: $a_n = \log_{n+1}(n+2)$
51. a. The sequence of salaries is: \$28000, \$29120, \$30284.80, \$31,496.19, \$32756.04
    b. The general term is: $a_n = 28000(1.04)^{n-1}$
53. a. The sequence of values is: 16 ft, 48 ft, 80 ft, 112 ft, 144 ft
    b. The sum of the values is 400 feet.
    c. No, since the sum is less than 420 feet.
55. a. The distances traveled are: 10 ft, 8 ft, 6.4 ft
    b. The general term is: $a_n = 10\left(\frac{4}{5}\right)^{n-1}$
    c. Substituting $n = 10$: $a_{10} = 10\left(\frac{4}{5}\right)^9 \approx 1.34$ feet
57. Simplifying: $9-27+81-243=-180$
59. Simplifying: $-4+8-16+32-64=-44$
61. Simplifying: $(1-3)+(9+1)+(16+1)+(25+1)+(36+1)=-2+10+17+26+37=88$

63. Simplifying: $\frac{1}{2}+\frac{2}{3}+\frac{3}{4}+\frac{4}{5}=\frac{30}{60}+\frac{40}{60}+\frac{45}{60}+\frac{48}{60}=\frac{163}{60}$

65. Simplifying: $\frac{1}{16}+\frac{1}{32}+\frac{1}{64}=\frac{4}{64}+\frac{2}{64}+\frac{1}{64}=\frac{7}{64}$

9.2 Series

1. Expanding the sum: $\sum_{i=1}^{4}(2i+4)=6+8+10+12=36$

3. Expanding the sum: $\sum_{i=2}^{3}(i^2-1)=3+8=11$

5. Expanding the sum: $\sum_{i=1}^{4}(i^2-3)=-2+1+6+13=18$

7. Expanding the sum: $\sum_{i=1}^{4}\frac{i}{1+i}=\frac{1}{2}+\frac{2}{3}+\frac{3}{4}+\frac{4}{5}=\frac{30}{60}+\frac{40}{60}+\frac{45}{60}+\frac{48}{60}=\frac{163}{60}$

9. Expanding the sum: $\sum_{i=1}^{4}(-3)^i=-3+9-27+81=60$

11. Expanding the sum: $\sum_{i=3}^{6}(-2)^i=-8+16-32+64=40$

13. Expanding the sum: $\sum_{i=2}^{6}(-2)^i=4-8+16-32+64=44$

15. Expanding the sum: $\sum_{i=1}^{5}\left(-\frac{1}{2}\right)^i=-\frac{1}{2}+\frac{1}{4}-\frac{1}{8}+\frac{1}{16}-\frac{1}{32}=-\frac{16}{32}+\frac{8}{32}-\frac{4}{32}+\frac{2}{32}-\frac{1}{32}=-\frac{11}{32}$

17. Expanding the sum: $\sum_{i=2}^{5}\frac{i-1}{i+1}=\frac{1}{3}+\frac{1}{2}+\frac{3}{5}+\frac{2}{3}=1+\frac{5}{10}+\frac{6}{10}=\frac{21}{10}$

19. Expanding the sum: $\sum_{i=1}^{5}(x+i)=(x+1)+(x+2)+(x+3)+(x+4)+(x+5)=5x+15$

21. Expanding the sum: $\sum_{i=1}^{4}(x-2)^i=(x-2)+(x-2)^2+(x-2)^3+(x-2)^4$

23. Expanding the sum: $\sum_{i=1}^{5}\frac{x+i}{x-1}=\frac{x+1}{x-1}+\frac{x+2}{x-1}+\frac{x+3}{x-1}+\frac{x+4}{x-1}+\frac{x+5}{x-1}$

25. Expanding the sum: $\sum_{i=3}^{8}(x+i)^i=(x+3)^3+(x+4)^4+(x+5)^5+(x+6)^6+(x+7)^7+(x+8)^8$

27. Expanding the sum: $\sum_{i=3}^{6}(x-2i)^{i+3}=(x-6)^6+(x-8)^7+(x-10)^8+(x-12)^9$

29. Writing with summation notation: $2+4+8+16=\sum_{i=1}^{4}2^i$

31. Writing with summation notation: $4+8+16+32+64=\sum_{i=2}^{6} 2^i$

33. Writing with summation notation: $5+9+13+17+21=\sum_{i=1}^{5}(4i+1)$

35. Writing with summation notation: $-4+8-16+32=\sum_{i=2}^{5} -(-2)^i$

37. Writing with summation notation: $\frac{3}{4}+\frac{4}{5}+\frac{5}{6}+\frac{6}{7}+\frac{7}{8}=\sum_{i=3}^{7}\frac{i}{i+1}$

39. Writing with summation notation: $\frac{1}{3}+\frac{2}{5}+\frac{3}{7}+\frac{4}{9}=\sum_{i=1}^{4}\frac{i}{2i+1}$

41. Writing with summation notation: $(x-2)^6+(x-2)^7+(x-2)^8+(x-2)^9=\sum_{i=6}^{9}(x-2)^i$

43. Writing with summation notation: $\left(1+\frac{1}{x}\right)^2+\left(1+\frac{2}{x}\right)^3+\left(1+\frac{3}{x}\right)^4+\left(1+\frac{4}{x}\right)^5=\sum_{i=1}^{4}\left(1+\frac{i}{x}\right)^{i+1}$

45. Writing with summation notation: $\frac{x}{x+3}+\frac{x}{x+4}+\frac{x}{x+5}=\sum_{i=3}^{5}\frac{x}{x+i}$

47. Writing with summation notation: $x^2(x+2)+x^3(x+3)+x^4(x+4)=\sum_{i=2}^{4}x^i(x+i)$

49. **a.** Writing as a series: $\frac{1}{3}=0.3+0.03+0.003+0.0003+\ldots$

b. Writing as a series: $\frac{2}{9}=0.2+0.02+0.002+0.0002+\ldots$

c. Writing as a series: $\frac{3}{11}=0.27+0.0027+0.000027+\ldots$

51. The sequence of values he falls is: 16, 48, 80, 112, 144, 176, 208
During the seventh second he falls 208 feet, and the total he falls is 784 feet.

53. **a.** The series is $16+48+80+112+144$.

b. Writing in summation notation: $\sum_{i=1}^{5}(32i-16)$

55. Simplifying: $2+9(8)=2+72=74$

57. Simplifying: $\frac{10}{2}\left(\frac{1}{2}+5\right)=5\left(\frac{1}{2}+\frac{10}{2}\right)=5\left(\frac{11}{2}\right)=\frac{55}{2}$

59. Simplifying: $3+(n-1)(2)=3+2n-2=2n+1$

61. Multiplying the first equation by –1:

$$-x-2y=-7$$
$$x+7y=17$$

Adding yields:

$$5y=10$$
$$y=2$$

Substituting into the first equation:

$$x+2(2)=7$$
$$x+4=7$$
$$x=3$$

The solution is (3,2).

9.3 Arithmetic Sequences

1. The sequence is arithmetic: $d=1$

3. The sequence is not arithmetic.

5. The sequence is arithmetic: $d=-5$

7. The sequence is not arithmetic.

9. The sequence is arithmetic: $d=\frac{2}{3}$

11. Finding the general term: $a_n=3+(n-1)\bullet 4=3+4n-4=4n-1$

Therefore: $a_{24}=4\bullet 24-1=96-1=95$

13. Finding the required term: $a_{10}=6+(10-1)\bullet(-2)=6-18=-12$

Finding the sum: $S_{10}=\frac{10}{2}(6-12)=5(-6)=-30$

15. Writing out the equations:

$$a_6=a_1+5d \quad a_{12}=a_1+11d$$
$$17=a_1+5d \quad 29=a_1+11d$$

The system of equations is:

$$a_1+11d=29$$
$$a_1+5d=17$$

Subtracting yields:

$$6d=12$$
$$d=2$$
$$a_1=7$$

Finding the required term: $a_{30}=7+29\bullet 2=7+58=65$

17. Writing out the equations:

$$a_3=a_1+2d \quad a_8=a_1+7d$$
$$16=a_1+2d \quad 26=a_1+7d$$

The system of equations is:

$$a_1+7d=26$$
$$a_1+2d=16$$

Subtracting yields:

$$5d=10$$
$$d=2$$
$$a_1=12$$

Finding the required term: $a_{20}=12+19\bullet 2=12+38=50$

Finding the sum: $S_{20}=\frac{20}{2}(12+50)=10\bullet 62=620$

19. Finding the required term: $a_{20} = 3 + 19 \bullet 4 = 3 + 76 = 79$

Finding the sum: $S_{20} = \frac{20}{2}(3 + 79) = 10 \bullet 82 = 820$

21. Writing out the equations:

$$a_4 = a_1 + 3d \quad a_{10} = a_1 + 9d$$

$$14 = a_1 + 3d \quad 32 = a_1 + 9d$$

The system of equations is:

$$a_1 + 9d = 32$$

$$a_1 + 3d = 14$$

Subtracting yields:

$$6d = 18$$

$$d = 3$$

$$a_1 = 5$$

Finding the required term: $a_{40} = 5 + 39 \bullet 3 = 5 + 117 = 122$

Finding the sum: $S_{40} = \frac{40}{2}(5 + 122) = 20 \bullet 127 = 2540$

23. Using the summation formula:

$$S_6 = \frac{6}{2}(a_1 + a_6)$$

$$-12 = 3(a_1 - 17)$$

$$a_1 - 17 = -4$$

$$a_1 = 13$$

Now find d:

$$a_6 = 13 + 5 \bullet d$$

$$-17 = 13 + 5d$$

$$5d = -30$$

$$d = -6$$

25. Using $a_1 = 14$ and $d = -3$: $a_{85} = 14 + 84 \bullet (-3) = 14 - 252 = -238$

27. Using the summation formula:

$$S_{20} = \frac{20}{2}(a_1 + a_{20})$$

$$80 = 10(-4 + a_{20})$$

$$-4 + a_{20} = 8$$

$$a_{20} = 12$$

Now finding d:

$$a_{20} = a_1 + 19d$$
$$12 = -4 + 19d$$
$$16 = 19d$$
$$d = \frac{16}{19}$$

Finding the required term: $a_{39} = -4 + 38\left(\frac{16}{19}\right) = -4 + 32 = 28$

29. Using $a_1 = 5$ and $d = 4$: $a_{100} = a_1 + 99d = 5 + 99 \bullet 4 = 5 + 396 = 401$

Now finding the required sum: $S_{100} = \frac{100}{2}(5 + 401) = 50 \bullet 406 = 20,300$

31. Using $a_1 = 12$ and $d = -5$: $a_{35} = a_1 + 34d = 12 + 34 \bullet (-5) = 12 - 170 = -158$

33. Using $a_1 = \frac{1}{2}$ and $d = \frac{1}{2}$: $a_{10} = a_1 + 9d = \frac{1}{2} + 9 \bullet \frac{1}{2} = \frac{10}{2} = 5$

Finding the sum: $S_{10} = \frac{10}{2}\left(\frac{1}{2} + 5\right) = \frac{10}{2} \bullet \frac{11}{2} = \frac{55}{2}$

35. **a.** The first five terms are: \$18,000, \$14,700, \$11,400, \$8,100, \$4,800

b. The common difference is –\$3,300.

c. Constructing a line graph:

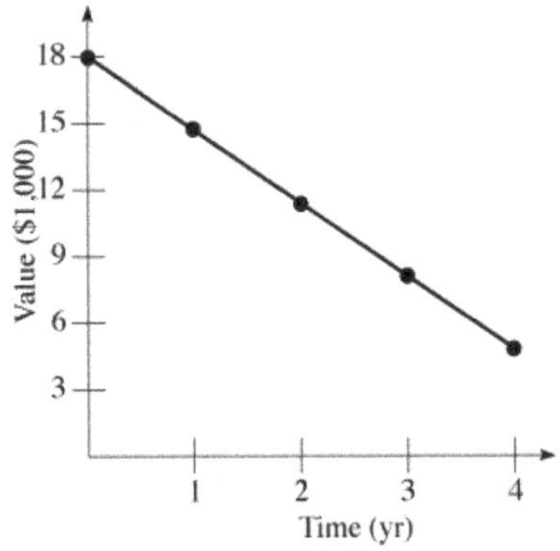

d. The value is approximately \$9,750.

e. The recursive formula is: $a_0 = 18000; a_n = a_{n-1} - 3300$ for $n \geq 1$

37. **a.** The sequence of values is: 1500 ft, 1460 ft, 1420 ft, 1380 ft, 1340 ft, 1300 ft

b. It is arithmetic because the same amount is subtracted from each succeeding term.

c. The general term is: $a_n = 1500 + (n-1) \bullet (-40) = 1500 - 40n + 40 = 1540 - 40n$

39. **a.** The first 15 triangular numbers is: 1,3,6,10,15,21,28,36,45,55,66,78,91,105,120

b. The recursive formula is: $a_1 = 1; a_n = n + a_{n-1}$ for $n \geq 2$

c. It is not arithmetic because the same amount is not added to each term.

41. **a.** The general term is: $a_n = 16 + (n-1) \bullet 32 = 16 + 32n - 32 = 32n - 16$

b. Substituting $n = 10$: $a_{10} = 32 \bullet 10 - 16 = 304$ feet

c. Finding the sum: $S_{10} = \frac{10}{2}(16 + 304) = 5 \bullet 320 = 1600$ feet

43. Simplifying: $\frac{1}{8}\left(\frac{1}{2}\right)=\frac{1}{16}$

45. Simplifying: $\frac{3\sqrt{3}}{3}=\sqrt{3}$

47. Simplifying: $2\bullet 2^{n-1}=2^{1+n-1}=2^n$

49. Simplifying: $\frac{ar^6}{ar^3}=r^{6-3}=r^3$

51. Simplifying: $\frac{\frac{1}{5}}{1-\frac{1}{2}}=\frac{\frac{1}{5}}{1-\frac{1}{2}}\bullet\frac{10}{10}=\frac{2}{10-5}=\frac{2}{5}$

53. Simplifying: $\frac{3\left[(-2)^8-1\right]}{-2-1}=\frac{3(256-1)}{-2-1}=\frac{3(255)}{-3}=-255$

9.4 Geometric Sequences

1. The sequence is geometric: $r=5$

3. The sequence is geometric: $r=\frac{1}{3}$

5. The sequence is not geometric.

7. The sequence is geometric: $r=-2$

9. The sequence is not geometric.

11. Finding the general term: $a_n=4\bullet 3^{n-1}$

13. Finding the term: $a_6=-2\left(-\frac{1}{2}\right)^{6-1}=-2\left(-\frac{1}{2}\right)^5=-2\left(-\frac{1}{32}\right)=\frac{1}{16}$

15. Finding the term: $a_{20}=3(-1)^{20-1}=3(-1)^{19}=-3$

17. Finding the sum: $S_{10}=\frac{10\left(2^{10}-1\right)}{2-1}=10\bullet 1023=10,230$

19. Finding the sum: $S_{20}=\frac{1\left((-1)^{20}-1\right)}{-1-1}=\frac{1\bullet 0}{-2}=0$

21. Using $a_1=\frac{1}{5}$ and $r=\frac{1}{2}$, the term is: $a_8=\frac{1}{5}\bullet\left(\frac{1}{2}\right)^{8-1}=\frac{1}{5}\bullet\left(\frac{1}{2}\right)^7=\frac{1}{5}\bullet\frac{1}{128}=\frac{1}{640}$

23. Using $a_1=-\frac{1}{2}$ and $r=\frac{1}{2}$, the sum is: $S_5=\frac{-\frac{1}{2}\left(\left(\frac{1}{2}\right)^5-1\right)}{\frac{1}{2}-1}=\frac{-\frac{1}{2}\left(\frac{1}{32}-1\right)}{-\frac{1}{2}}=\frac{1}{32}-1=-\frac{31}{32}$

25. Using $a_1=\sqrt{2}$ and $r=\sqrt{2}$, the term is: $a_{10}=\sqrt{2}\left(\sqrt{2}\right)^9=\left(\sqrt{2}\right)^{10}=2^5=32$

The sum is: $S_{10}=\frac{\sqrt{2}\left(\left(\sqrt{2}\right)^{10}-1\right)}{\sqrt{2}-1}=\frac{\sqrt{2}(32-1)}{\sqrt{2}-1}=\frac{31\sqrt{2}}{\sqrt{2}-1}\bullet\frac{\sqrt{2}+1}{\sqrt{2}+1}=\frac{62+31\sqrt{2}}{2-1}=62+31\sqrt{2}$

27. Using $a_1 = 100$ and $r = 0.1$, the term is: $a_6 = 100(0.1)^5 = 10^2(10^{-5}) = 10^{-3} = \frac{1}{1000}$

The sum is: $S_6 = \frac{100((0.1)^6 - 1)}{0.1 - 1} = \frac{100(10^{-6} - 1)}{-0.9} = \frac{-99.9999}{-0.9} = 111.111$

29. Since $a_4 \bullet r \bullet r = a_6$, we have the equation:

$$a_4 r^2 = a_6$$
$$40r^2 = 160$$
$$r^2 = 4$$
$$r = \pm 2$$

31. Since $a_1 = -3$ and $r = -2$, the values are:

$$a_8 = -3(-2)^7 = -3(-128) = 384$$
$$S_8 = \frac{-3((-2)^8 - 1)}{-2 - 1} = \frac{-3(256 - 1)}{-3} = 255$$

33. Since $a_7 \bullet r \bullet r \bullet r = a_{10}$, we have the equation:

$$a_7 r^3 = a_{10}$$
$$13r^3 = 104$$
$$r^3 = 8$$
$$r = 2$$

35. Using $a_1 = \frac{1}{2}$ and $r = \frac{1}{2}$ in the sum formula: $S = \frac{\frac{1}{2}}{1 - \frac{1}{2}} = \frac{\frac{1}{2}}{\frac{1}{2}} = 1$

37. Using $a_1 = 4$ and $r = \frac{1}{2}$ in the sum formula: $S = \frac{4}{1 - \frac{1}{2}} = \frac{4}{\frac{1}{2}} = 8$

39. Using $a_1 = 2$ and $r = \frac{1}{2}$ in the sum formula: $S = \frac{2}{1 - \frac{1}{2}} = \frac{2}{\frac{1}{2}} = 4$

41. Using $a_1 = \frac{4}{3}$ and $r = -\frac{1}{2}$ in the sum formula: $S = \frac{\frac{4}{3}}{1 + \frac{1}{2}} = \frac{\frac{4}{3}}{\frac{3}{2}} = \frac{4}{3} \bullet \frac{2}{3} = \frac{8}{9}$

43. Using $a_1 = \frac{2}{5}$ and $r = \frac{2}{5}$ in the sum formula: $S = \frac{\frac{2}{5}}{1 - \frac{2}{5}} = \frac{\frac{2}{5}}{\frac{3}{5}} = \frac{2}{5} \bullet \frac{5}{3} = \frac{2}{3}$

45. Using $a_1 = \frac{3}{4}$ and $r = \frac{1}{3}$ in the sum formula: $S = \frac{\frac{3}{4}}{1 - \frac{1}{3}} = \frac{\frac{3}{4}}{\frac{2}{3}} = \frac{3}{4} \bullet \frac{3}{2} = \frac{9}{8}$

47. Interpreting the decimal as an infinite sum with $a_1 = 0.4$ and $r = 0.1$: $S = \frac{0.4}{1 - 0.1} = \frac{0.4}{0.9} \bullet \frac{10}{10} = \frac{4}{9}$

49. Interpreting the decimal as an infinite sum with $a_1 = 0.27$ and $r = 0.01$:

$$S=\frac{0.27}{1-0.01}=\frac{0.27}{0.99}\bullet\frac{100}{100}=\frac{27}{99}=\frac{3}{11}$$

51. **a.** The first five terms are: \$450,000, \$315,000, \$220,500, \$154,350, \$108,045

b. The common ratio is 0.7.

c. Constructing a line graph:

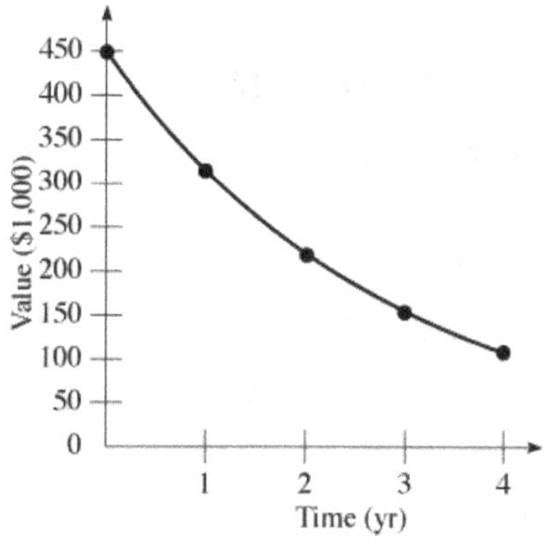

d. The value is approximately \$90,000.

e. The recursive formula is: $a_0=450000; a_n=0.7a_{n-1}$ for $n\geq 1$

53. **a.** Using $a_1=\frac{1}{3}$ and $r=\frac{1}{3}$ in the sum formula: $S=\dfrac{\frac{1}{3}}{1-\frac{1}{3}}=\dfrac{\frac{1}{3}}{\frac{2}{3}}=\frac{1}{2}$

b. Finding the sum: $S_6=\dfrac{\frac{1}{3}\left(\left(\frac{1}{3}\right)^6-1\right)}{\frac{1}{3}-1}=\dfrac{\frac{1}{3}\left(\frac{1}{729}-1\right)}{-\frac{2}{3}}=\dfrac{\frac{1}{3}\left(-\frac{728}{729}\right)}{-\frac{2}{3}}=-\frac{1}{2}\left(-\frac{728}{729}\right)=\frac{364}{729}$

c. Finding the difference of these two answers: $S-S_6=\frac{1}{2}-\frac{364}{729}=\frac{729}{1458}-\frac{728}{1458}=\frac{1}{1458}$

55. **a.** The pile is now $2(0.002)=0.004$ inches.

b. Using $a_1=0.002$ and $r=2$, find the term: $a_5=0.002(2)^5=0.064$ inches

c. Using $a_1=0.002$ and $r=2$, find the term: $a_{25}=0.002(2)^{25}=67{,}108.864$ inches

57. **a.** The sequence of incomes is: \$60,000, \$64,200, \$68,694, \$73,503, \$78,648

b. The general term is: $a_n=60000(1.07)^{n-1}$

c. Finding the sum: $S_{10}=\dfrac{60000\left(1.07^{10}-1\right)}{1.07-1}=\$828{,}987$

59. Expanding: $(x+y)^1=x+y$

61. Expanding: $(x+y)^3=(x+y)(x+y)^2=(x+y)\left(x^2+2xy+y^2\right)=x^3+3x^2y+3xy^2+y^3$

63. Simplifying: $\dfrac{7\bullet6\bullet5\bullet4\bullet3\bullet2\bullet1}{(5\bullet4\bullet3\bullet2\bullet1)(2\bullet1)}=\dfrac{7\bullet6}{2\bullet1}=\dfrac{42}{2}=21$

9.5 The Binomial Expansion

1. Using the binomial formula:

$$(x+2)^4=\binom{4}{0}x^4+\binom{4}{1}x^3(2)+\binom{4}{2}x^2(2)^2+\binom{4}{3}x(2)^3+\binom{4}{4}(2)^4$$
$$=x^4+4\bullet 2x^3+6\bullet 4x^2+4\bullet 8x+16$$
$$=x^4+8x^3+24x^2+32x+16$$

3. Using the binomial formula:

$$(x+y)^6=\binom{6}{0}x^6+\binom{6}{1}x^5y+\binom{6}{2}x^4y^2+\binom{6}{3}x^3y^3+\binom{6}{4}x^2y^4+\binom{6}{5}xy^5+\binom{6}{6}y^6$$
$$=x^6+6x^5y+15x^4y^2+20x^3y^3+15x^2y^4+6xy^5+y^6$$

5. Using the binomial formula:

$$(2x+1)^5=\binom{5}{0}(2x)^5+\binom{5}{1}(2x)^4(1)+\binom{5}{2}(2x)^3(1)^2+\binom{5}{3}(2x)^2(1)^3+\binom{5}{4}(2x)(1)^4+\binom{5}{5}(1)^5$$
$$=32x^5+5\bullet 16x^4+10\bullet 8x^3+10\bullet 4x^2+5\bullet 2x+1$$
$$=32x^5+80x^4+80x^3+40x^2+10x+1$$

7. Using the binomial formula:

$$(x-2y)^5=\binom{5}{0}x^5+\binom{5}{1}x^4(-2y)+\binom{5}{2}x^3(-2y)^2+\binom{5}{3}x^2(-2y)^3+\binom{5}{4}x(-2y)^4+\binom{5}{5}(-2y)^5$$
$$=x^5-5\bullet 2x^4y+10\bullet 4x^3y^2-10\bullet 8x^2y^3+5\bullet 16xy^4-32y^5$$
$$=x^5-10x^4y+40x^3y^2-80x^2y^3+80xy^4-32y^5$$

9. Using the binomial formula:

$$(3x-2)^4=\binom{4}{0}(3x)^4+\binom{4}{1}(3x)^3(-2)+\binom{4}{2}(3x)^2(-2)^2+\binom{4}{3}(3x)(-2)^3+\binom{4}{4}(-2)^4$$
$$=81x^4-4\bullet 54x^3+6\bullet 36x^2-4\bullet 24x+16$$
$$=81x^4-216x^3+216x^2-96x+16$$

11. Using the binomial formula:

$$(4x-3y)^3=\binom{3}{0}(4x)^3+\binom{3}{1}(4x)^2(-3y)+\binom{3}{2}(4x)(-3y)^2+\binom{3}{3}(-3y)^3$$
$$=64x^3-3\bullet 48x^2y+3\bullet 36xy^2-27y^3$$
$$=64x^3-144x^2y+108xy^2-27y^3$$

13. Using the binomial formula:

$$(x^2+2)^4=\binom{4}{0}(x^2)^4+\binom{4}{1}(x^2)^3(2)+\binom{4}{2}(x^2)^2(2)^2+\binom{4}{3}(x^2)(2)^3+\binom{4}{4}(2)^4$$
$$=x^8+4\bullet 2x^6+6\bullet 4x^4+4\bullet 8x^2+16$$
$$=x^8+8x^6+24x^4+32x^2+16$$

15. Using the binomial formula:

$$\left(x^2+y^2\right)^3=\binom{3}{0}\left(x^2\right)^3+\binom{3}{1}\left(x^2\right)^2\left(y^2\right)+\binom{3}{2}\left(x^2\right)\left(y^2\right)^2+\binom{3}{3}\left(y^2\right)^3=x^6+3x^4y^2+3x^2y^4+y^6$$

17. Using the binomial formula:

$$(2x+3y)^4=\binom{4}{0}(2x)^4+\binom{4}{1}(2x)^3(3y)+\binom{4}{2}(2x)^2(3y)^2+\binom{4}{3}(2x)(3y)^3+\binom{4}{4}(3y)^4$$
$$=16x^4+4\bullet 24x^3y+6\bullet 36x^2y^2+4\bullet 54xy^3+81y^4$$
$$=16x^4+96x^3y+216x^2y^2+216xy^3+81y^4$$

19. Using the binomial formula:

$$\left(\frac{x}{2}+\frac{y}{3}\right)^3=\binom{3}{0}\left(\frac{x}{2}\right)^3+\binom{3}{1}\left(\frac{x}{2}\right)^2\left(\frac{y}{3}\right)+\binom{3}{2}\left(\frac{x}{2}\right)\left(\frac{y}{3}\right)^2+\binom{3}{3}\left(\frac{y}{3}\right)^3$$
$$=\frac{x^3}{8}+3\bullet\frac{x^2y}{12}+3\bullet\frac{xy^2}{18}+\frac{y^3}{27}$$
$$=\frac{x^3}{8}+\frac{x^2y}{4}+\frac{xy^2}{6}+\frac{y^3}{27}$$

21. Using the binomial formula:

$$\left(\frac{x}{2}-4\right)^3=\binom{3}{0}\left(\frac{x}{2}\right)^3+\binom{3}{1}\left(\frac{x}{2}\right)^2(-4)+\binom{3}{2}\left(\frac{x}{2}\right)(-4)^2+\binom{3}{3}(-4)^3$$
$$=\frac{x^3}{8}-3\bullet x^2+3\bullet 8x-64$$
$$=\frac{x^3}{8}-3x^2+24x-64$$

23. Using the binomial formula:

$$\left(\frac{x}{3}+\frac{y}{2}\right)^4=\binom{4}{0}\left(\frac{x}{3}\right)^4+\binom{4}{1}\left(\frac{x}{3}\right)^3\left(\frac{y}{2}\right)+\binom{4}{2}\left(\frac{x}{3}\right)^2\left(\frac{y}{2}\right)^2+\binom{4}{3}\left(\frac{x}{3}\right)\left(\frac{y}{2}\right)^3+\binom{4}{4}\left(\frac{y}{2}\right)^4$$
$$=\frac{x^4}{81}+4\bullet\frac{x^3y}{54}+6\bullet\frac{x^2y^2}{36}+4\bullet\frac{xy^3}{24}+\frac{y^4}{16}$$
$$=\frac{x^4}{81}+\frac{2x^3y}{27}+\frac{x^2y^2}{6}+\frac{xy^3}{6}+\frac{y^4}{16}$$

25. Writing the first four terms:

$$\binom{9}{0}x^9+\binom{9}{1}x^8(2)+\binom{9}{2}x^7(2)^2+\binom{9}{3}x^6(2)^3$$
$$=x^9+9\bullet 2x^8+36\bullet 4x^7+84\bullet 8x^6$$
$$=x^9+18x^8+144x^7+672x^6$$

27. Writing the first four terms:

$$\binom{10}{0}x^{10}+\binom{10}{1}x^9(-y)+\binom{10}{2}x^8(-y)^2+\binom{10}{3}x^7(-y)^3=x^{10}-10x^9y+45x^8y^2-120x^7y^3$$

29. Writing the first four terms:

$$\binom{25}{0}x^{25}+\binom{25}{1}x^{24}(3)+\binom{25}{2}x^{23}(3)^2+\binom{25}{3}x^{22}(3)^3$$
$$=x^{25}+25\bullet 3x^{24}+300\bullet 9x^{23}+2300\bullet 27x^{22}$$
$$=x^{25}+75x^{24}+2700x^{23}+62100x^{22}$$

31. Writing the first four terms:

$$\binom{60}{0}x^{60}+\binom{60}{1}x^{59}(-2)+\binom{60}{2}x^{58}(-2)^2+\binom{60}{3}x^{57}(-2)^3$$
$$=x^{60}-60\bullet 2x^{59}+1770\bullet 4x^{58}-34220\bullet 8x^{57}$$
$$=x^{60}-120x^{59}+7080x^{58}-273760x^{57}$$

33. Writing the first four terms:

$$\binom{18}{0}x^{18}+\binom{18}{1}x^{17}(-y)+\binom{18}{2}x^{16}(-y)^2+\binom{18}{3}x^{15}(-y)^3=x^{18}-18x^{17}y+153x^{16}y^2-816x^{15}y^3$$

35. Writing the first three terms: $\binom{15}{0}x^{15}+\binom{15}{1}x^{14}(1)+\binom{15}{2}x^{13}(1)^2=x^{15}+15x^{14}+105x^{13}$

37. Writing the first three terms: $\binom{12}{0}x^{12}+\binom{12}{1}x^{11}(-y)+\binom{12}{2}x^{10}(-y)^2=x^{12}-12x^{11}y+66x^{10}y^2$

39. Writing the first three terms:

$$\binom{20}{0}x^{20}+\binom{20}{1}x^{19}(2)+\binom{20}{2}x^{18}(2)^2=x^{20}+20\bullet 2x^{19}+190\bullet 4x^{18}=x^{20}+40x^{19}+760x^{18}$$

41. Writing the first two terms: $\binom{100}{0}x^{100}+\binom{100}{1}x^{99}(2)=x^{100}+100\bullet 2x^{99}=x^{100}+200x^{99}$

43. Writing the first two terms: $\binom{50}{0}x^{50}+\binom{50}{1}x^{49}y=x^{50}+50x^{49}y$

45. Finding the required term: $\binom{12}{8}(2x)^4(3y)^8=495\bullet 2^4\bullet 3^8x^4y^8=51{,}963{,}120x^4y^8$

47. Finding the required term: $\binom{10}{4}x^6(-2)^4=210\bullet 16x^6=3{,}360x^6$

49. Finding the required term: $\binom{12}{5}x^7(-2)^5=-792\bullet 32x^7=-25{,}344x^7$

51. Finding the required term: $\binom{25}{2}x^{23}(-3y)^2=300\bullet 9x^{23}y^2=2{,}700x^{23}y^2$

53. Finding the required term: $\binom{20}{11}(2x)^9(5y)^{11}=\dfrac{20!}{11!9!}(2x)^9(5y)^{11}$

55. Writing the first three terms:

$$\binom{10}{0}\left(x^2y\right)^{10}+\binom{10}{1}\left(x^2y\right)^9(-3)+\binom{10}{2}\left(x^2y\right)^8(-3)^2$$
$$=x^{20}y^{10}-10\bullet 3x^{18}y^9+45\bullet 9x^{16}y^8$$
$$=x^{20}y^{10}-30x^{18}y^9+405x^{16}y^8$$

57. Finding the third term: $\binom{7}{2}\left(\frac{1}{2}\right)^5\left(\frac{1}{2}\right)^2=21\bullet\frac{1}{128}=\frac{21}{128}$

59. Solving the equation:

$$5^x=7$$
$$\log 5^x=\log 7$$
$$x\log 5=\log 7$$
$$x=\frac{\log 7}{\log 5}\approx 1.21$$

61. Solving the equation:

$$8^{2x+1}=16$$
$$2^{6x+3}=2^4$$
$$6x+3=4$$
$$6x=1$$
$$x=\frac{1}{6}$$

63. Using the compound interest formula:

$$400\left(1+\frac{0.10}{4}\right)^{4t}=800$$
$$(1.025)^{4t}=2$$
$$\ln(1.025)^{4t}=\ln 2$$
$$4t\ln 1.025=\ln 2$$
$$t=\frac{\ln 2}{4\ln 1.025}\approx 7.02$$

It will take 7.02 years.

65. Evaluating the logarithm: $\log_4 20=\frac{\log 20}{\log 4}\approx 2.16$

67. Evaluating the logarithm: $\ln 576\approx 6.36$

69. Solving for t:

$$A=10e^{5t}$$
$$\frac{A}{10}=e^{5t}$$
$$5t=\ln\left(\frac{A}{10}\right)$$
$$t=\frac{1}{5}\ln\left(\frac{A}{10}\right)$$

Chapter 9 Test

See www.mathtv.com for video solutions to all problems in this chapter test.

Chapter 10
Conic Sections

10.1 The Circle

1. Using the distance formula: $d=\sqrt{(6-3)^2+(3-7)^2}=\sqrt{9+16}=\sqrt{25}=5$

3. Using the distance formula: $d=\sqrt{(5-0)^2+(0-9)^2}=\sqrt{25+81}=\sqrt{106}$

5. Using the distance formula: $d=\sqrt{(-2-3)^2+(1+5)^2}=\sqrt{25+36}=\sqrt{61}$

7. Using the distance formula: $d=\sqrt{(-10+1)^2+(5+2)^2}=\sqrt{81+49}=\sqrt{130}$

9. Solving the equation:

$$\begin{aligned}\sqrt{(x-1)^2+(2-5)^2}&=\sqrt{13}\\(x-1)^2+9&=13\\(x-1)^2&=4\\x-1&=\pm 2\\x-1&=-2,2\\x&=-1,3\end{aligned}$$

11. Solving the equation:

$$\begin{aligned}\sqrt{(x-3)^2+(5-9)^2}&=5\\(x-3)^2+16&=25\\(x-3)^2&=9\\x-3&=\pm 3\\x-3&=-3,3\\x&=0,6\end{aligned}$$

13. Solving the equation:

$$\begin{aligned}\sqrt{(2x+1-x)^2+(6-4)^2}&=6\\(x+1)^2+4&=36\\(x+1)^2&=32\\x+1&=\pm\sqrt{32}\\x+1&=\pm 4\sqrt{2}\\x&=-1\pm 4\sqrt{2}\end{aligned}$$

15. The equation is $(x-3)^2+(y+2)^2=9$.

17. The equation is $(x+5)^2+(y+1)^2=5$.

19. The equation is $x^2+(y+5)^2=1$.

21. The equation is $x^2+y^2=4$.

23. The center is (0,0) and the radius is 2.

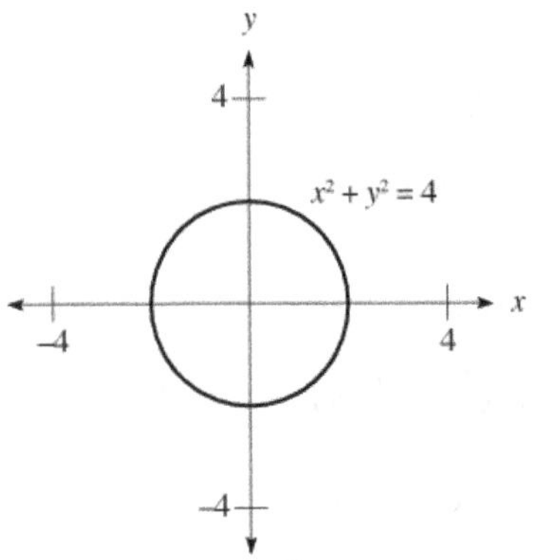

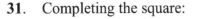

25. The center is (1,3) and the radius is 5.

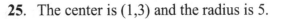

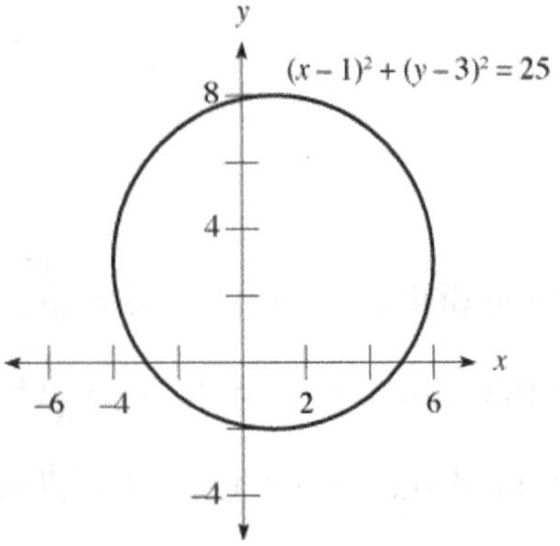

27. The center is (–2,4) and the radius is $2\sqrt{2}$.

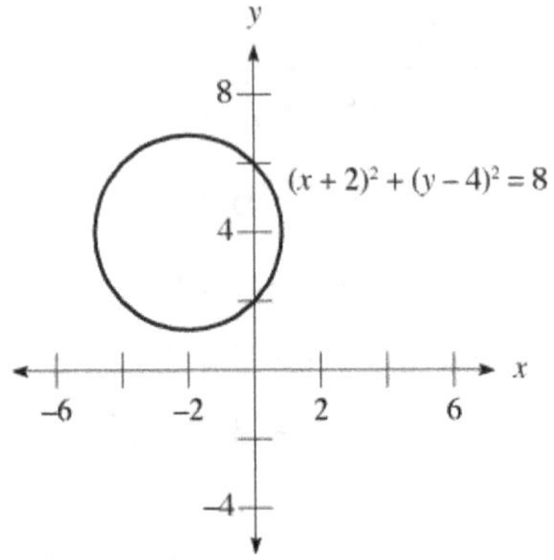

29. The center is (–2,4) and the radius is $\sqrt{17}$.

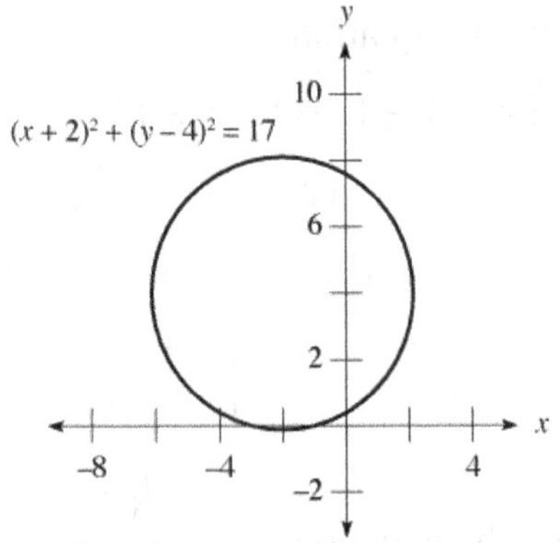

31. Completing the square:

$$x^2 + y^2 + 2x - 4y = 4$$
$$(x^2 + 2x + 1) + (y^2 - 4y + 4) = 4 + 1 + 4$$
$$(x+1)^2 + (y-2)^2 = 9$$

The center is (–1,2) and the radius is 3.

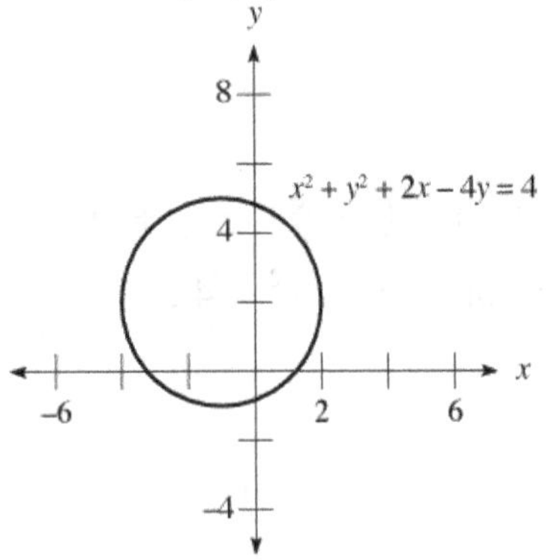

33. Completing the square:

$$x^2 + y^2 - 6y = 7$$
$$x^2 + (y^2 - 6y + 9) = 7 + 9$$
$$x^2 + (y-3)^2 = 16$$

The center is (0,3) and the radius is 4.

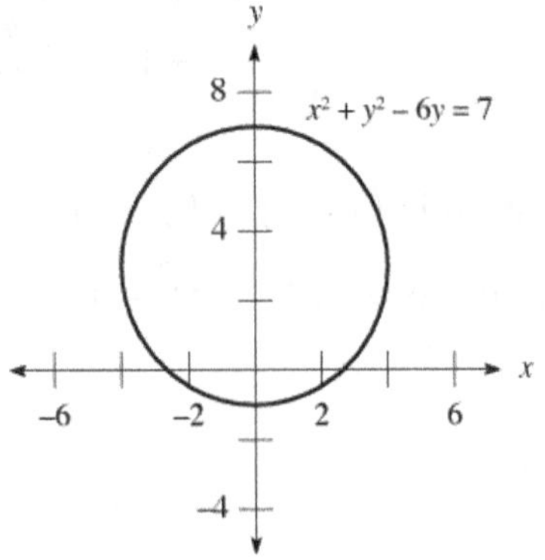

35. Completing the square:

$$x^2+y^2+2x=1$$
$$\left(x^2+2x+1\right)+y^2=1+1$$
$$(x+1)^2+y^2=2$$

The center is (–1,0) and the radius is $\sqrt{2}$.

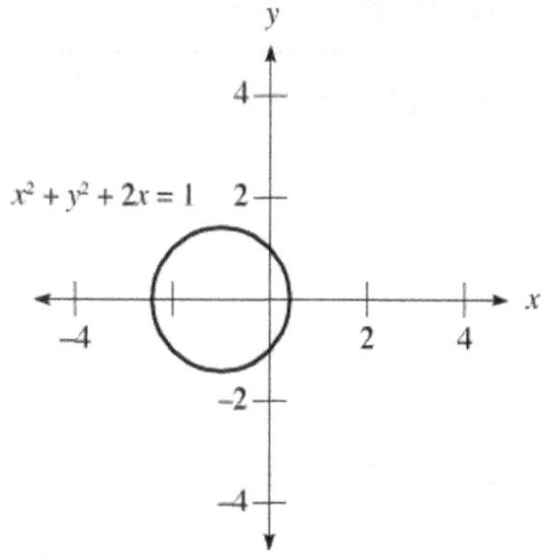

37. Completing the square:

$$x^2+y^2-4x-6y=-4$$
$$\left(x^2-4x+4\right)+\left(y^2-6y+9\right)=-4+4+9$$
$$(x-2)^2+(y-3)^2=9$$

The center is (2,3) and the radius is 3.

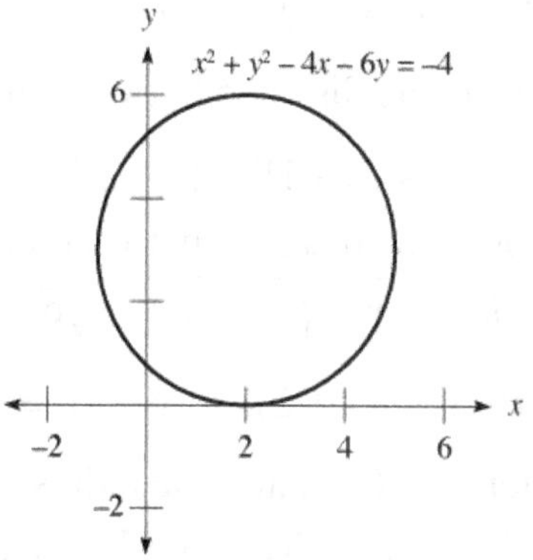

39. Completing the square:

$$x^2+y^2+2x+y=\frac{11}{4}$$
$$\left(x^2+2x+1\right)+\left(y^2+y+\frac{1}{4}\right)=\frac{11}{4}+1+\frac{1}{4}$$
$$(x+1)^2+\left(y+\frac{1}{2}\right)^2=4$$

The center is $\left(-1,-\frac{1}{2}\right)$ and the radius is 2.

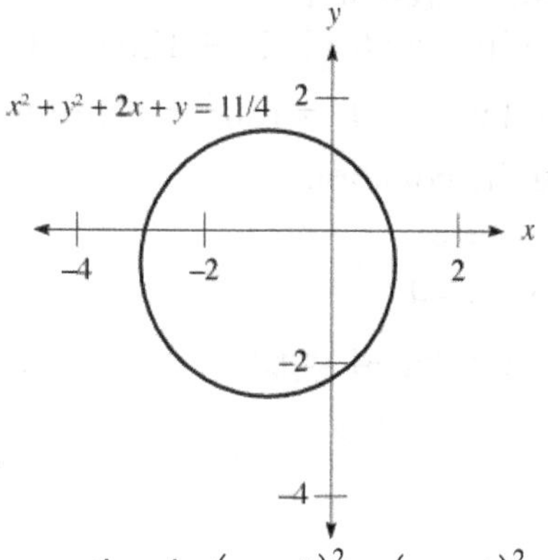

41. Completing the square:

$$4x^2+4y^2-4x+8y=11$$
$$x^2+y^2-x+2y=\frac{11}{4}$$
$$\left(x^2-x+\frac{1}{4}\right)+\left(y^2+2y+1\right)=\frac{11}{4}+\frac{1}{4}+1$$
$$\left(x-\frac{1}{2}\right)^2+(y+1)^2=4$$

The center is $\left(\frac{1}{2},-1\right)$ and the radius is 2.

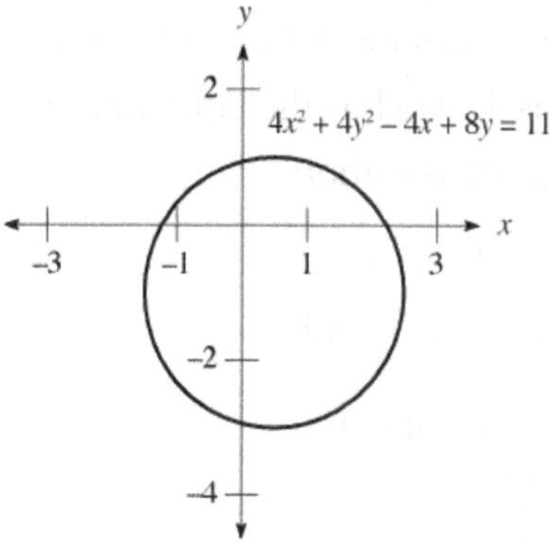

43. The equation is $(x-3)^2+(y-4)^2=25$.

45. The equations are:

$$\text{A: } \left(x-\frac{1}{2}\right)^2+(y-1)^2=\frac{1}{4}$$

$$\text{B: } (x-1)^2+(y-1)^2=1$$

$$\text{C: } (x-2)^2+(y-1)^2=4$$

47. The equation is $x^2+y^2=25$.

49. The equation is $x^2+y^2=9$.

51. The radius is the distance between these two points, which is: $r=\sqrt{(-1-4)^2+(3-3)^2}=\sqrt{25}=5$

The equation is $(x+1)^2+(y-3)^2=25$.

53. The radius is the distance between these two points, which is:

$$r=\sqrt{(1+2)^2+(-3-5)^2}=\sqrt{9+64}=\sqrt{73}$$

The equation is $(x+2)^2+(y-5)^2=73$.

55. The center is at (0,2) and the radius is 4, so the equation is $x^2+(y-2)^2=16$.

57. Since the radius is $\sqrt{18}=3\sqrt{2}$, the circumference and area are:

$$C=2\pi\left(3\sqrt{2}\right)=6\pi\sqrt{2} \qquad A=\pi\left(3\sqrt{2}\right)^2=18\pi$$

59. First complete the square:

$$x^2+y^2+4x+2y=20$$

$$\left(x^2+4x+4\right)+\left(y^2+2y+1\right)=20+4+1$$

$$(x+2)^2+(y+1)^2=25$$

Since the radius is 5, the circumference and area are:

$$C=2\pi(5)=10\pi \qquad A=\pi(5)^2=25\pi$$

61. His distance from home is: $d=\sqrt{5^2+3^2}=\sqrt{25+9}=\sqrt{34}\approx 5.8$

Yes, he was about 5.8 blocks from home, which is within the search area.

63. The x-coordinate of the center is $x = 500$, the y-coordinate of the center is $12 + 120 = 132$, and the radius is 120. Thus the equation of the circle is $(x-500)^2+(y-132)^2=120^2=14{,}400$.

65. Solving the equation:

$$y^2=9$$

$$y=\pm\sqrt{9}=\pm 3$$

67. Solving the equation:

$$-y^2=4$$

$$y^2=-4$$

$$y=\pm\sqrt{-4}=\pm 2i$$

69. Solving the equation:

$$\frac{-x^2}{9}=1$$

$$x^2=-9$$

$$x=\pm\sqrt{-9}=\pm 3i$$

71. Dividing: $\dfrac{4x^2+9y^2}{36}=\dfrac{4x^2}{36}+\dfrac{9y^2}{36}=\dfrac{x^2}{9}+\dfrac{y^2}{4}$

73. To find the x-intercept, let $y = 0$:

$$\begin{aligned} 3x - 4(0) &= 12 \\ 3x - 0 &= 12 \\ 3x &= 12 \\ x &= 4 \end{aligned}$$

To find the y-intercept, let $x = 0$:

$$\begin{aligned} 3(0) - 4y &= 12 \\ 0 - 4y &= 12 \\ -4y &= 12 \\ y &= -3 \end{aligned}$$

75. Substituting $x = 3$:

$$\begin{aligned} \frac{3^2}{25} + \frac{y^2}{9} &= 1 \\ \frac{9}{25} + \frac{y^2}{9} &= 1 \\ \frac{y^2}{9} &= \frac{16}{25} \\ y^2 &= \frac{144}{25} \\ y &= \pm\sqrt{\frac{144}{25}} = \pm\frac{12}{5} = \pm 2.4 \end{aligned}$$

10.2 Ellipses and Hyperbolas

1. Graphing the ellipse:

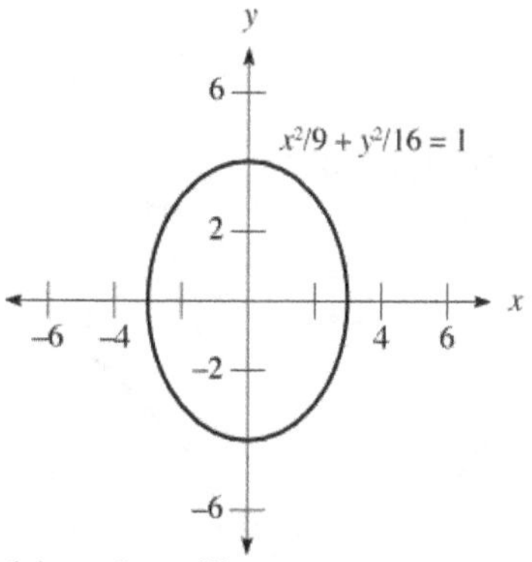

3. Graphing the ellipse:

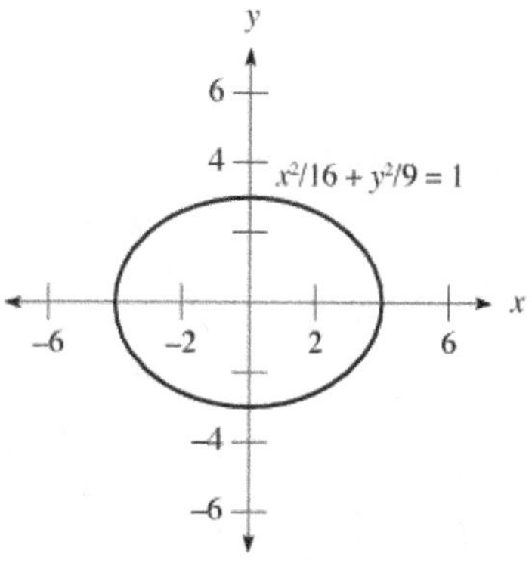

5. Graphing the ellipse:

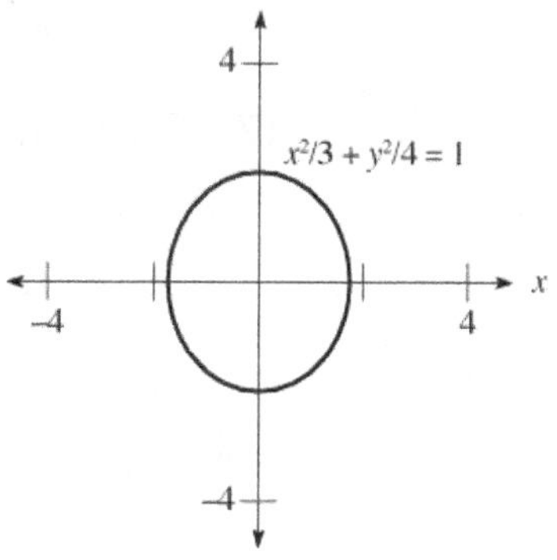

7. The standard form is $\frac{x^2}{25}+\frac{y^2}{4}=1$. Graphing the ellipse:

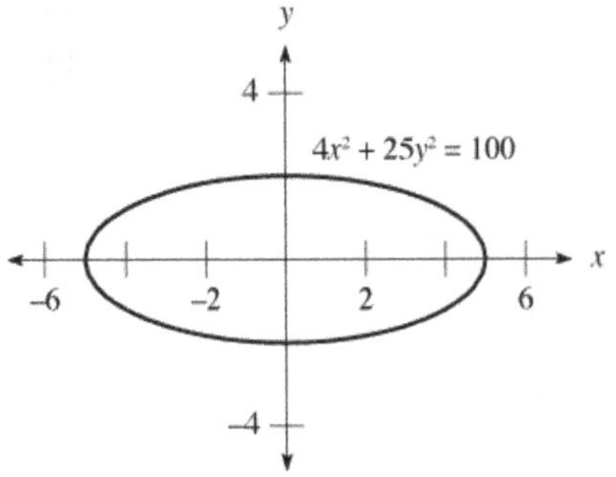

9. The standard form is $\frac{x^2}{16}+\frac{y^2}{2}=1$. Graphing the ellipse:

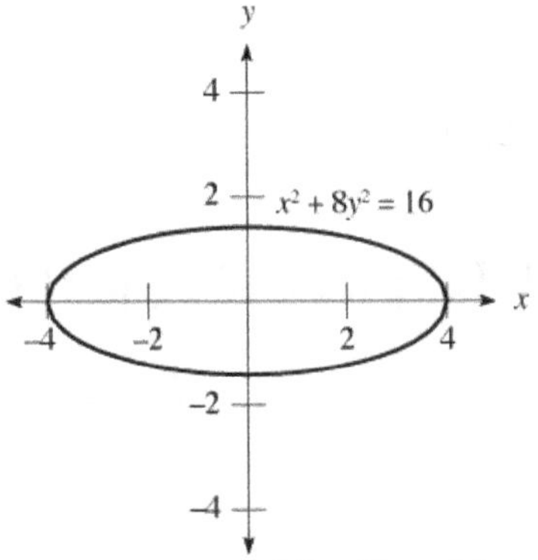

11. Graphing the hyperbola:

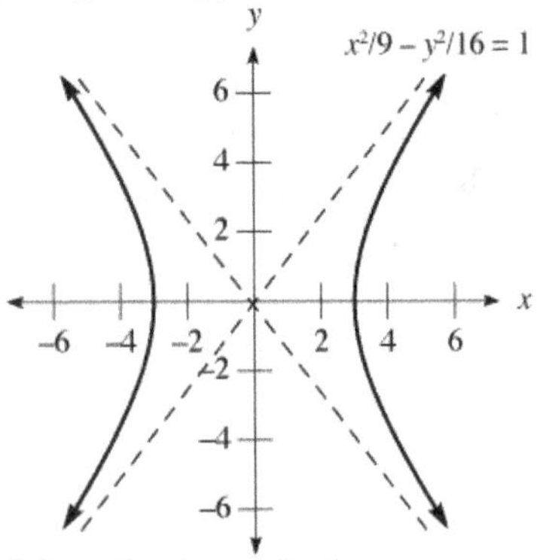

13. Graphing the hyperbola:

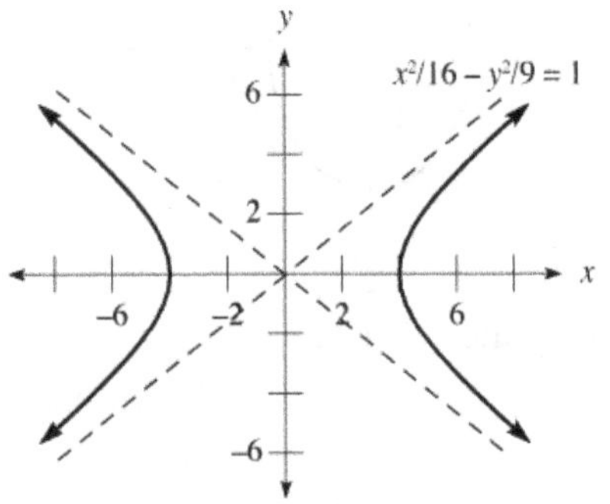

15. Graphing the hyperbola:

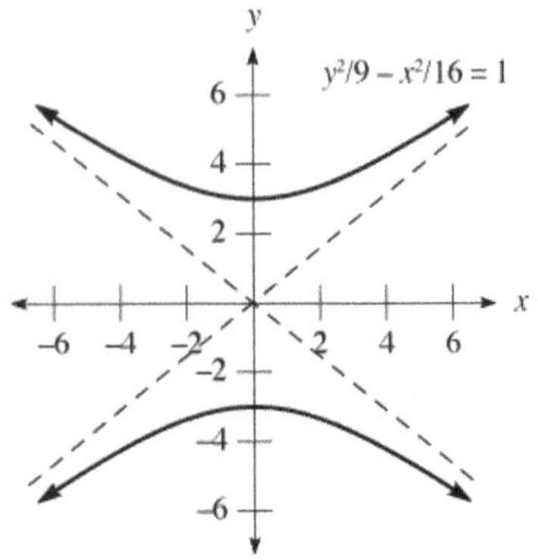

17. Graphing the hyperbola:

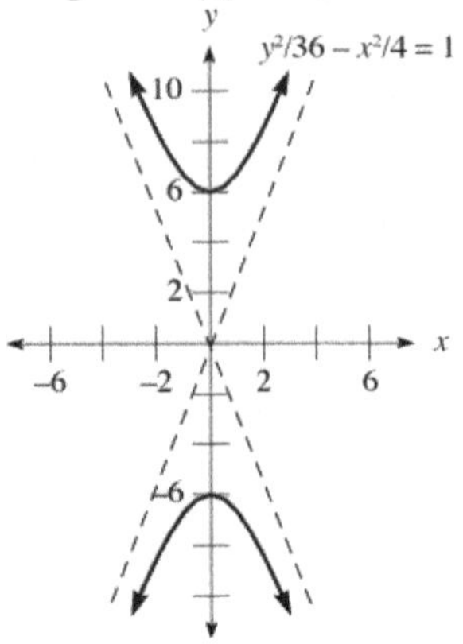

19. The standard form is $\frac{x^2}{4}-\frac{y^2}{1}=1$. Graphing the hyperbola:

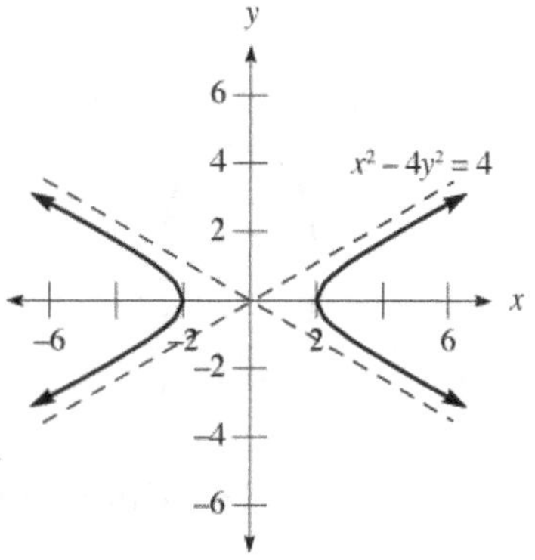

21. The standard form is $\frac{y^2}{9}-\frac{x^2}{16}=1$. Graphing the hyperbola:

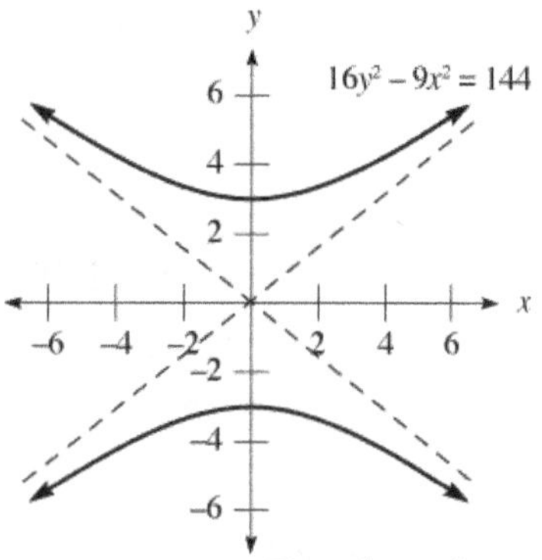

23. For the x-intercepts, set $y = 0$:

$$0.4x^2 = 3.6$$
$$x^2 = 9$$
$$x = \pm 3$$

For the y-intercepts, set $x = 0$:

$$0.9y^2 = 3.6$$
$$y^2 = 4$$
$$y = \pm 2$$

25. For the x-intercepts, set $y = 0$:

$$\frac{x^2}{0.04} = 1$$
$$x^2 = 0.04$$
$$x = \pm 0.2$$

For the y-intercepts, set $x = 0$:

$$-\frac{y^2}{0.09} = 1$$
$$y^2 = -0.09$$

There are no y-intercepts.

27. For the x-intercepts, set $y = 0$:

$$\frac{25x^2}{9} = 1$$
$$x^2 = \frac{9}{25}$$
$$x = \pm\frac{3}{5}$$

For the y-intercepts, set $x = 0$:

$$\frac{25y^2}{4} = 1$$
$$y^2 = \frac{4}{25}$$
$$y = \pm\frac{2}{5}$$

29. Graphing the ellipse:

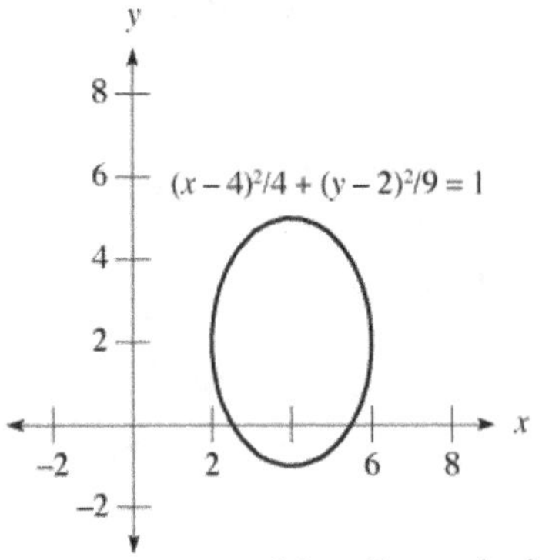

31. Completing the square:

$$4x^2 + y^2 - 4y - 12 = 0$$
$$4x^2 + \left(y^2 - 4y + 4\right) = 12 + 4$$
$$4x^2 + (y-2)^2 = 16$$
$$\frac{x^2}{4} + \frac{(y-2)^2}{16} = 1$$

Graphing the ellipse:

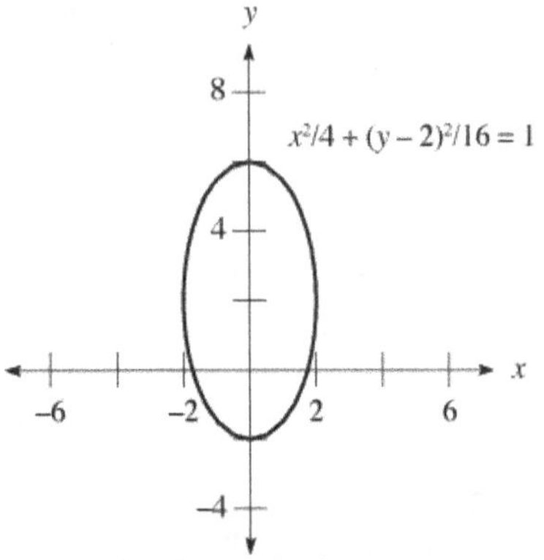

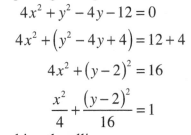

33. Completing the square:

$$x^2 + 9y^2 + 4x - 54y + 76 = 0$$
$$\left(x^2 + 4x + 4\right) + 9\left(y^2 - 6y + 9\right) = -76 + 4 + 81$$
$$(x+2)^2 + 9(y-3)^2 = 9$$
$$\frac{(x+2)^2}{9} + \frac{(y-3)^2}{1} = 1$$

Graphing the ellipse:

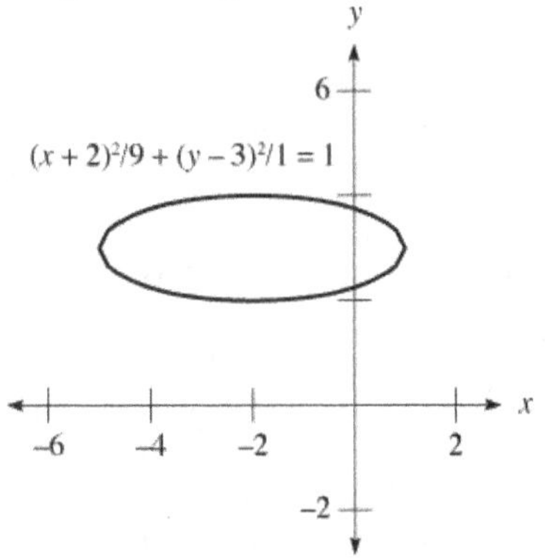

35. Graphing the hyperbola:

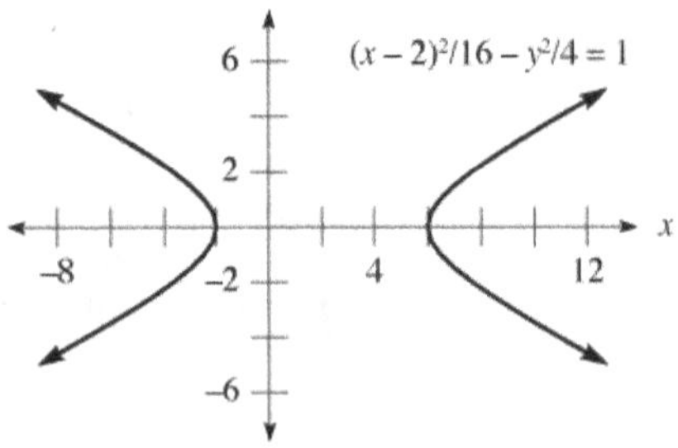

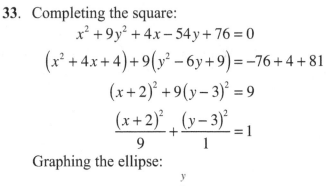

37. Completing the square:

$$9y^2 - x^2 - 4x + 54y + 68 = 0$$
$$9\left(y^2 + 6y + 9\right) - \left(x^2 + 4x + 4\right) = -68 + 81 - 4$$
$$9(y+3)^2 - (x+2)^2 = 9$$
$$\frac{(y+3)^2}{1} - \frac{(x+2)^2}{9} = 1$$

Graphing the hyperbola:

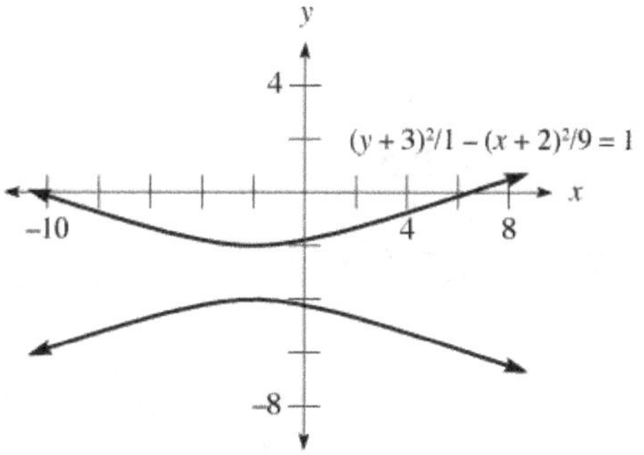

39. Completing the square:

$$4y^2 - 9x^2 - 16y + 72x - 164 = 0$$
$$4\left(y^2 - 4y + 4\right) - 9\left(x^2 - 8x + 16\right) = 164 + 16 - 144$$
$$4(y-2)^2 - 9(x-4)^2 = 36$$
$$\frac{(y-2)^2}{9} - \frac{(x-4)^2}{4} = 1$$

Graphing the hyperbola:

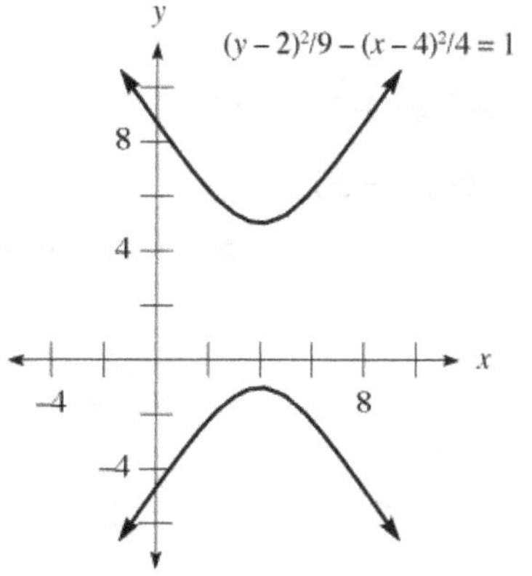

41. Substituting $y = 4$:

$$\frac{x^2}{25}+\frac{4^2}{16}=1$$
$$\frac{x^2}{25}+1=1$$
$$\frac{x^2}{25}=0$$
$$x=0$$

43. Substituting $x = -3$:

$$\frac{(-3)^2}{9}+\frac{y^2}{16}=1$$
$$1+\frac{y^2}{16}=1$$
$$\frac{y^2}{16}=0$$
$$y=0$$

45. The length of the major axis is 8.

47. The equation of the ellipse is $\frac{x^2}{307.5^2}+\frac{y^2}{255^2}=1$.

49. The equation for the path is $\frac{x^2}{229^2}+\frac{y^2}{195^2}=1$.

51. Substituting $a = 4$ and $c = 3$:

$$4^2=b^2+3^2$$
$$16=b^2+9$$
$$b^2=7$$
$$b=\sqrt{7}\approx 2.65$$

The width should be approximately 2(2.65) = 5.3 feet wide.

53. Since $4^2+0^2=16$ and $0^2+5^2=25$, while $0^2+0^2=0$, only (0,0) is a solution.

55. Multiplying: $(2y+4)^2=(2y)^2+2(2y)(4)+4^2=4y^2+16y+16$

57. Solving for x:

$$x-2y=4$$
$$x=2y+4$$

59. Simplifying: $x^2-2(x^2-3)=x^2-2x^2+6=-x^2+6$

61. Factoring: $5y^2+16y+12=(5y+6)(y+2)$

63. Solving the equation:

$$y^2=4$$
$$y=\pm\sqrt{4}=\pm 2$$

65. Solving the equation:

$$-x^2+6=2$$
$$-x^2=-4$$
$$x^2=4$$
$$x=\pm\sqrt{4}=\pm 2$$

10.3 Second-Degree Inequalities and Nonlinear Systems

1. Graphing the inequality:

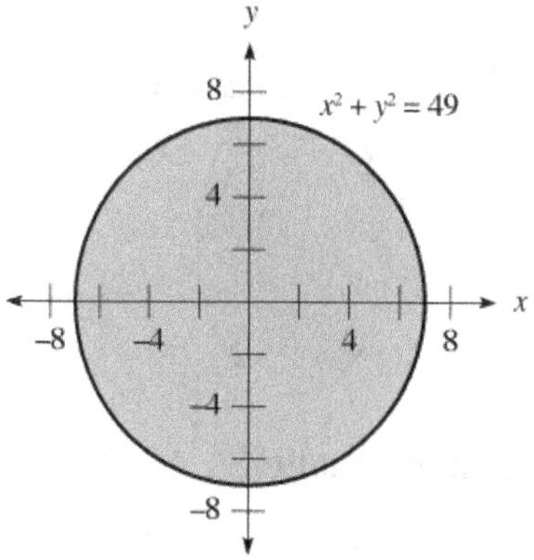

3. Graphing the inequality:

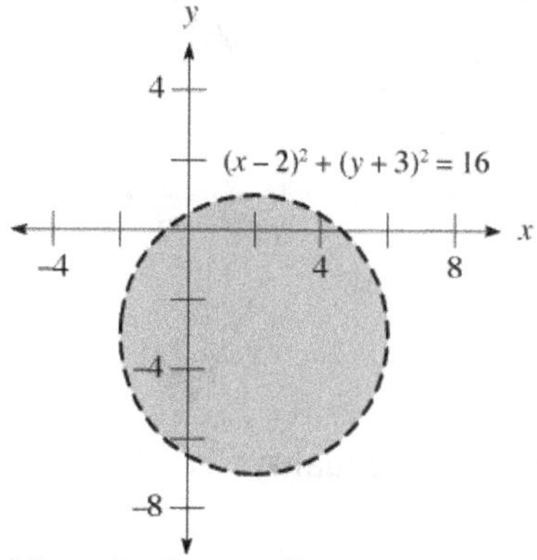

5. Graphing the inequality:

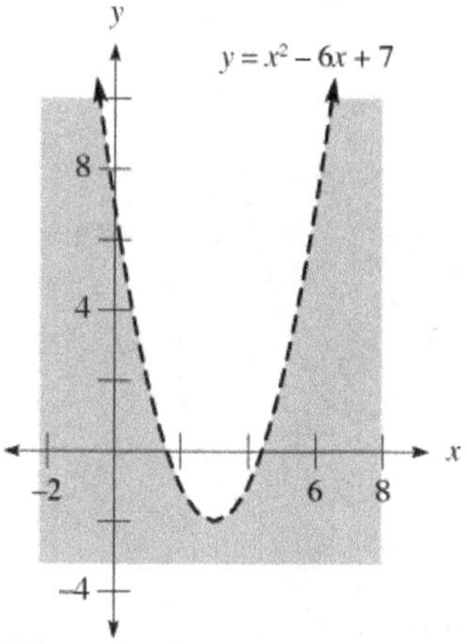

7. Graphing the inequality:

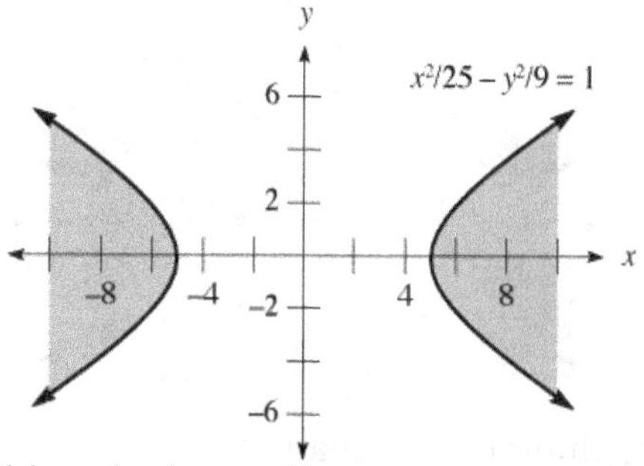

9. Graphing the inequality:

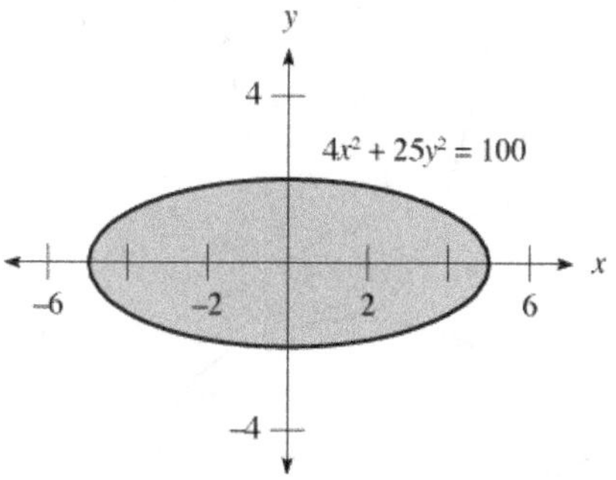

11. Graphing the inequality:

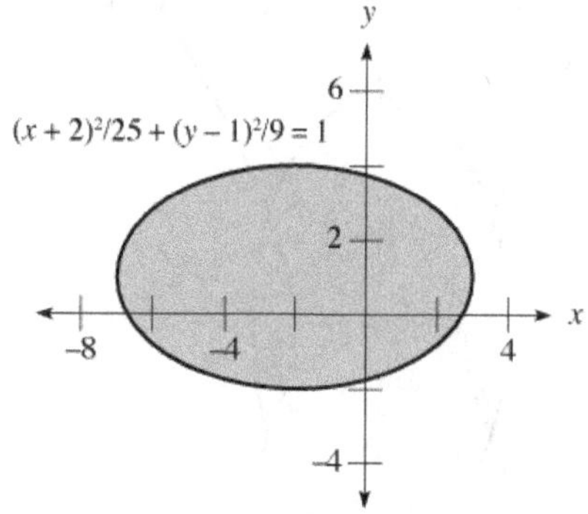

13. Graphing the inequality:

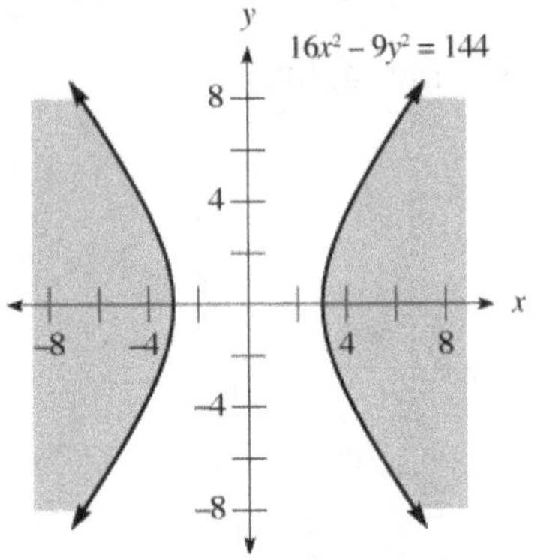

15. Graphing the inequality:

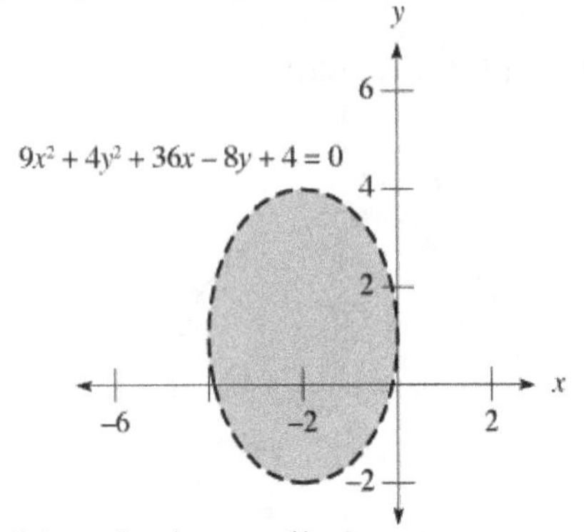

17. Graphing the inequality:

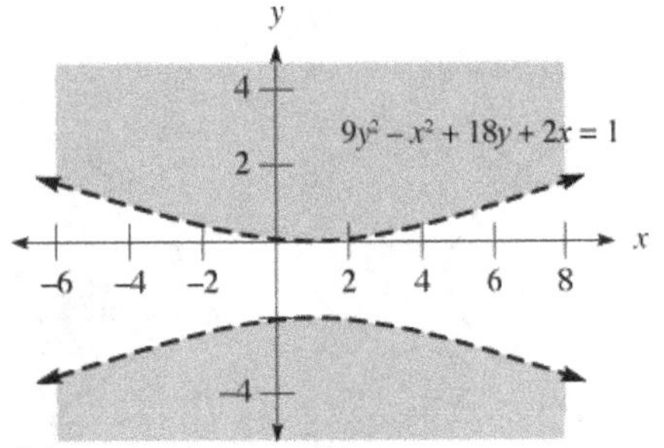

19. Graphing the inequality:

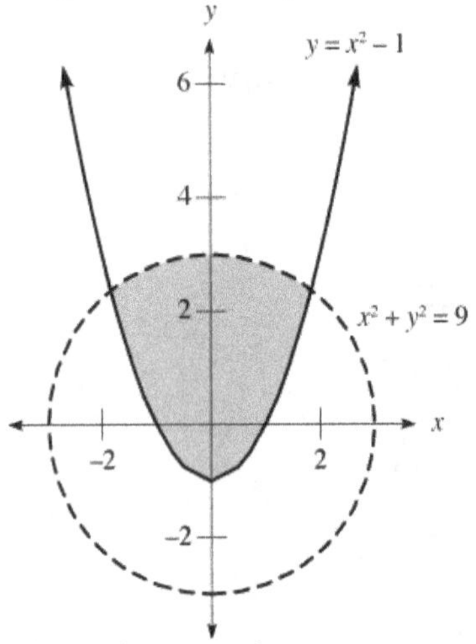

21. Graphing the inequality:

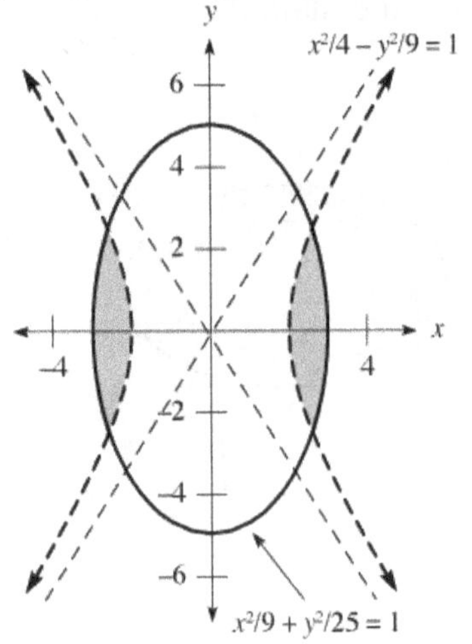

23. There is no intersection.

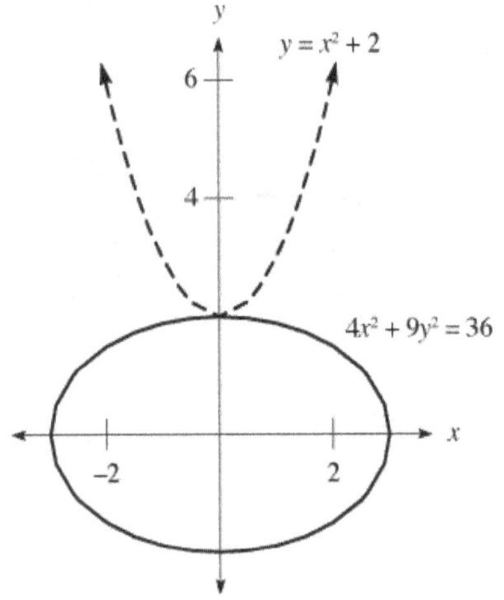

25. Graphing the inequality:

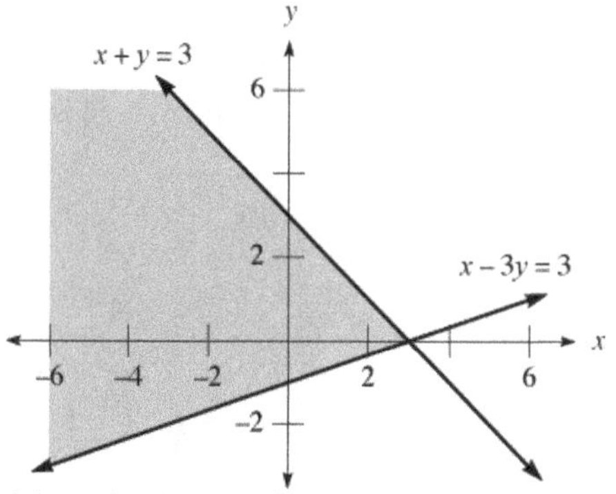

27. Graphing the inequality:

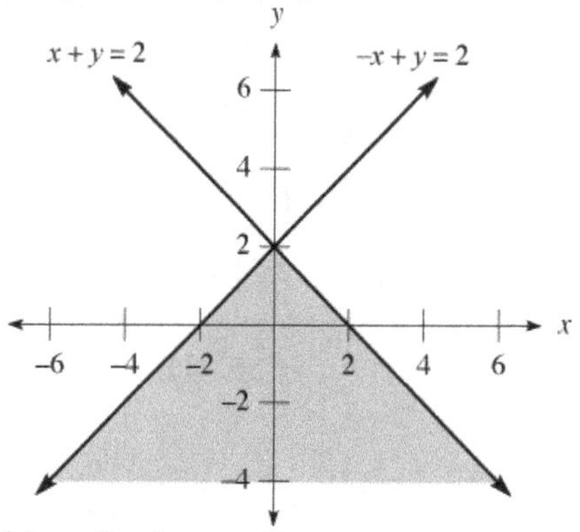

29. Graphing the inequality:

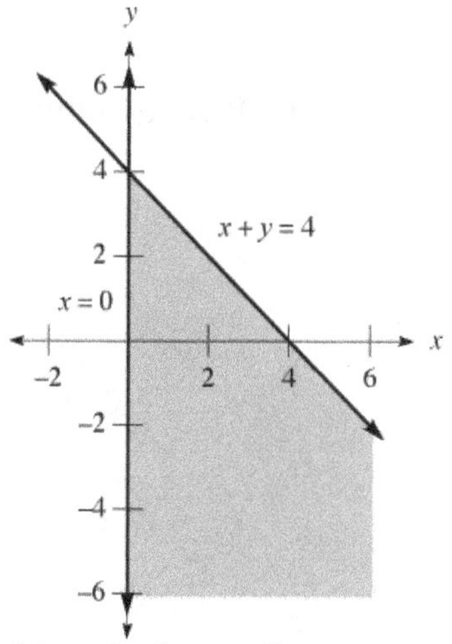

31. Graphing the inequality:

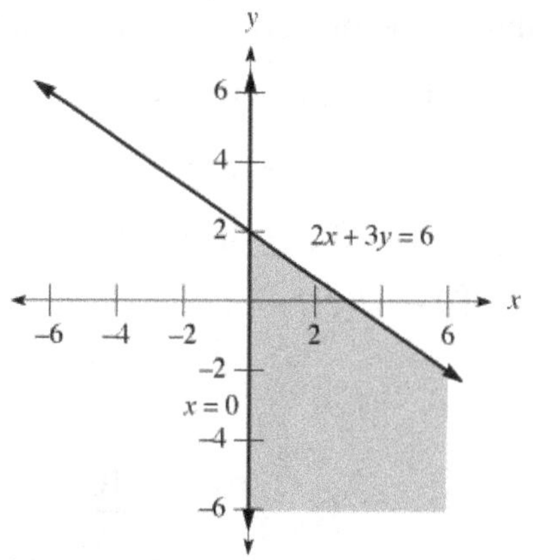

33. Graphing the inequality:

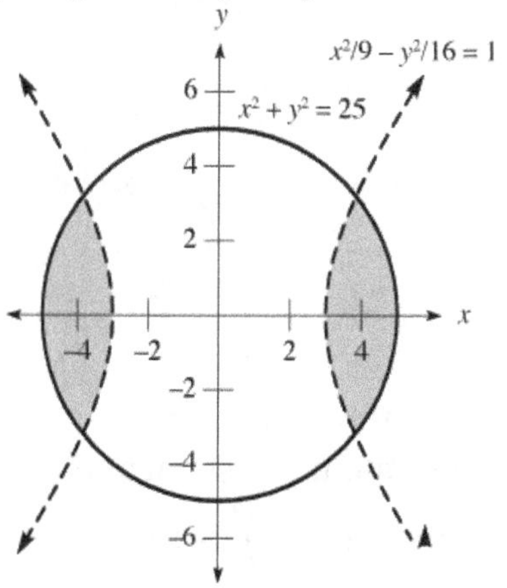

35. Graphing the inequality:

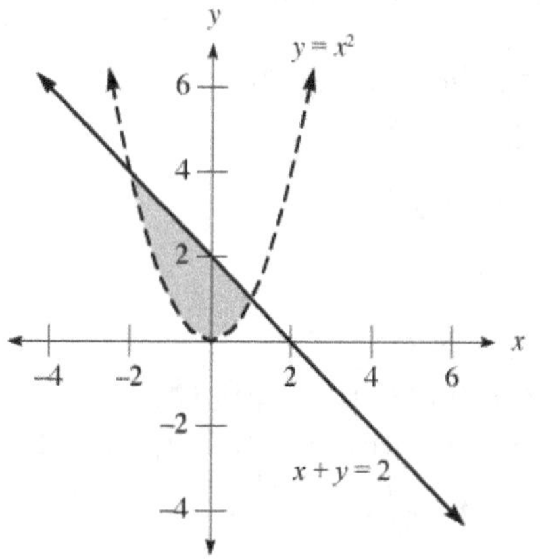

37. Solving the second equation for *y* yields $y = 3 - 2x$. Substituting into the first equation:

$$\begin{aligned} x^2 + (3-2x)^2 &= 9 \\ x^2 + 9 - 12x + 4x^2 &= 9 \\ 5x^2 - 12x &= 0 \\ x(5x-12) &= 0 \\ x &= 0, \frac{12}{5} \\ y &= 3, -\frac{9}{5} \end{aligned}$$

The solutions are $(0,3), \left(\frac{12}{5}, -\frac{9}{5}\right)$.

39. Solving the second equation for *x* yields $x = 8 - 2y$. Substituting into the first equation:

$$\begin{aligned} (8-2y)^2 + y^2 &= 16 \\ 64 - 32y + 4y^2 + y^2 &= 16 \\ 5y^2 - 32y + 48 &= 0 \\ (y-4)(5y-12) &= 0 \\ y &= 4, \frac{12}{5} \\ x &= 0, \frac{16}{5} \end{aligned}$$

The solutions are $(0,4), \left(\frac{16}{5}, \frac{12}{5}\right)$.

41. Adding the two equations yields:

$$\begin{aligned} 2x^2 &= 50 \\ x^2 &= 25 \\ x &= -5, 5 \\ y &= 0 \end{aligned}$$

The solutions are (–5,0), (5,0).

43. Substituting into the first equation:

$$\begin{aligned} x^2 + \left(x^2 - 3\right)^2 &= 9 \\ x^2 + x^4 - 6x^2 + 9 &= 9 \\ x^4 - 5x^2 &= 0 \\ x^2\left(x^2 - 5\right) &= 0 \\ x &= 0, -\sqrt{5}, \sqrt{5} \\ y &= -3, 2, 2 \end{aligned}$$

The solutions are $(0,-3), \left(-\sqrt{5}, 2\right), \left(\sqrt{5}, 2\right)$.

45. Substituting into the first equation:

$$\begin{aligned} x^2 + \left(x^2 - 4\right)^2 &= 16 \\ x^2 + x^4 - 8x^2 + 16 &= 16 \\ x^4 - 7x^2 &= 0 \\ x^2\left(x^2 - 7\right) &= 0 \\ x &= 0, -\sqrt{7}, \sqrt{7} \\ y &= -4, 3, 3 \end{aligned}$$

The solutions are $(0,-4), \left(-\sqrt{7}, 3\right), \left(\sqrt{7}, 3\right)$.

47. Substituting into the first equation:

$$3x+2(x^2-5)=10$$
$$3x+2x^2-10=10$$
$$2x^2+3x-20=0$$
$$(x+4)(2x-5)=0$$
$$x=-4,\frac{5}{2}$$
$$y=11,\frac{5}{4}$$

The solutions are $(-4,11),\left(\frac{5}{2},\frac{5}{4}\right)$.

49. Substituting into the first equation:

$$-x+1=x^2+2x-3$$
$$x^2+3x-4=0$$
$$(x+4)(x-1)=0$$
$$x=-4,1$$
$$y=5,0$$

The solutions are (–4,5), (1,0).

51. Substituting into the first equation:

$$x-5=x^2-6x+5$$
$$x^2-7x+10=0$$
$$(x-2)(x-5)=0$$
$$x=2,5$$
$$y=-3,0$$

The solutions are (2,–3), (5,0).

53. Adding the two equations yields:

$$8x^2=72$$
$$x^2=9$$
$$x=\pm 3$$
$$y=0$$

The solutions are (–3,0), (3,0).

55. Solving the first equation for x yields $x=y+4$. Substituting into the second equation:

$$(y+4)^2+y^2=16$$
$$y^2+8y+16+y^2=16$$
$$2y^2+8y=0$$
$$2y(y+4)=0$$
$$y=0,-4$$
$$x=4,0$$

The solutions are (0,–4), (4,0).

57. Adding the two equations yields:

$$3x^2=8$$
$$x^2=\frac{8}{3}$$
$$x=\pm\sqrt{\frac{8}{3}}$$
$$x=\pm\frac{\sqrt{24}}{\sqrt{9}}=\pm\frac{2\sqrt{6}}{3}$$

Substituting into the second equation:

$$\frac{8}{3}+y=7$$

$$y=\frac{13}{3}$$

The solutions are $\left(-\frac{2\sqrt{6}}{3},\frac{13}{3}\right),\left(\frac{2\sqrt{6}}{3},\frac{13}{3}\right)$.

59. Substituting into the second equation:

$$x^2-3=x^2-2x-1$$

$$-3=-2x-1$$

$$-2x=-2$$

$$x=1$$

$$y=-2$$

The solution is (1,–2).

61. Adding the two equations yields:

$$8x^2=80$$

$$x^2=10$$

$$x=\pm\sqrt{10}$$

$$y=0$$

The solutions are $\left(-\sqrt{10},0\right),\left(\sqrt{10},0\right)$.

63. The system of equations is:

$$x^2+y^2=89$$

$$x^2-y^2=39$$

Adding the two equations yields:

$$2x^2=128$$

$$x^2=64$$

$$x=\pm 8$$

$$y=\pm 5$$

The numbers are either 8 and 5, 8 and –5, –8 and 5, or –8 and –5.

65. **a.** Subtracting the two equations yields:

$$(x+8)^2-x^2=0$$

$$x^2+16x+64-x^2=0$$

$$16x=-64$$

$$x=-4$$

Substituting to find y:

$$(-4)^2+y^2=64$$

$$y^2+16=64$$

$$y^2=48$$

$$y=\pm\sqrt{48}=\pm 4\sqrt{3}$$

The intersection points are $\left(-4,-4\sqrt{3}\right)$ and $\left(-4,4\sqrt{3}\right)$.

b. Subtracting the two equations yields:

$$x^2-(x-8)^2=0$$
$$x^2-x^2+16x-64=0$$
$$16x=64$$
$$x=4$$

Substituting to find y:

$$4^2+y^2=64$$
$$y^2+16=64$$
$$y^2=48$$
$$y=\pm\sqrt{48}=\pm4\sqrt{3}$$

The intersection points are $\left(4,-4\sqrt{3}\right)$ and $\left(4,4\sqrt{3}\right)$.

67. Expanding using the binomial theorem:

$$(x+2)^4=\binom{4}{0}x^4+\binom{4}{1}x^3(2)+\binom{4}{2}x^2(2)^2+\binom{4}{3}x(2)^3+\binom{4}{4}(2)^4$$
$$=x^4+8x^3+24x^2+32x+16$$

69. Expanding using the binomial theorem:

$$(2x+y)^3=\binom{3}{0}(2x)^3+\binom{3}{1}(2x)^2y+\binom{3}{2}(2x)y^2+\binom{3}{3}y^3=8x^3+12x^2y+6xy^2+y^3$$

71. The first two terms are: $\binom{50}{0}x^{50}+\binom{50}{1}x^{49}(3)=x^{50}+150x^{49}$

Chapter 10 Test

See www.mathtv.com for video solutions to all problems in this chapter test.

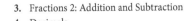